Air Pollution at the Urban and Regional Level: Sources, Sinks, and Transportation

Air Pollution at the Urban and Regional Level: Sources, Sinks, and Transportation

Editors

Elena Hristova
Manousos Ioannis Manousakas
Anikó Angyal
Maria Gini

MDPI

Basel • Beijing • Wuhan • Barcelona • Belgrade • Novi Sad • Cluj • Manchester

Editors

Elena Hristova
Department Meteorology
National Institute of
Meteorology and Hydrology
Sofia
Bulgaria

Manousos Ioannis
Manousakas
Laboratory of Atmospheric
Chemistry
Paul Scherrer Institute
Villigen
Switzerland

Anikó Angyal
Institute of Nuclear Research
Hungarian Academy
of Sciences
Debrecen
Hungary

Maria Gini
INRASTES Institute
NCSR "Demokritos"
Athens
Greece

Editorial Office
MDPI
St. Alban-Anlage 66
4052 Basel, Switzerland

This is a reprint of articles from the Special Issue published online in the open access journal *Atmosphere* (ISSN 2073-4433) (available at: www.mdpi.com/journal/atmosphere/special_issues/Air_Pollution_Sources_Sinks_Transportation).

For citation purposes, cite each article independently as indicated on the article page online and as indicated below:

Lastname, A.A.; Lastname, B.B. Article Title. *Journal Name* **Year**, *Volume Number*, Page Range.

ISBN 978-3-0365-8731-8 (Hbk)
ISBN 978-3-0365-8730-1 (PDF)
doi.org/10.3390/books978-3-0365-8730-1

Contents

About the Editors

Elena Hristova

Elena Hristova is an associate professor at the National Institute of Meteorology and Hydrology, Bulgaria. She holds a PhD in Chemical Engineering (University of Chemical Technology and Metallurgy, Sofia, Bulgaria, 2012). Her research interests are in the field of chemical composition of atmospheric aerosols, the identification of major source groups of PM pollution in urban environments with receptor model, the measurement of Black carbon concentrations in ambient air, and the chemical composition of atmospheric deposition in Bulgaria. She has led and participated in a number of scientific and applied research projects and tasks at NIMH, from different national and international funding programs (International Atomic Energy Agency, Transboundary Cooperation with Turkey, European Regional Development Fund through the Innovative Actions for Urban Development initiative, and Bulgarian National Science Found).

Manousos Ioannis Manousakas

Manousos-Ioannis Manousakas received a Ph.D. in Environmental Monitoring and Analysis in 2014 at the Chemistry Department of the University of Patras, Greece. After his studies, he has worked as Collaborating Researcher in NCSR Demokritos and the Institute of Chemical Engineering Sciences in Greece, and at Paul Scherrer Institute in Switzerland. His key research interests include atmospheric pollution, and specifically particulate-related pollution, chemical analysis, and source apportionment. He has a strong international network of collaborators, shown by his published research papers in Scientific Journals, and participated in several Scientific Projects from different funding programs. He serves as a source apportionment expert in many projects worldwide organized by the International Atomic Energy Agency (IAEA). He is a reviewer board member at five journals, while until 2021 has served as Guest Editor in three journals.

Anikó Angyal

Anikó Angyal received her Ph.D. in Environmental Sciences in 2016 at Pál Juhász-Nagy Doctoral School of the University of Debrecen, Hungary. After finishing her studies, she work as a Research Associate in Institute for Nuclear Research in Debrecen, Hungary. Her main research interests include atmospheric pollution, in particular chemical analysis and source identification. Another important scientific field is cultural heritage, where the main focus is the analysis of artefacts using ion beam analysis. She has been involved in several national and international scientific projects under various funding programs. She has been Guest Editor of three journals.

Maria Gini

Maria Gini is a collaborating researcher at the Institute of Nuclear and Radiological Sciences and Technology, Energy and Safety, NCSR "Demokritos" (NCSRD). She holds a PhD in Environmental Physics (National and Kapodistrian University of Athens, Greece, Physics Department, 2015). Her research focuses on aerosol microphysical, optical and chemical characterization with respect to air quality and climate change, aerosol-based nanotechnology and nano-particle metrology, GHGs, environmental radioactivity, and population exposure to aerosol contaminants. She has actively participated in several European and National projects (i.e. Research and Innovation, Metrology, Climate Change Adaptation).

Preface

Air pollution poses a significant risk to human health. Emissions from industrial activities, energy production, transport, waste management activities, and natural sources contribute to the many air pollution-induced problems, such as reduced visibility, adverse health effects, and global climate change. Air pollution is not a local concern. The residence time of pollutants in the atmosphere can extend from several days to months, and the corresponding spatial transportation scales are proportionally large, ranging from local to continental. Even though there is no region not affected by air pollution at some level, the situation is significantly worse in urban areas.

The aim of this Special Issue was to gather up-to-date research knowledge aiming at assessing air pollution at the urban and regional level, including both experimental and monitoring studies and mathematical/numerical modeling studies. The publications of the issue covered the subjects of air pollution with particulates and gaseous pollutants, Particulate Matter (PM) sources and source apportionment, Carbonaceous species (Organic and Elemental or Black Carbon) in the atmosphere, and Ozon-related pollution.

Elena Hristova, Manousos Ioannis Manousakas, Anikó Angyal, and Maria Gini
Editors

atmosphere

Editorial

Special Issue: Air Pollution at the Urban and Regional Level: Sources, Sinks, and Transportation

Elena Svilenova Hristova [1,*] and Manousos Ioannis Manousakas [2]

[1] Department of Meteorology, National Institute of Meteorology and Hydrology, 1784 Sofia, Bulgaria
[2] Laboratory of Atmospheric Chemistry, Paul Scherrer Institute (PSI), 5232 Villigen, Switzerland
* Correspondence: elena.hristova@meteo.bg

1. Introduction

Air pollution poses a significant risk to human health. Emissions from industrial activities, energy production, transport, waste management activities, and natural sources contribute to the many air-pollution-induced problems, such as reduced visibility, adverse health effects, and global climate change. Air pollution is not a local concern. The residence time of pollutants in the atmosphere can extend from several days to months, and the corresponding spatial transportation scales are proportionally large, ranging from local to continental. Even though there is no region not affected by air pollution at some level, the situation is significantly worse in urban areas. Urban areas are hotspots of air pollution, especially the large and densely populated metropolitan areas. Air pollutants can vary in type and characteristics, can be of gaseous or particulate form, and can either be directly emitted or formed in the atmosphere from their precursor molecules.

The aim of this Special Issue was to gather up-to-date research knowledge aiming at assessing air pollution at the urban and regional level, including both experimental and monitoring studies and mathematical/numerical modeling studies.

2. Results

The publications of the issue cover the subjects of air pollution with particulates and gaseous pollutants (five), Particulate Matter (PM) sources and source apportionment (three), Carbonaceous species (Organic and Elemental or Black Carbon) in the atmosphere (two), and Ozone-related pollution (one). The relatively large geographical coverage of the studies included in this issue provides a good overview of particulate-related pollution on an intercontinental level. The Special Issue contains ten published studies referring to different regions around the world: Europe (five), Asia (three), and the Middle East (two). A brief overview of the main finding and conclusions of the studies included in the Special Issue will be presented below in chronological order of publication.

In the first publication, the authors used modeling approaches to study the atmospheric composition in Sofia, Bulgaria [1]. The models used were WRF as a meteorological pre-processor, CMAQ as a chemical transport model, and SMOKE as the emission pre-processor of Models-3 system. Based on the findings of the study, it was identified that the daily concentration variation of the two main air pollution species—NO_2 and $PM_{2.5}$—have different magnitudes, and the effect of different emission sources on the relative contributions to the concentration of the species is highly variant. The results produced by the CMAQ "integrated process rate analysis" demonstrate the complex behavior and interaction of the different processes. Further analysis of these processes; their spatial, diurnal, and seasonal variability; and their interactions can be helpful for an explanation of the overall picture and origin of the pollution in the considered region.

The second paper tackled the subject of the effect of vehicular emissions on air pollution in Warsaw, Poland, and specifically the impact a modernization of the fleet composition could have in reducing pollutant concentrations in the region [2]. Using the Calpuff model,

Citation: Hristova, E.S.; Manousakas, M.I. Special Issue: Air Pollution at the Urban and Regional Level: Sources, Sinks, and Transportation. *Atmosphere* **2023**, *14*, 132. https://doi.org/10.3390/atmos14010132

Received: 2 December 2022
Accepted: 4 January 2023
Published: 7 January 2023

simulations of the yearly averaged concentrations of NOx, CO, PM_{10}, and $PM_{2.5}$ were estimated, together with an assessment of the population's exposure to individual pollutants. The simulations indicated that a fleet modernization following the latest European standards would reduce NOx concentrations due to reduced emissions from passenger cars, heavy-duty vehicles, and public transport. On the other hand, improving air quality in terms of CO concentration is based exclusively upon the modernization of gasoline-powered cars. Finally, it is suggested that despite the substantial contribution of traffic to the overall PM concentrations in the area, modernization of the fleet leads to only minor effects because PM pollution in Warsaw originates mainly from the municipal sector and transboundary inflow.

The authors of the third publication investigated the concentrations of organic (OC) and elemental carbon (EC) between May 2018 and March 2019 in Amman, Jordan [3]. The results show that the OC and EC annual mean concentrations in $PM_{2.5}$ samples were 5.9 ± 2.8 μg m^{-3} and 1.7 ± 1.1 μg m^{-3}, respectively. It was found that OC and EC concentrations were mostly in the fine fraction of PM. During sand and dust storm (SDS) episodes, OC and EC concentrations were higher than the annual means. Based on this observation, the SDS episodes were identified as being responsible for an increased carbonaceous aerosol content in addition to the overall increase in $PM_{2.5}$ and PM_{10} concentrations, which may have direct implications on human health.

The fourth publication was focused on the determination of black carbon (BC) in $PM_{2.5}$, using both experimental and modeling approaches in two Bulgarian cities—Sofia and Burgas—during October 2020 and January 2022 [4]. For the experimental evaluation of BC, the Multi-Wavelength Absorption Black Carbon Instrument (MABI) was used, while for the modeled ones, data from the chemical transport models (CTM) of the European (regional) air quality system established at the Copernicus Atmosphere Monitoring Service (CAMS) were utilized. The BC and $PM_{2.5}$ concentrations were higher in January than in October for both cities. It was identified that in October, the model underestimated the observed BC concentrations (Sofia 2.44 μg m^{-3}, Burgas 1.63 μg m^{-3}) by 17% and 51%, respectively. In January 2021, the observed monthly BC concentrations were higher (Sofia 3.62 μg m^{-3}, Burgas 1.75 μg m^{-3}), and the model's bias was less than in October, with an observed overestimation of 22% for Sofia. Regarding $PM_{2.5}$, the relative bias in October (17% for Sofia, and 6% for Burgas) was less than the relative bias in January when the model underestimated $PM_{2.5}$ monthly mean concentrations by 20% (Sofia) and 42% (Burgas).

In the fifth publication, the authors investigated the effect of cooking processes in Dammam City, Saudi Arabia, by directly measuring the emissions from the chimneys of different types of restaurants and the surrounding ambient air [5]. Five air pollutants were measured simultaneously. The highest mean levels of CO (64.8 ± 44.3 ppm), CO2 (916.7 ± 463.4 ppm), VOCs (105.1 ± 61.3 ppm), NO_2 (4.2 ± 2.4 ppm), and SO_2 (8.0 ± 7.4 ppm) were recorded in chimneys of grilling restaurants. Similarly, the highest pollutant concentrations were recorded in the areas adjacent to the grilling restaurants compared to other restaurants using different cooking processes.

The authors of the sixth publication analyze the relationships between the particulate matter concentration and land use changes in the Beijing–Tianjin–Hebei region, China, from 2015 to 2018 [6]. The obtained results are sumarazed in three main conclusions: (1) an improved sine function model can suitably fit the periodic changes in the particulate matter concentration, with the average R2 value increasing to 0.65 from the traditional model value of 0.49, while each model coefficient effectively estimates the change characteristics of each stage; (2) among all land use types, the particulate matter concentrations in construction land and farmland are high, with a large annual difference between high and low values; (3) the landscape pattern of land use exerts a significant influence on the particulate matter concentration.

The characteristics of the difference between the concentration of resuspended dust (Ci) and the background concentration of roads (CBg) and the background of city atmosphere (Bg) concentration measured were compared with the effects of traffic and weather conditions discussed in the seventh publication [7]. The PM reduction measures are being implemented according to the occurrence of high concentrations of PM10 and PM2.5 provided by the city Bg observations in South Korea. The work mainly focuses on the following four topics: (1) the increased level of resuspended dust according to vehicle speed and silt loading (sL) level; (2) the difference between atmospheric pollution concentration at adjacent monitoring stations and background concentration levels on roads due to atmospheric weather changes; (3) the correlation between traffic and weather factors with resuspended dust levels; (4) the evaluation of resuspended dust levels by road section. The results of this study suggested the need for an efficient alternative considering the effect of yellow dust over time, because, due to the occurrence of yellow dust, Bg can differ from the resuspended dust concentration. Additionally, the concentration of resuspended dust on roads may differ significantly from that of the adjacent Bg observation caused by vehicles driving on roads. Therefore, it is suggested that a more frequent occurrence of high levels of resuspended dust may occur, and research should be continued to quantify the influence of dust collected through mobile measurements and to provide accurate forecasts.

The authors of the eighth publication studied the O_3 pollution situation in Tai'an, China, from May to September in the period from 2016 to 2021 [8]. They found that the pollution during this period was mainly light pollution, accounting for 69% to 100% of the total polluted days, with June being the most polluted month. High temperature (>30 °C), low relative humidity (20~40%), and low wind speed (1~3 m/s) provided favourable meteorological conditions for the generation of O_3, especially with a prevailing southerly wind, resulting in light and air pollution in Tai'an. The results of backward trajectory analysis showed that in the summer pollution trajectories, the pollution trajectories in the southwest direction accounted for the highest proportion of all kinds of trajectories. When the southerly winds dominated, the accumulated O_3 concentrations were significantly higher than those of the surrounding cities due to the blocking of the northern mountains. The comparison of the average O_3 concentration obtained for three monitoring stations in Tai'an (Renkou School station closer to the mountain, and the Dianli College and the Shandong First Medical University stations relatively far away from the mountains) indicated 13~15 $\mu g/m^3$ higher O_3 concentration at the Renkou School station than at other stations, indicating that the Renkou School station was more affected by the obstructions of the mountains. The application of WRF-CMAQ in this study to quantify the blocking effect of the mountains indicated that the average O_3 concentrations in Tai'an decreased by about 1.7~7.5 $\mu g/m^3$ after reducing the terrain height of Mount Tai in the model.

The study presented in the ninth publication is an analysis the consequences of implementing the official scenario prepared for air quality improvement in Warsaw, Poland, particularly in terms of population exposure and the associated health risks at the end of the present decade. The main tool used was a Gaussian system, CALPUFF, for modelling atmospheric pollution dispersion [9]. Four pollutants, NOx, PM10, PM2.5, and BaP, all of which presently exceed environmental limits in Warsaw, were considered. The results show a reduction in population exposure attributed to the specific pollutants and scenarios being implemented. The final reduction in population exposure to NOx is about 28% which means that the base avoidable mortality assigned to this pollutant is 743 avoidable deaths, which would be reduced by about 204 cases. The analogous result for PM2.5 was a more significant (~30%) reduction in population exposure would reduce the number of avoidable yearly deaths by 607 cases from the initial value of 2023 avoidable deaths. The reduction in exposure to PM10 was similar, amounting to about 28%, while for carcinogenic BaP, it is as much as approximately 50%.

The last publication in this Special Issue is on the application of the EPA PMF 5.0 model to the PM2.5 chemical composition and hygroscopicity for the period August 2016–July 2017 in Athens, Greece [10]. Source apportionment analysis identified six major sources, including four anthropogenic sources (vehicle exhaust and no exhaust, heavy oil combustion, and a mixed source of secondary aerosol formation and biomass combustion) and two natural sources (mineral dust and aged sea salt). The authors found that the mixed source is the main contributor to PM2.5 levels (44%), followed by heavy oil combustion (26%) and automotive emissions from road traffic and non-exhaust emissions (15%). The aerosol hygroscopic growth factor is mainly related to mixed source (36%) and heavy oil combustion (24%), and to a lower extent with vehicle exhaust (by 19%), aged sea salt (by 14%), and unburned vehicle exhaust (c 6%).

3. Conclusions

Urban air quality is determined by many factors: air masses, characteristic local flow and meteorology, low dispersion ability in built-up environments, the concentration of emission sources of different types, and various chemical processes. The publications in this Special Issue address the topics of air pollution with particulates (PM10, PM2.5) and gaseous pollutants (NO_x, NO_2, SO_2, CO, VOCs, O_3); PM sources and source apportionment; and Carbonaceous species such as Organic and Elemental or Black Carbon. Both experimental and modelling approaches are applied. The relatively wide geographic range of the studies provides a good overview of particulate matter pollution at the intercontinental level. All ten published studies cover different regions of the world: Europe (Poland, Greece, Bulgaria), Asia (South Korea, China), and the Middle East (Saudi Arabia, Jordan).

Since the chemical composition of PM remains a reflection of complex, site-specific processes, research aimed at better understanding the sources of PM, and the processes they undergo in the atmosphere remains very relevant. Therefore, this Special Issue will continue to summarise the research in this field (https://www.mdpi.com/journal/atmosphere/special_issues/0Q62Z8ZMZQ, 6 January 2023).

Author Contributions: Conceptualization, E.S.H. and M.I.M.; methodology, E.S.H.; writing—original draft preparation, E.S.H. and M.I.M.; All authors have read and agreed to the published version of the manuscript.

Funding: This research received no external funding.

Institutional Review Board Statement: Not applicable.

Informed Consent Statement: Not applicable.

Data Availability Statement: Not applicable.

Acknowledgments: The editor would like to thank the authors for their contributions to this Special Issue and the reviewers for their constructive and helpful comments to improve the manuscripts.

Conflicts of Interest: The authors declare no conflict of interest.

References

1. Ivanov, V.; Georgieva, I. Basic facts about numerical simulations of atmospheric composition in the city of Sofia. *Atmosphere* **2021**, *12*, 1450. [CrossRef]
2. Holnicki, P.; Nahorski, Z.; Kałuszko, A. Impact of vehicle fleet modernization on the traffic-originated air pollution in an urban area—A case study. *Atmosphere* **2021**, *12*, 1581. [CrossRef]
3. Hussein, T.; Li, X.; Bakri, Z.; Alastuey, A.; Arar, S.; Al-Hunaiti, A.; Viana, M.; Petäjä, T. Organic and Elemental Carbon in the Urban Background in an Eastern Mediterranean City. *Atmosphere* **2022**, *13*, 197. [CrossRef]
4. Hristova, E.; Georgieva, E.; Veleva, B.; Neykova, N.; Naydenova, S.; Gonsalvesh-Musakova, L.; Neykova, R.; Petrov, A. Black Carbon in Bulgaria—Observed and Modelled Concentrations in Two Cities for Two Months. *Atmosphere* **2022**, *13*, 213. [CrossRef]
5. ElSharkawy, M.F.; Ibrahim, O.A. Impact of the Restaurant Chimney Emissions on the Outdoor Air Quality. *Atmosphere* **2022**, *13*, 261. [CrossRef]
6. Zhai, H.; Yao, J.; Wang, G.; Tang, X. Impact of Land Use on Atmospheric Particulate Matter Concentrations: A Case Study of the Beijing-Tianjin-Hebei Region, China. *Atmosphere* **2022**, *13*, 391. [CrossRef]
7. Hong, S.; Yoo, H.; Cho, J.; Yeon, G.; Kim, I. Characteristics of Resuspended Road Dust with Traffic and Atmospheric Environment in South Korea. *Atmosphere* **2022**, *13*, 1215. [CrossRef]
8. Ge, Y.; Chen, C.; Jiang, Y.; Yang, T.; Kang, H.; Li, J.; Zhao, X.; Zhang, Y.; Li, M.; Hou, T.; et al. Characteristics of Ozone Pollution in Tai'an and Topographic Effects of Mount Tai. *Atmosphere* **2022**, *13*, 1299. [CrossRef]

9. Holnicki, P.; Kałuszko, A.; Nahorski, Z. Scenario Analysis of Air Quality Improvement in Warsaw, Poland, by the End of the Current Decade. *Atmosphere* **2022**, *13*, 1613. [CrossRef]
10. Diapouli, E.; Fetfatzis, P.; Panteliadis, P.; Spitieri, C.; Gini, M.I.; Papagiannis, S.; Vasilatou, V.; Eleftheriadis, K. $PM_{2.5}$ Source Apportionment and Implications for Particle Hygroscopicity at an Urban Background Site in Athens, Greece. *Atmosphere* **2022**, *13*, 1685. [CrossRef]

atmosphere

MDPI

Article

$PM_{2.5}$ Source Apportionment and Implications for Particle Hygroscopicity at an Urban Background Site in Athens, Greece

Evangelia Diapouli *, Prodromos Fetfatzis, Pavlos Panteliadis, Christina Spitieri, Maria I. Gini, Stefanos Papagiannis, Vasiliki Vasilatou and Konstantinos Eleftheriadis

Institute of Nuclear & Radiological Science & Technology, Energy & Safety, National Centre of Scientific Research "Demokritos", 15341 Athens, Greece
* Correspondence: ldiapouli@ipta.demokritos.gr

Abstract: Aerosol hygroscopicity is a key aerosol property, influencing a number of other physical properties, and the impacts of PM pollution on the environment, climate change, and health. The present work aims to provide insight into the contribution of major PM sources to aerosol hygroscopicity, focusing on an urban background site, with a significant impact from both primary and secondary sources. The EPA PMF 5.0 model was applied to $PM_{2.5}$ chemical composition and hygroscopicity data collected from August 2016 to July 2017 in Athens, Greece. Source apportionment analysis resulted in six major sources, including four anthropogenic sources (vehicular exhaust and non-exhaust, heavy oil combustion, and a mixed source of secondary aerosol formation and biomass burning) and two natural sources (mineral dust and aged sea salt). The mixed source was found to be the main contributor to $PM_{2.5}$ levels (44%), followed by heavy oil combustion (26%) and vehicular traffic exhaust and non-exhaust emissions (15%). The aerosol hygroscopic growth factor (GF) was found to be mainly associated with the mixed source (by 36%) and heavy oil combustion (by 24%) and, to a lesser extent, with vehicle exhaust (by 19%), aged sea salt (by 14%), and vehicle non-exhaust (by 6%).

Keywords: $PM_{2.5}$; hygroscopicity; source apportionment; positive matrix factorization; urban background

Citation: Diapouli, E.; Fetfatzis, P.; Panteliadis, P.; Spitieri, C.; Gini, M.I.; Papagiannis, S.; Vasilatou, V.; Eleftheriadis, K. $PM_{2.5}$ Source Apportionment and Implications for Particle Hygroscopicity at an Urban Background Site in Athens, Greece. *Atmosphere* **2022**, *13*, 1685. https://doi.org/10.3390/atmos13101685

Academic Editor: László Bencs

Received: 22 August 2022
Accepted: 11 October 2022
Published: 14 October 2022

Publisher's Note: MDPI stays neutral with regard to jurisdictional claims in published maps and institutional affiliations.

1. Introduction

Suspended particulate matter (PM) remains an atmospheric pollutant of great concern, especially in urban centres, where anthropogenic emission sources are concentrated and a large share of the global population resides. Particulate pollution contributes to climate change and visibility degradation; impacts the atmospheric chemical processes, affecting the concentration of gaseous pollutants; and displays documented adverse effects on human health [1,2]. In order to fully understand the different impacts of particulate pollution, a comprehensive characterization of the atmospheric aerosol is needed, including its physical and chemical properties.

Aerosol hygroscopicity, i.e., the ability of particles to absorb water in response to increasing relative humidity, is a key aerosol property, influencing a number of other physical properties, and the impacts of PM pollution on the environment, climate, and health. The water content of particles affects their size and surface area, as well as the total aerosol mass, acidity, chemical reactivity, and atmospheric lifetime [3,4]. Particle hygroscopicity also affects the aerosol optical properties, with significant implications for visibility and radiative forcing [5–7]. In addition, hygroscopic growth is directly associated with aerosol cloud droplet activation properties, which impact climate change [8]. Hygroscopicity is also strongly associated with particle health impacts. Hygroscopic particles grow significantly in the near supersaturated environment of the respiratory tract; this may substantially alter particle deposition (both with respect to lung deposition probability and place of deposition), as well as the metabolic pathway after deposition, in comparison with non-soluble particles [9,10].

The atmospheric aerosol ability to absorb water depends on its chemical composition and mixing state [11,12]. Fossil fuel combustion processes usually emit hydrophobic or slightly hygroscopic particles [9], while biomass burning emissions may also contain hygroscopic particles [13]. Secondary inorganic aerosols (e.g., ammonium sulfate and nitrate) are hydrophilic. With respect to primary natural particles, sea salt particles are hygroscopic, while soil dust particles are normally insoluble [4]. Among the different aerosol species, the less characterized in terms of hygroscopic growth is organic aerosol, owing to the wide range of organic compounds found in the atmosphere. It is generally expected that aged and more oxidized organic aerosol will be more hygroscopic [14]; nevertheless, it has been shown that hygroscopic growth is not always correlated with the O=C ratio, demonstrating that other factors also play an important role (such as the different functional groups, carbon chain length, and so on) [15]. These uncertainties related to organic aerosol hygroscopic properties hamper a better understanding of the role of primary and secondary organic matter in the atmospheric environment and climate [16].

Several studies have attempted to link aerosol hygroscopic behavior with chemical composition [15,17–21] or to characterize hygroscopic properties of various aerosol types, based on air mass back trajectories or specific pollution episodes [22–24]. In addition, the hygroscopic properties of particles emitted by typical combustion processes, such as diesel vehicle exhaust [25,26] and biomass burning emissions [27–30], have been assessed. Nevertheless, while it has been acknowledged that hygroscopicity is highly dependent on both aerosol emission sources and atmospheric transformations, the hygroscopic properties of the various sources in receptor sites, where freshly emitted and aged aerosol are mixed, have been scarcely explored. To the best of our knowledge, only two very recent studies have attempted to evaluate the hygroscopicity of major PM sources identified by receptor modelling. Li et al. performed a source apportionment study combining $PM_{2.5}$ chemical composition data with hygroscopicity measurements from a 3-month period, during 2019–2020, in an urban site of Guangzhou, China [31]. Vu et al. applied receptor modelling for source apportionment of sub-micrometer particles, using particle number size distribution data together with hygroscopicity data, collected at an urban background site in London, United Kingdom, during two 2-month campaigns in 2012, covering cold and warm season [32]. Both studies confirmed the good potential of PMF in characterizing the hygroscopic properties of particles originating from different primary and secondary sources. The most hydrophilic sources were related to secondary inorganic aerosol and mixed secondary organic and inorganic aerosol; the less-hygroscopic particles were associated with primary combustion emissions and soil dust [31,32].

The present work aims to provide further insight into the contribution of PM sources to aerosol hygroscopicity, focusing on an urban background site, with a significant impact from both primary and secondary sources. A year-long measurement campaign was performed during 2016–2017 at the Demokritos (DEM) urban background station, in Athens, Greece, including $PM_{2.5}$ chemical speciation and hygroscopicity measurements. Limited hygroscopicity data are available for Southern Europe, and Greece in particular. Hygroscopicity measurements were performed at a regional background site in the island of Crete [22] and a background site in the island of Lemnos [33]. Data on urban aerosol hygroscopicity in Greece have been reported only for Athens, during a short campaign (11–26 June 2003) at a suburban background site, and at DEM station for the year-long campaign of 2016–2017 [34,35]. In the present work, these latter data were used together with concurrent $PM_{2.5}$ chemical composition data in order to apply receptor modelling for PM and hygroscopicity source apportionment. Source apportionment analysis resulted in six major sources, including four anthropogenic sources (vehicular exhaust and non-exhaust, heavy oil combustion, and a mixed source of secondary aerosol formation and biomass burning) and two natural sources (mineral dust and aged sea salt). The aerosol hygroscopic growth factor (GF) was found mainly associated with the mixed source (by 36%) and heavy oil combustion (by 24%) and, to a lesser extent, with vehicle exhaust (by 19%), aged sea salt (by 14%), and vehicle non-exhaust (by 6%).

2. Materials and Methods

2.1. PM Sampling and Analysis

A year-long measurement campaign was conducted from August 2016 to July 2017 at the Demokritos (DEM) urban background station, in Athens, Greece (37.995° N 23.816° E, at 270 m above sea level (a.s.l.)), a member of GAW and part of the ACTRIS and PANACEA infrastructures (Figure 1). The station is away from direct emissions; the site is partially influenced by the urban area (i.e., under most atmospheric conditions) and partially by the incoming regional aerosol (i.e., under northern or southern winds), and can be considered as typical for a suburban Mediterranean atmospheric environment [36]. The climate is characterized by hot dry summers and wet mild winters. During the measurement campaign, the ambient temperature was measured to be equal to 10.9 ± 4.4 °C in the cold period (15 October 2016–14 April 2017) and 23.5 ± 4.7 °C in the warm period (1 August–14 October 2016 and 15 April–31 July 2017).

Figure 1. The Demokritos (DEM) urban background station in Ag. Paraskevi, Athens, Greece (37.995° N 23.816° E, at 270 m above sea level (a.s.l.)).

In this research, 24 h $PM_{2.5}$ samples were collected by a low-volume reference sampler (Sequential 47/50-CD, Sven Leckel GmbH, Berlin, Germany) at a flow rate of 2.3 m^3/h on Teflon filters and were analyzed gravimetrically for the determination of PM mass concentrations, according to EN12341. The samples were also analyzed by X-ray fluorescence (XRF) with the use of an energy dispersive x-ray fluorescence spectrometer (ED-XRF) laboratory instrument (Epsilon 5, PANalytical, Malvern, UK). The following major and trace elements were determined: Na, Mg, Al, Si, S, Cl, K, Ca, Ti, V, Cr, Mn, Fe, Ni, Cu, Zn, Br, Ba, and Pb. Detection limits ranged from 0.2 ng/m^3 (for Cr) to 16.2 ng/m^3 (for Ba). Details on the method may be found in Manousakas et al. [37].

Near-real time elemental (EC) and organic carbon (OC) concentrations in $PM_{2.5}$ were recorded on a 3 h basis, using the thermo-optical transmittance (TOT) method, using a semi-continuous OC-EC field analyzer (Model-4, Sunset Laboratory, Inc., Tigard, OR, U.S.A.), equipped with an in-line parallel carbon denuder for the removal of organic gases. The EUSAAR2 protocol was employed for analysis. The output of the analyzer includes the different OC fractions (OC1 to OC4) corresponding to the four temperature steps of the

EUSAAR2 protocol (at 200 °C, 300 °C, 450 °C, and 650 °C); the pyrolytic organic carbon (POC), formed through pyrolysis during the thermal volatilization of OC; and the elemental carbon (EC). The OC fractions provide insight into the volatility of the different organic compounds, with OC1 representing the more volatile OC fractions and OC4 and POC the less volatile fractions [38]. Details on the method may be found in Panteliadis et al. [39].

In addition, a custom-built humidified tandem differential mobility analyzer (HTDMA) was used to measure the hygroscopic growth factor distributions of ambient aerosol particles with selected dry diameters (D0) of 30 nm, 50 nm, 80 nm, and 250 nm (at a relative humidity of 30 ± 3%), exposed to a relative humidity (RH) of 90 ± 2%. Owing to water uptake, the diameter of the humidified particles (D(RH)) increased and the ratio between the humidified (D(RH)) and dry particle diameter (D0) was defined as the growth factor (GF):

$$\mathrm{GF} = \frac{\mathrm{D(RH)}}{\mathrm{D0}} \tag{1}$$

The particle concentration at the HTDMA outlet as a function of the growth factor is referred to as the measurement distribution function (MDF). An inversion algorithm was applied to the measured MDF to retrieve the actual growth factor probability density functions (GF-PDFs), which describe the probability that a particle with a defined dry size exhibits a certain GF at the specified relative humidity. Details on the HTDMA system and the inversion algorithm applied may be found in Spitieri et al. [35].

2.2. Source Apportionment by Positive Matrix Factorization

Source apportionment was performed by means of receptor modeling, specifically the positive matrix factorization (PMF) model. PMF is a widely used receptor model, which allows for the identification of the major PM sources, based on their chemical profiles, as well as the quantification of their contribution to the measured PM concentration levels [40]. The model is based on the following mass balance equation:

$$c_{ij} = \sum\nolimits_{k=1}^{p} g_{ik} \cdot f_{kj} + e_{ij} \tag{2}$$

where c_{ij} is the concentration of chemical element j measured in sample i, p is the number of factors (sources) that contribute to the measured concentrations, g_{ik} is the contribution of source k to sample i, f_{kj} is the concentration of the chemical element j in source k, and e_{ij} is the residual (the difference between the measured value and the value fitted by the model) for chemical element j in sample i.

The model solves Equation (2) by minimizing the sum of squared residuals, as shown in Equation (3) below:

$$Q = \sum\nolimits_{i=1}^{n} \sum\nolimits_{j=1}^{m} \left(\frac{c_{ij} - \sum\nolimits_{k=1}^{p} g_{ik} \cdot f_{kj}}{u_{ij}} \right)^2 = \sum\nolimits_{i=1}^{n} \sum\nolimits_{j=1}^{m} \left(\frac{e_{ij}}{u_{ij}} \right)^2 \tag{3}$$

where u_{ij} is the uncertainty of the concentration value c_{ij}. PMF produces two matrices (F—factor profiles and G—factor contributions) under the constraint of non-negativity for matrix F and non-significant negativity for matrix G.

In the present work, the EPA PMF 5.0 model was applied to the $PM_{2.5}$ chemical composition database in order to quantify the contribution of the different sources to the observed $PM_{2.5}$ concentration levels; at a second stage, the model was applied to the combined chemical composition and hygroscopicity data in order to explore the hygroscopicity of the major PM sources identified. The hygroscopic growth factor (GF) of particles with dry diameters (D0) of 250 nm was used in this analysis. This particle size fraction corresponds to the accumulation mode and may be thus considered representative of the $PM_{2.5}$ hygroscopic properties. The OC/EC and hygroscopicity data were averaged to match the 24 h $PM_{2.5}$ concentration data. Overall, the database comprised 265 sampling days and 26 input variables, including 24 variables defined as "strong": EC, OC1, OC2, OC3, OC4, Na, Mg,

Al, Si, S, Cl, K, Ca, Ti, V, Cr, Mn, Fe, Ni, Cu, Br, Ba, Pb, and GF, and two variables defined as "weak": POC and Zn. $PM_{2.5}$ mass concentration was included as the total variable in the initial analysis (without GF as an input variable). The expanded uncertainty (including sampling and analytical uncertainties) was calculated for all input data and ranged from 10 to 40% for values above the detection limit. For values below the detection limit, the uncertainty was set equal to 5/6 of the detection limit. Missing values in the datasets were substituted by the median of the respective time series, calculated separately for each season. The uncertainty for these values was set equal to four times the concentration value. An extra modeling uncertainty of 5% was added to account for modeling errors. The model was run for different numbers of sources, ranging from 4 to 10 sources. The best solution was identified through the use of key performance indicators, including the lowest Q values, distribution of scaled residuals, and fit of measured PM concentrations, as well as by assessing the physical meaning of the obtained source profiles and contributions. The final solutions were based on 100 runs, while the uncertainty of the solutions was assessed by implementing the bootstrapping and displacement error estimation tools provided by EPA PMF 5.0 [41].

3. Results

3.1. $PM_{2.5}$ Chemical Composition and Hygroscopicity

The mean annual $PM_{2.5}$ concentration was measured equal to 10.5 ± 4.0 μg/m^3, below the EU annual limit value of 25 μg/m^3. Carbonaceous particles accounted for a large fraction of the PM mass. OC and EC were measured on average equal to 2.1 ± 1.0 μg/m^3 and 0.4 ± 0.3 μg/m^3, respectively. The average OC-to-EC concentration ratio (OC/EC) was calculated equal to 5.4 ± 2.5, indicative of the urban background character of the site [42]. Assuming an organic mass (OM) to organic carbon (OC) ratio of 1.7 [43], carbonaceous aerosol (OM + EC) was calculated to comprise on average 42 ± 22% of the total $PM_{2.5}$ mass. The most abundant species, following carbonaceous matter, was S, with an average concentration of 0.5 ± 0.5 μg/m^3.

The seasonal variability of all measured concentrations is provided in Table 1. $PM_{2.5}$, as well as many PM components, displayed a significant difference (at $p = 0.01$) between cold and warm period concentrations. Carbonaceous aerosol was higher during the cold period, pointing towards an additional carbon source, related to residential heating. The OC/EC ratio, however, did not display a significant difference between heating (cold) and non-heating (warm) periods, suggesting a limited contribution from residential wood burning. The concentrations of S and crustal components (Al, Si, K, Ca, Ti, and Fe) were higher during the warm period. Sulphate, which is the main form of S in the atmosphere, is formed photochemically, and thus is expected to be present in higher concentrations during the warm months, when solar radiation is more intense. The high levels of crustal components during this season are also expected as a result of the dry conditions that promote soil dust resuspension.

Table 1. Seasonal variability of $PM_{2.5}$ and measured chemical components' concentrations. $PM_{2.5}$ concentrations are provided in μg/m^3 and all components' concentrations in ng/m^3. Species that display statistically significant difference in concentrations between warm and cold periods (at $p = 0.01$) are marked in bold.

	Cold Period			Warm Period		
	Median	Range		Median	Range	
$PM_{2.5}$	9.4	2.9	18.8	10.5	4.8	24.7
OC	2280	436	6582	1762	737	4067
EC	469	99	1725	327	BDL [1]	1096
OC/EC	4.5	2.0	12.5	4.6	1.9	17.0

Table 1. *Cont.*

	Cold Period			Warm Period		
	Median	Range		Median	Range	
Na	36.4	BDL	594	31.1	BDL	369
Mg	3.6	BDL	319	2.2	BDL	252
Al	4.8	BDL	517	34.7	BDL	567
Si	5.4	BDL	1295	76.1	BDL	1238
S	109	12.7	1595	526	23.2	1812
Cl	1.2	BDL	638	0.8	BDL	53.5
K	16.4	3.3	342	78.5	1.6	345
Ca	11.8	0.7	431	82.5	2.6	437
Ti	0.8	BDL	40.5	3.3	BDL	39.2
V	0.5	BDL	6.9	0.6	BDL	7.2
Cr	0.4	BDL	4.1	0.3	BDL	3.5
Mn	1.0	BDL	5.4	1.0	BDL	7.2
Fe	10.5	1.0	369	63.3	3.0	361
Ni	0.8	BDL	12.4	0.6	BDL	11.2
Cu	0.6	BDL	15.8	0.6	BDL	13.4
Zn	2.8	BDL	36.0	3.5	BDL	49.5
Br	1.1	BDL	7.6	1.1	BDL	6.3
Ba	13.5	BDL	13.5	8.1	BDL	18.4
Pb	1.9	BDL	17.6	1.4	BDL	18.4

[1] Value below the detection limit.

The median value of the hygroscopic growth factor, at 90% relative humidity, for particles with a dry diameter at 250 nm, was measured equal to 1.21 ± 0.07. The cold and warm period values were 1.22 ± 0.06 and 1.18 ± 0.07, respectively.

3.2. *$PM_{2.5}$ and Hygroscopicity Source Apportionment*

The assessment of the different PMF solutions, in terms of the goodness of fit of input data and physical meaning of the obtained factors, led to the identification of the six-factor solution as the one that best described the input data, for both cases of model application (including and without the hygroscopicity as an input parameter). The source profiles obtained for the full dataset (including the hygroscopic growth factor) are presented in Figure 2. The apportionment of different species to each factor/source is presented in Figure 3. The initial PMF analysis, without hygroscopicity data, provided almost identical source profiles to the ones displayed in Figure 3, in terms of the relative mass contribution of the different chemical components to the profiles. The six major $PM_{2.5}$ sources identified were as follows: vehicle exhaust, vehicle non-exhaust, heavy oil combustion, mixed source related to secondary aerosol formation and biomass burning, mineral dust, and aged sea salt.

The vehicle exhaust profile accounted for most of the carbonaceous species' mass (60% of EC and 75–86% of OC1–OC3 fractions). OC1–OC3 have been associated with combustion sources, including traffic, as emissions from these sources are rich in semi-volatile organic compounds [44]. The vehicle exhaust profile displayed high loadings of EC and OC1–OC4, while POC's contribution was negligible, in agreement with diesel and gasoline vehicle emission profiles reported in the literature [45,46]. The OC/EC ratio in the profile was calculated equal to 4.6, which suggests a mixture of fresh and aged exhaust emissions, in agreement with the urban background character of the site. Vehicle exhaust emissions contributed 16% (1.5 μg/m^3) during the cold period and 10% (1.1 μg/m^3) during the warm period.

The vehicle non-exhaust profile accounted for most of Cu (88%) and Pb (50%) mass. Cu and Pb are emitted during brake wear, while Pb has been also linked to asphalt wear [47]. Non-exhaust emissions' contribution to $PM_{2.5}$ was low (2% during the cold period and 3% during the warm period), as expected, as they are related to coarse rather than fine particles.

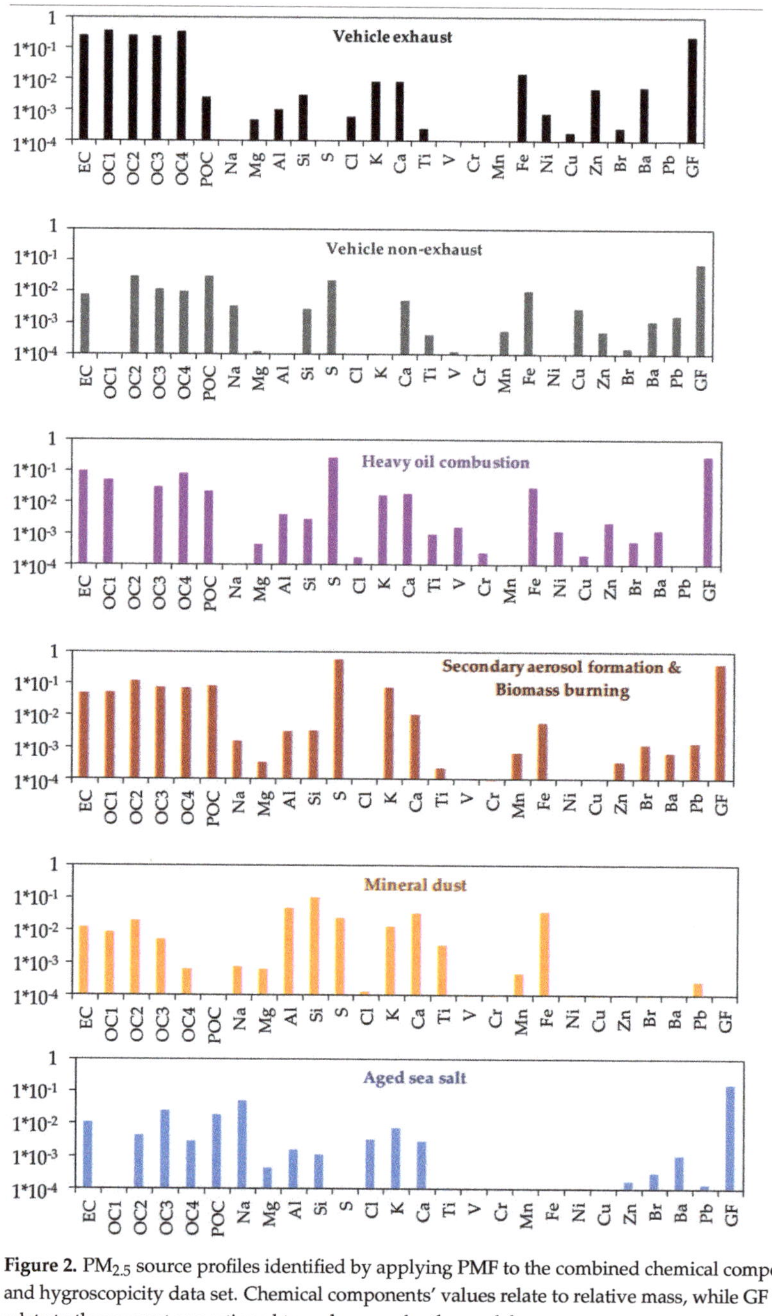

Figure 2. $PM_{2.5}$ source profiles identified by applying PMF to the combined chemical composition and hygroscopicity data set. Chemical components' values relate to relative mass, while GF values relate to the amount apportioned to each source by the model.

The heavy oil combustion profile was identified by V and Ni. Most of V mass (89%) and a large fraction of Ni mass (58%) were associated with this source. The profile displayed high loadings of carbonaceous species and S, as well as Fe, Ca, and K, in accordance with heavy oil combustion profiles found in the literature [48,49]. Heavy oil combustion emissions were associated with less volatile OC fractions in comparison with vehicle exhaust [45]. This source contributed 23% (2.2 μg/m^3) during the cold period and 29% (3.4 μg/m^3) during the warm period.

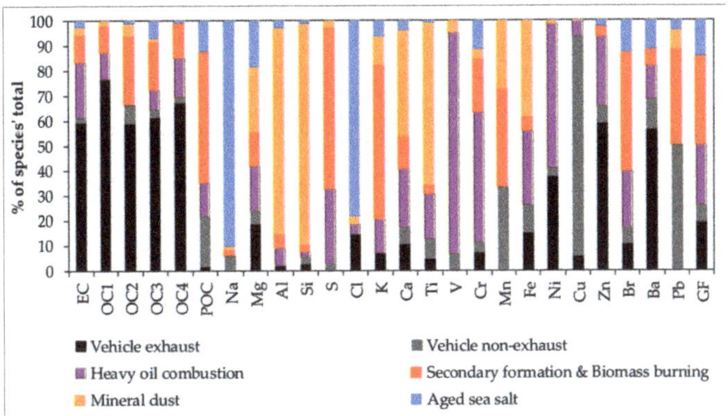

Figure 3. Relative amount of species (%) apportioned to each factor/source, obtained by applying PMF to the combined chemical composition and hygroscopicity dataset.

The fourth source identified was a mixed source of secondary aerosol formation and biomass burning. The source profile was characterized by high loadings of S and OC and accounted for 65% of total S mass, 53% of POC mass, and 62% of K mass. POC has been linked to both secondary organic aerosol formation and biomass burning emissions [44,50]; on the other hand, the biomass burning profiles reported by Liu et al. [51] demonstrate large amounts of OC1–OC4 fractions. The lack of key secondary ionic species (i.e., nitrate and ammonium) in the $PM_{2.5}$ chemical speciation database did not allow for the identification of a separate source related to secondary aerosol formation. The mixed secondary aerosol formation and biomass burning source did not display significant seasonal variability (Figure S1, Supplementary Materials), contributing 46% to total $PM_{2.5}$ concentrations (4.5 μg/m^3) during the cold period and 41% (4.7 μg/m^3) during the warm period.

The mineral dust profile was identified by the high loadings of crustal elements (Al, Si, K, Ca, Ti, and Fe). This profile accounted for more than 80% of the total mass of Al and Si and 65% of Ti. Mineral dust contributed 3% (0.3 μg/m^3) during the cold period and 8% (0.9 μg/m^3) during the warm period.

The aged sea salt profile was identified by the high loading of Na, while it accounted for most of Na and Cl total mass (90% and 79%, respectively). The Cl/Na mass ratio in the profile was much lower than the ratio reported for seawater, suggesting Cl depletion due to transformation of fresh sea salt aerosol through mixing with urban gaseous emissions. Aged sea salt profiles, containing high loadings of nitrate, have been previously reported for the monitoring site [43,52]. Aged sea salt contributed 5% (0.4 μg/m^3) during the cold period and 2% (0.2 μg/m^3) during the warm period.

According to PMF analysis, the aerosol hygroscopic growth factor (GF) was mainly associated with the mixed source of secondary aerosol formation and biomass burning (36%) and heavy oil combustion (24%). To a lesser extent, GF was also associated with vehicle exhaust (19%), aged sea salt (14%), and vehicle non-exhaust (6%). Mineral dust was found to be hydrophobic.

PMF analysis performed only on the chemical composition database, including $PM_{2.5}$ as a total variable, allowed for the quantification of source contributions. The average source contributions over the whole year are presented in Figure 4. The mixed source related to secondary aerosol formation and biomass burning was found to be the main contributor to $PM_{2.5}$ levels (44%), followed by heavy oil combustion (26%) and vehicular traffic exhaust and non-exhaust emissions (15%). Contributions from the two natural sources (mineral dust and aged sea salt) were much lower, owing to the coarse nature of these particles.

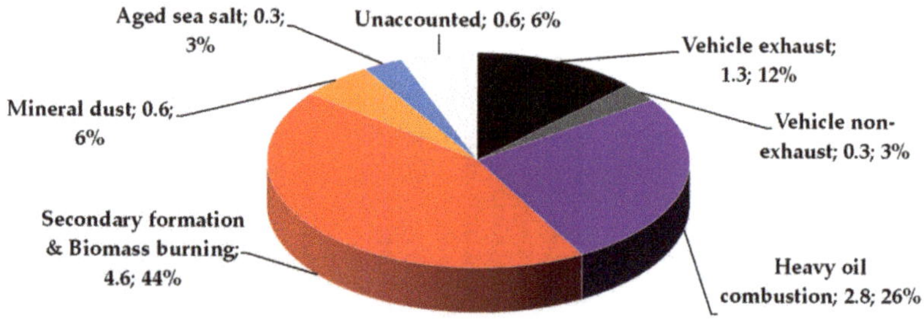

Figure 4. Average contribution (in μg/m^3 and as % of total $PM_{2.5}$ concentration) of the different sources obtained by PMF to the $PM_{2.5}$ concentration levels measured during August 2016–July 2017.

4. Discussion

The measured concentration levels of $PM_{2.5}$ and chemical constituents were generally in agreement with previous studies [43,52,53]. Over the last decade, $PM_{2.5}$ concentrations at the site were relatively low in comparison with the EU annual limit value of 25 μg/m^3. Nevertheless, the levels are higher than the WHO guidelines, which recommend an annual limit value of 5 μg/m^3 and a 24 h limit value of 15 μg/m^3. Carbonaceous aerosol comprised a significant fraction of total $PM_{2.5}$, but a modest decrease in the total carbon (TC) concentrations was observed, with TC accounting for 24% of $PM_{2.5}$, in comparison with 27–30% during 2011–2013. This decrease was related to OC rather than EC, leading to a significant reduction in the observed OC/EC ratio (5.4 on average in this study, as opposed to 7.6–10.5 in previous years). K concentration levels were also lower in this study and the warm period levels were higher than those observed during the cold period. These results suggest that residential biomass burning was not so prominent during 2016–2017, in comparison with previous years.

The $PM_{2.5}$ chemical composition data collected during a full year (August 2016–July 2017) were combined with concurrent hygroscopicity data in order to apply PMF for source apportionment. Source apportionment of $PM_{2.5}$ concentrations resulted in six major emission sources, including four anthropogenic sources (vehicular exhaust and non-exhaust, heavy oil combustion, and a mixed source of secondary aerosol formation and biomass burning) and two natural sources (mineral dust and aged sea salt). The contribution of the natural sources to the observed $PM_{2.5}$ levels was limited (less than 10%); fine particles are known to originate mainly from anthropogenic sources.

The contribution from vehicular traffic (for both exhaust and non-exhaust emissions) was found to be slightly lower than in 2013–2014 (1.6 μg/m^3 in comparison with 2.1–2.3 μg/m^3). The contribution of heavy oil combustion, on the contrary, was found to be elevated (2.8 μg/m^3 in the present study, versus 2.0 μg/m^3 in 2014 and 0.8 μg/m^3 in 2013). This source is considered to be related to shipping emissions, as well as industrial emissions from oil refineries located in the west of the Athens metropolitan area [54]. The increase in the heavy oil combustion contribution over the last years may be related to enhanced economic activities due to the gradual recovery of the Greek economy following the financial crisis.

The fourth anthropogenic source found in the present study was a mixed source of secondary aerosol formation and biomass burning. Previous studies have found separate sources related to secondary sulphates and nitrates and to biomass burning. The use of ionic species in some of these studies allowed for a better characterization of the impact of secondary aerosol formation on PM concentration levels [43,52]. In the present study, a secondary aerosol formation source with high loadings of S and organics was identified and may be associated with regional pollution [52]; however, it was not possible to identify locally produced secondary aerosol, which is usually traced by high loadings of nitrate [55].

Regarding the impact from biomass burning, as already discussed, the low OC/EC ratios measured during the cold period (average and median values equal to 5.0 and 4.5, respectively) and the low K levels suggest that biomass burning emissions during the study period were mainly related to the transport of wildfire smoke plumes, rather than local biomass burning for residential heating. According to the JRC Annual Fire Reports, Greece experienced increased levels of fire activity during the 2016 and 2017 fire seasons, which may have resulted in frequent transport events of aged biomass burning plumes, mixing with the regional pollution reaching the measurement site. The contribution of the mixed source (secondary aerosol formation and biomass burning) did not display significant temporal variability; specific peak contributions, though, may be associated with known wildfire events, such as the intense wildfires in Evia in August 2016 (at a distance of less than 100 km from the measurement site) (Figure S1, Supplementary Materials).

The inclusion of the hygroscopic growth factor (GF) in PMF analysis provided insight into the source apportionment of aerosol hygroscopicity, i.e., the contribution of the different particle sources to the total hygroscopity of ambient aerosol. The mixed source related to secondary aerosol formation and biomass burning was found to be the most hydrophilic source, in agreement with previous source apportionment studies [31,32]. This source displayed high contributions by S and organics, as well as EC and K in lower loadings. Secondary inorganic aerosol, such as sulphate, is known to contribute significantly to atmospheric aerosol hygroscopicity [56]. In addition, highly oxidized secondary organic species have been found to play a major role in organic aerosol hygroscopicity [15,57]. On the other hand, the hygroscopic properties of biomass burning smoke plumes reported in the literature vary significantly, depending on biofuel characteristics, combustion conditions, plume's age (freshly emitted or aged plumes), and the presence of inorganic constituents [28]. Mochida and Kawamura [58] demonstrated that biomass burning emissions contain oxygenated organics that may significantly increase aerosol hygroscopicity. In addition, atmospheric ageing of biomass burning plumes is expected to lead to enhanced hygroscopic properties [9].

The hygroscopic growth rate was also associated with heavy oil combustion (24%) and vehicular exhaust emissions (19%). Li et al. have also found a small part of aerosol hygroscopicity associated with vehicular traffic and ship emissions [31]. The nearly hydrophobic particles emitted from fossil fuel combustion may be transformed into less-hygroscopic particles during atmospheric ageing processes, such as through condensation of secondary organic aerosols formed by photo-oxidation of volatile organic compounds emitted with exhaust gases [59,60]. The high S loading in the heavy oil combustion profile further promotes hygroscopicity.

Aged sea salt was also associated with hygroscopic growth (14%). Sea spray particles are inherently hygroscopic. Zieger et al. have demonstrated that the hygroscopic growth factor of sea salt particles is significantly lower than NaCl [61]. The presence of organics, also evident in the aged sea salt profile obtained by PMF, may further decrease the hygroscopicity of the ambient sea spray particles.

Vehicle non-exhaust emissions contributed very little to aerosol hygroscopicity, while mineral dust was the only source with no contribution to the hydroscopic GF, in agreement with Li et al. [31]. Given that major crustal components, such as Al- and Si-based compounds, are non-hygroscopic, mineral dust is generally considered insoluble and may supress aerosol hygroscopicity [62,63].

The application of PMF analysis on combined chemical composition and hygroscopicity data may enhance our understanding of the hygroscopic properties of ambient aerosol in different types of environments, by identifying the various kinds of hygroscopic particles emitted by anthropogenic and natural sources or formed during atmospheric ageing. In addition, the inclusion in PMF analysis of OC carbon fractions (as provided by TOT method) may provide further insight with respect to the role of organics in aerosol hygroscopicity. The results of this study highlight the influence of the different source emissions and atmospheric ageing on the chemical composition and hygroscopic properties of atmospheric aerosol. Good knowledge of the hygroscopic properties of ambient aerosol is important for air quality and visibility management, as well as climate research. Aerosol hygroscopicity has also been identified as a key parameter in terms of exposure to airborne particles and related health impacts. Taking into account that a large fraction of urban populations resides in suburban and urban background areas, understanding how the transformed emissions reaching the receptor site affect local aerosol hygroscopicity is crucial when developing mitigation measures for air quality and the protection of public health.

Supplementary Materials: The following supporting information can be downloaded at: https://www.mdpi.com/article/10.3390/atmos13101685/s1, Figure S1: Temporal variability of source contributions, obtained through the application of PMF model on $PM_{2.5}$ chemical speciation data.

Author Contributions: E.D. and K.E. contributed to conceptualization and methodology; C.S., M.I.G., P.F., P.P., S.P., V.V. and E.D. contributed to data curation; P.F., P.P. and E.D. contributed to formal analysis, investigation, writing—original draft preparation, and visualization; all contributed to writing—review and editing; E.D. contributed to supervision; E.D., P.F., P.P. and M.I.G. contributed to funding acquisition. All authors have read and agreed to the published version of the manuscript.

Funding: This research is co-financed by Greece and the European Union (European Social Fund—ESF) through the Operational Programme «Human Resources Development, Education, and Lifelong Learning 2014–2020» in the context of the project "Study of the impact of emission sources and environmental conditions on the atmospheric aerosol mixture in the Athens Basin, with emphasis on particle hygroscopicity and related human health effects" (MIS 5047792).

Data Availability Statement: Data will be available by request to the authors.

Conflicts of Interest: The authors declare no conflict of interest. The funders had no role in the design of the study; in the collection, analyses, or interpretation of data; in the writing of the manuscript; or in the decision to publish the results.

References

1. Myhre, G.; Shindell, D.; Breon, F.-M.; Collins, W.; Fuglestvedt, J.; Huang, J.; Koch, D.; Lamarque, J.-F.; Lee, D.S.; Mendoza, B.; et al. Anthropogenic and natural radiative forcing. In *Climate Change 2013: The Physical Science Basis, Contribution of Working Group I to the Fifth Assessment Report of the Intergovernmental Panel on Climate Change*; Stocker, T.F., Qin, D., Plattner, G.-K., Tignor, M.M.B., Allen, S.K., Boschung, J., Nauels, A., Xia, Y., Bex, V., Midgley, P.M., Eds.; Cambridge University Press: New York, NY, USA, 2013; pp. 659–740.
2. Beelen, R.; Hoek, G.; Raaschou-Nielsen, O.; Stafoggia, M.; Jovanovic Andersen, J.; Weinmayr, G.; Hoffmann, B.; Wolf, K.; Samoli, E.; Fischer, P.H.; et al. Natural-cause mortality and long-term exposure to particle components: An analysis of 19 European cohorts within the multi-center ESCAPE project. *Environ. Health Perspect.* **2015**, *123*, 525–533. [CrossRef]
3. Gasparini, R.; Li, R.; Collins, D.R. Integration of size distributions and size-resolved hygroscopicity measured during the Houston Supersite for compositional categorization of the aerosol. *Atmos. Environ.* **2004**, *38*, 3285–3303. [CrossRef]
4. Kerminen, V.-M. The effects of particle chemical character and atmospheric processes on particle hygroscopic properties. *J. Aerosol Sci.* **1997**, *28*, 121–132. [CrossRef]
5. Lai, L.W. Poor Visibility in Winter Due to Synergistic Effect Related to Fine Particulate Matter and Relative Humidity in the Taipei Metropolis, Taiwan. *Atmosphere* **2022**, *13*, 270. [CrossRef]
6. Massoli, P.; Bates, T.S.; Quinn, P.K.; Lack, D.A.; Baynard, T.; Lerner, B.M.; Tucker, S.C.; Brioude, J.; Stohl, A.; Williams, E.J. Aerosol optical and hygroscopic properties during TexAQS-GoMACCS 2006 and their impact on aerosol direct radiative forcing. *J. Geophys. Res. Atmos.* **2009**, *114*, D00F07. [CrossRef]
7. Malm, W.C.; Day, D.E. Estimates of aerosol species scattering characteristics as a function of relative humidity. *Atmos. Environ.* **2001**, *35*, 2845–2860. [CrossRef]
8. Väisänen, O.; Ruuskanen, A.; Ylisirniö, A.; Miettinen, P.; Portin, H.; Hao, L.; Leskinen, A.; Komppula, M.; Romakkaniemi, S.; Lehtinen, K.E.J.; et al. In-cloud measurements highlight the role of aerosol hygroscopicity in cloud droplet formation. *Atmos. Chem. Phys.* **2016**, *16*, 10385–10398. [CrossRef]
9. Vu, T.V.; Delgado-Saborit, J.M.; Harrison, R.M. A review of hygroscopic growth factors of submicron aerosols from different sources and its implication for calculation of lung deposition efficiency of ambient aerosols. *Air Qual. Atmos. Health* **2015**, *8*, 429–440. [CrossRef]
10. Broday, D.M.; Georgopoulos, P.G. Growth and deposition of hygroscopic particulate matter in the human lungs. *Aerosol Sci. Technol.* **2001**, *34*, 144–159. [CrossRef]
11. Krieger, U.K.; Marcolli, C.; Reid, J.P. Exploring the complexity of aerosol particle properties and processes using single particle techniques. *Chem. Soc. Rev.* **2012**, *41*, 6631–6662.

12. Choi, M.Y.; Chan, C.K. The effects of organic species on the hygroscopic behaviors of inorganic aerosols. *Environ. Sci. Technol.* **2002**, *36*, 2422–2428. [CrossRef] [PubMed]
13. Simoneit, B.R.T.; Rogge, W.F.; Mazurek, M.A.; Standley, L.J.; Cass, G.R. Lignan pyrolysis products, lignans, and renin acids as specified tracers of plant classes in emissions from biomass combustion. *Environ. Sci. Technol.* **1993**, *27*, 2533–2541. [CrossRef]
14. Jimenez, J.L.; Canagaratna, M.R.; Donahue, N.M.; Prevot, A.S.H.; Zhang, Q.; Kroll, J.H.; DeCarlo, P.F.; Allan, J.D.; Coe, H.; Ng, N.L.; et al. Evolution of organic aerosols in the atmosphere. *Science* **2009**, *326*, 1525–1529. [CrossRef] [PubMed]
15. Kuang, Y.; Huang, S.; Xue, B.; Luo, B.; Song, Q.; Chen, W.; Hu, W.; Li, W.; Zhao, P.; Cai, M.; et al. Contrasting effects of secondary organic aerosol formations on organic aerosol hygroscopicity. *Atmos. Chem. Phys.* **2021**, *21*, 10375–10391. [CrossRef]
16. Kanakidou, M.; Seinfeld, J.H.; Pandis, S.N.; Barnes, I.; Dentener, F.J.; Facchini, M.C.; Van Dingenen, R.; Ervens, B.; Nenes, A.; Nielsen, C.J.; et al. Organic aerosol and global climate modelling: A review. *Atmos. Chem. Phys.* **2005**, *5*, 1053–1123. [CrossRef]
17. Kong, L.; Hu, M.; Tan, Q.; Feng, M.; Qu, Y.; An, J.; Zhang, Y.; Liu, X.; Cheng, N. Aerosol optical properties under different pollution levels in the Pearl River Delta (PRD) region of China. *J. Environ. Sci.* **2020**, *87*, 49–59. [CrossRef]
18. Mandariya, A.K.; Tripathi, S.N.; Gupta, T.; Mishra, G. Wintertime hygroscopic growth factors (HGFs) of accumulation mode particles and their linkage to chemical composition in a heavily polluted urban atmosphere of Kanpur at the Centre of IGP, India: Impact of ambient relative humidity. *Sci. Total Environ.* **2020**, *704*, 135363. [CrossRef]
19. Wang, X.; Ye, X.; Chen, H.; Chen, J.; Yang, X.; Gross, D.S. Online hygroscopicity and chemical measurement of urban aerosol in Shanghai, China. *Atmos. Environ.* **2014**, *95*, 318–326. [CrossRef]
20. Liu, X.; Zhang, Y.; Jung, J.; Gu, J.; Li, Y.; Guo, S.; Chang, S.-Y.; Yue, D.; Lin, P.; Kim, Y.J.; et al. Research on the hygroscopic properties of aerosols by measurement and modeling during CAREBeijing-2006. *J. Geophys. Res.* **2009**, *114*, D00G16. [CrossRef]
21. Carrico, C.M.; Kreidenweis, S.M.; Malm, W.C.; Day, D.E.; Lee, T.; Carrillo, J.; McMeeking, G.R.; Collett, J.L., Jr. Hygroscopic growth behavior of a carbon-dominated aerosol in Yosemite National Park. *Atmos. Environ.* **2005**, *39*, 1393–1404. [CrossRef]
22. Stock, M.; Cheng, Y.F.; Birmili, W.; Massling, A.; Wehner, B.; Muller, T.; Leinert, S.; Kalivitis, N.; Mihalopoulos, N.; Wiedensohler, A. Hygroscopic properties of atmospheric aerosol particles over the Eastern Mediterranean: Implications for regional direct radiative forcing under clean and polluted conditions. *Atmos. Chem. Phys.* **2011**, *11*, 4251–4271. [CrossRef]
23. Gysel, M.; Crosier, J.; Topping, D.O.; Whitehead, J.D.; Bower, K.N.; Cubison, M.J.; Williams, P.I.; Flynn, M.J.; McFiggans, G.B.; Coe, H. Closure study between chemical composition and hygroscopic growth of aerosol particles during TORCH2. *Atmos. Chem. Phys.* **2007**, *7*, 6131–6144. [CrossRef]
24. Kreidenweis, S.M.; Remer, L.A.; Bruintjes, R.; Dubovik, O. Smoke aerosol from biomass burning in Mexico: Hygroscopic smoke optical model. *J. Geophys. Res.* **2001**, *106*, 4831–4844.
25. Grose, M.; Sakurai, H.; Savstrom, J.; Stolzenburg, M.R.; Watts, W.F.; Morgan, C.G.; Murray, I.P.; Twigg, M.V.; Kittelson, D.B.; McMurry, P.H. Chemical and physical properties of ultrafine diesel exhaust particles sampled downstream of a catalytic trap. *Environ. Sci. Technol.* **2006**, *40*, 5502–5507. [CrossRef] [PubMed]
26. Weingartner, E.; Burtscher, H.; Baltensperger, U. Hygroscopic properties of carbon and diesel soot particles. *Atmos. Environ.* **1997**, *31*, 2311–2327. [CrossRef]
27. Li, C.; Hu, Y.; Chen, J.; Ma, Z.; Ye, X.; Yang, X.; Wang, L.; Wang, X.; Mellouki, A. Physiochemical properties of carbonaceous aerosol from agricultural residue burning: Density, volatility, and hygroscopicity. *Atmos. Environ.* **2016**, *140*, 94–105. [CrossRef]
28. Carrico, C.M.; Petters, M.D.; Kreidenweis, S.M.; Sullivan, A.P.; McMeeking, G.R.; Levin, E.J.T.; Engling, G.; Malm, W.C.; Collett, J.L., Jr. Water uptake and chemical composition of fresh aerosols generated in open burning of biomass. *Atmos. Chem. Phys.* **2010**, *10*, 5165–5178. [CrossRef]
29. Lewis, K.A.; Arnott, W.P.; Moosmuller, H.; Chakrabarty, R.K.; Carrico, C.M.; Kreidenweis, S.M.; Day, D.E.; Malm, W.C.; Laskin, A.; Jimenez, J.L.; et al. Reduction in biomass burning aerosol light absorption upon humidification: Roles of inorganically-induced hygroscopicity, particle collapse, and photoacoustic heat and mass transfer. *Atmos. Chem. Phys.* **2009**, *9*, 8949–8966. [CrossRef]
30. Rissler, J.; Pagels, J.; Swietlicki, E.; Wierzbicka, A.; Strand, M.; Lillieblad, L.; Sanati, M.; Bohgard, M. Hygroscopic behaviour of aerosol particles emitted from biomass fired grate boilers. *Aerosol. Sci. Technol.* **2005**, *39*, 919–930. [CrossRef]
31. Li, J.; Zhang, Z.; Wu, Y.; Tao, J.; Xia, Y.; Wang, C.; Zhang, R. Effects of chemical compositions in fine particles and their identified sources on hygroscopic growth factor during dry season in urban Guangzhou of South China. *Sci. Total Environ.* **2021**, *801*, 149749. [CrossRef]
32. Vu, T.V.; Shi, Z.; Harrison, R.M. Estimation of hygroscopic growth properties of source-related sub-micrometre particle types in a mixed urban aerosol. *NPJ Clim. Atmos. Sci.* **2021**, *4*, 21. [CrossRef]
33. Bezantakos, S.; Barmpounis, K.; Giamarelou, M.; Bossioli, E.; Tombrou, M.; Mihalopoulos, N.; Eleftheriadis, K.; Kalogiros, J.; Allan, J.D.; Bacak, A.; et al. Chemical composition and hygroscopic properties of aerosol particles over the Aegean Sea. *Atmos. Chem. Phys.* **2013**, *13*, 11595–11608. [CrossRef]
34. Petaja, T.; Kerminen, V.-M.; Dal Maso, M.; Junninen, H.; Koponen, I.K.; Hussein, T.; Aalto, P.P.; Andronopoulos, S.; Robin, D.; Hameri, K.; et al. Sub-micron atmospheric aerosols in the surroundins of Marseille and Athens: Physical characterization and new particle formation. *Atmos. Chem. Phys.* **2007**, *7*, 2705–2720. [CrossRef]
35. Spitieri, C.; Gini, M.; Gysel-Beer, M.; Eleftheriadis, K. Annual cycle of hygroscopic properties and mixing state of the suburban aerosol in Athens, Greece. *Atmos. Chem. Phys. Discuss.* **2022**. *preprint*. [CrossRef]
36. Eleftheriadis, K.; Gini, M.I.; Diapouli, E.; Vratolis, S.; Vasilatou, V.; Fetfatzis, P.; Manousakas, M.I. Aerosol microphysics and chemistry reveal the COVID19 lockdown impact on urban air quality. *Sci. Rep.* **2021**, *11*, 14477. [CrossRef] [PubMed]

37. Manousakas, M.; Diapouli, E.; Papaefthymiou, H.; Kantarelou, V.; Zarkadas, C.; Kalogridis, A.-C.; Karydas, A.-G.; Eleftheriadis, K. XRF characterization and source apportionment of PM_{10} samples collected in a coastal city. *X Ray Spectrom.* **2018**, *47*, 190–200. [CrossRef]
38. Ma, J.; Li, X.; Gu, P.; Dallmann, T.R.; Presto, A.A.; Donahue, N.M. Estimating ambient particulate organic carbon concentrations and partitioning using thermal optical measurements and the volatility basis set. *Aerosol Sci. Technol.* **2016**, *50*, 638–651. [CrossRef]
39. Panteliadis, P.; Hafkenscheid, T.; Cary, B.; Diapouli, E.; Fischer, A.; Favez, O.; Quincey, P.; Viana, M.; Hitzenberger, R.; Vecchi, R.; et al. ECOC comparison exercise with identical thermal protocols after temperature offset correction-Instrument diagnostics by in-depth evaluation of operational parameters. *Atmos. Meas. Tech.* **2015**, *8*, 779–792. [CrossRef]
40. Belis, C.A.; Pernigotti, D.; Pirovano, G.; Favez, O.; Jaffrezo, J.L.; Kuenen, J.; van Der Gon, H.D.; Reizer, M.; Riffault, V.; Alleman, L.Y.; et al. Evaluation of receptor and chemical transport models for PM10 source apportionment. *Atmos. Environ. X* **2020**, *5*, 100053. [CrossRef]
41. Manousakas, M.; Papaefthymiou, H.; Diapouli, E.; Migliori, A.; Karydas, A.G.; Bogdanovic-Radovic, I.; Eleftheriadis, K. Assessment of PM2.5 sources and their corresponding level of uncertainty in a coastal urban area using EPA PMF 5.0 enhanced diagnostics. *Sci. Total Environ.* **2017**, *574*, 155–164. [CrossRef]
42. Pio, C.; Cerqueira, M.; Harrison, R.M.; Nunes, T.; Mirante, F.; Alves, C.; Oliveira, C.; Sanchez de la Campa, A.; Artíñano, B.; Matos, M. OC/EC ratio observations in Europe: Re-thinking the approach for apportionment between primary and secondary organic carbon. *Atmos. Environ.* **2011**, *45*, 6121–6132. [CrossRef]
43. Amato, F.; Alastuey, A.; Karanasiou, A.; Lucarelli, F.; Nava, S.; Calzolai, G.; Severi, M.; Becagli, S.; Gianelle, V.L.; Colombi, C.; et al. AIRUSE-LIFE+: A harmonized PM speciation and source apportionment in five southern European cities. *Atmos. Chem. Phys.* **2016**, *16*, 3289–3309. [CrossRef]
44. Li, H.Z.; Dallmann, T.R.; Li, X.; Gu, P.; Presto, A.A. Urban Organic Aerosol Exposure: Spatial Variations in Composition and Source Impacts. *Environ. Sci. Technol.* **2018**, *52*, 415–426. [CrossRef]
45. Kim, E.; Hopke, P.K. Source apportionment of fine particles in Washington, DC, utilizing temperature-resolved carbon fractions. *J. Air Waste Manag. Assoc.* **2004**, *54*, 773–785. [CrossRef] [PubMed]
46. Kim, E.; Hopke, P.K.; Edgerton, E.S. Improving source identification of Atlanta aerosol using temperature resolved carbon fractions in positive matrix factorization. *Atmos. Environ.* **2004**, *38*, 3349–3362. [CrossRef]
47. Jeong, H.; Ryu, J.-S.; Ra, K. Characteristics of potentially toxic elements and multi-isotope signatures (Cu, Zn, Pb) in non-exhaust traffic emission sources. *Environ. Pollut.* **2022**, *292*, 118339. [CrossRef]
48. Bove, M.C.; Brotto, P.; Cassola, F.; Cuccia, E.; Massabò, D.; Mazzino, A.; Piazzalunga, A.; Prati, P. An integrated PM2.5 source apportionment study: Positive Matrix Factorisation vs. the chemical transport model CAMx. *Atmos. Environ.* **2014**, *94*, 274–286. [CrossRef]
49. Pey, J.; Pérez, N.; Cortés, J.; Alastuey, A.; Querol, X. Chemical fingerprint and impact of shipping emissions over a western Mediterranean metropolis: Primary and aged contributions. *Sci. Total Environ.* **2013**, *463–464*, 497–507. [CrossRef]
50. Kim, E.; Hopke, P.K. Improving Source Apportionment of Fine Particles in the Eastern United States Utilizing Temperature-Resolved Carbon Fractions. *J. Air Waste Manag. Assoc.* **2005**, *55*, 1456–1463. [CrossRef]
51. Liu, W.; Wang, Y.; Russell, A.; Edgerton, E.S. Enhanced source identification of southeast aerosols using temperature-resolved carbon fractions and gas phase components. *Atmos. Environ.* **2006**, *40*, S445–S466. [CrossRef]
52. Diapouli, E.; Manousakas, M.; Vratolis, S.; Vasilatou, V.; Maggos, T.; Saraga, D.; Grigoratos, T.; Argyropoulos, G.; Voutsa, D.; Samara, C.; et al. Evolution of air pollution source contributions over one decade, derived by PM10 and PM2.5 source apportionment in two metropolitan urban areas in Greece. *Atmos. Environ.* **2017**, *164*, 416–430. [CrossRef]
53. Almeida, S.M.; Manousakas, M.; Diapouli, E.; Kertesz, Z.; Samek, L.; Hristova, E.; Sega, K.; Padilla Alvarez, R.; Belis, C.A.; The IAEA European Region Study GROUP. Ambient particulate matter source apportionment using receptor modelling in European and Central Asia urban areas. *Environ. Pollut.* **2020**, *266*, 115199. [CrossRef] [PubMed]
54. Karanasiou, A.A.; Siskos, P.A.; Eleftheriadis, K. Assessment of source apportionment by Positive Matrix Factorization analysis on fine and coarse urban aerosol size fractions. *Atmos. Environ.* **2009**, *43*, 3385–3395. [CrossRef]
55. Manousakas, M.; Diapouli, E.; Belis, C.; Vasilatou, V.; Gini, M.; Lucarelli, F.; Querol, X.; Eleftheriadis, K. Quantitative assessment of the variability in chemical profiles from source apportionment analysis of PM10 and PM2.5 at different sites within a large metropolitan area. *Environ. Res.* **2021**, *192*, 110257. [CrossRef]
56. Zieger, P.; Fierz-Schmidhauser, R.; Poulain, L.; Müller, T.; Birmili, W.; Spindler, G.; Wiedensohler, A.; Baltensperger, U.; Weingartner, E. Influence of water uptake on the aerosol particle light scattering coefficients of the central european aerosol. *Tellus Ser. B Chem. Phys. Meteorol.* **2014**, *66*, 22716. [CrossRef]
57. Duplissy, J.; DeCarlo, P.F.; Dommen, J.; Alfarra, M.R.; Metzger, A.; Barmpadimos, I.; Prevot, A.S.H.; Weingartner, E.; Tritscher, T.; Gysel, M.; et al. Relating hygroscopicity and composition of organic aerosol particulate matter. *Atmos. Chem. Phys.* **2011**, *11*, 1155–1165. [CrossRef]
58. Mochida, M.; Kawamura, K. Hygroscopic properties of levoglucosan and related organic compounds characteristic to biomass burning aerosol particles. *J. Geophys. Res.* **2004**, *109*, D21202. [CrossRef]
59. Swietlicki, E.; Hansson, H.-C.; Hämeri, K.; Svenningsson, B.; Massling, A.; Mcfiggans, G.; Mcmurry, P.H.; Petäjä, T.; Tunved, P.; Gysel, M.; et al. Hygroscopic properties of submicrometer atmospheric aerosol particles measured with HTDMA instruments in various environments—A review. *Tellus B* **2008**, *60*, 432–469. [CrossRef]

60. Tritscher, T.; Jurányi, Z.; Martin, M.; Chirico, R.; Gysel, M.; Heringa, M.F.; DeCarlo, P.F.; Sierau, B.; Prévôt, A.S.H.; Weingartner, E.; et al. Changes of hygroscopicity and morphology during ageing of diesel soot. *Environ. Res. Lett.* **2011**, *6*, 034026. [CrossRef]
61. Zieger, P.; Vaisanen, O.; Corbin, J.C.; Partridge, D.G.; Bastelberger, S.; Mousavi-Fard, M.; Rosati, B.; Gysel, M.; Krieger, U.K.; Leck, C.; et al. Revising the hygroscopicity of inorganic sea salt particles. *Nat. Commun.* **2017**, *8*, 15883. [CrossRef]
62. Arub, Z.; Singh, G.; Habib, G.; Raman, R.S. Highly significant impact of mineral dust on aerosol hygroscopicity at New Delhi. *Atmos. Environ.* **2021**, *254*, 118375. [CrossRef]
63. Kelly, J.T.; Chuang, C.C.; Wexler, A.S. Influence of dust composition on cloud droplet formation. *Atmos. Environ.* **2007**, *41*, 2904–2916. [CrossRef]

atmosphere

Article

Scenario Analysis of Air Quality Improvement in Warsaw, Poland, by the End of the Current Decade

Piotr Holnicki [1,*], Andrzej Kałuszko [1] and Zbigniew Nahorski [1,2]

[1] Systems Research Institute, Polish Academy of Sciences, 01-447 Warsaw, Poland
[2] Warsaw School of Information Technology (WIT), 01-447 Warsaw, Poland
* Correspondence: holnicki@ibspan.waw.pl

Abstract: Very low air quality in the Warsaw conurbation, Poland, similarly to the case in many large European cities, poses a serious threat to the residents' health, being a significant source of premature mortality. Many results presented in earlier publications indicated local heating installations and car traffic as the main emission categories responsible for this adverse population exposure, where the dominant polluting compounds are NOx, PM_{10}, $PM_{2.5}$, and BaP. The last two mainly originate from individual household heating installations, both in the city of Warsaw and in its vicinity. To reduce the health risk of air pollution, the city authorities have recently made fundamental decisions, related to the individual housing sector, aimed at the radical decarbonization of all heating installations in Warsaw and its surroundings. On the other hand, the ongoing modernization of the city's car fleet (including individual and public transport), taking into account the restrictive EU emission standards, as well as the quickly growing share of electric and hybrid cars (BEVs and PHEVs), gives a good prospect of a fundamental improvement in air quality in Warsaw conurbation. The main subject of the paper is a quantitative assessment of the air quality improvement in the current decade (by 2030), resulting from the above modernization activities. The final results are expressed as the attributed reduction in population exposure, which was found to be 28–30% with respect to NOx and PM, and the associated health risk, i.e., 204 fewer avoidable deaths with respect to NOx and 607 fewer with respect to $PM_{2.5}$.

Keywords: urban air quality; emission abatement; emission Euro norms; BEV/PHEV cars; population exposure

Citation: Holnicki, P.; Kałuszko, A.; Nahorski, Z. Scenario Analysis of Air Quality Improvement in Warsaw, Poland, by the End of the Current Decade. *Atmosphere* **2022**, *13*, 1613. https://doi.org/10.3390/atmos13101613

Academic Editors: Elena Hristova, Manousos Ioannis Manousakas, Anikó Angyal and Maria Gini

Received: 5 September 2022
Accepted: 27 September 2022
Published: 2 October 2022

Publisher's Note: MDPI stays neutral with regard to jurisdictional claims in published maps and institutional affiliations.

1. Introduction

More than 70% of European citizens live in urban areas [1], where, due to high population densities and economic activities, they are exposed to high levels of air pollution [2–5]. The adverse concentrations often exceed the WHO guidelines, especially the currently established, more restrictive limits [6]. Exposure to particulate matter, ozone, nitrogen oxides, and other pollutants is associated with serious health problems [7], including premature mortality in European cities.

According to the above EEA reports, the most premature deaths related to nitrogen dioxide pollution in Europe were recorded in the main Western Europe conurbations, e.g., in Spain, Italy, France, and Belgium [8,9]. Although in Polish cities, particularly in Warsaw, the NO_2 concentrations are comparatively slightly lower (Warsaw's position is 24th in the NO_2 ranking list of 1000 most polluted European cities [9], with 743 avoidable deaths, estimated according to the current WHO recommendations [6]), the overall air quality in Warsaw is much worse compared to other European capitals, mainly due to extremely high $PM_{2.5}$ concentrations. A similar situation is also observed in most other Polish conurbations, which is the main reason for Poland suffering some of the lowest air quality [10–12] in Europe. This, in part, results from the burning of low-quality fuels in residential heating, as well as due to the country's reliance on coal.

The municipal and housing sectors in Poland mainly use coal fuel [13–15]. Emitters related to the individual heating of apartments and buildings are located in residential areas, and the related pollutions usually occur at a low height above the ground level. As a result, these emissions directly affect concentrations of the pollutants in places where people live and often exceed the air quality standards, especially for $PM_{2.5}$. In 2018, the emissions in Poland from this source category accounted for approximately 47% of total PM_{10} dust emissions and 52% of total $PM_{2.5}$ emissions. In the abovementioned list of 1000 most $PM_{2.5}$ polluted European cities [9], Warsaw's ranking position was 20 (with five other Polish cities having an even higher rank), with 2023 avoidable deaths according to WHO regulations [6]. It was also confirmed by [8,12] that the annual mean concentration of $PM_{2.5}$ particles in Poland was worst in all the EU, where 30 Polish cities were among 100 most polluted European ones.

This situation means that any effective scenario of air quality improvement in Warsaw must simultaneously take into account a reduction in both dominating pollution categories, the road traffic-induced NOx concentrations, as well as $PM_{2.5}$ and BaP (benzo[α]pyrene) pollutions, also emitted mainly by the municipal sector. The effectiveness of possible corrective actions in both cases will be directly dependent on the government's energy and climate policy implemented in the coming years.

The *Fit for 55* package adopted on 14 July 2021 by the European Commission [16] modifies the existing climate and energy legislation to meet the new EU objective of a minimum 55% reduction in greenhouse gas (GHG) emissions by 2030 (as compared with 1990 levels). Directly related to the Commission's *2030 Climate Target Plan* [17–19], it not only explicitly supports the 55% GHG target, but also indicates that 38–40% renewable energy sources (RES) share by 2030 is the minimum necessary to meet this goal.

As stated above, many Polish conurbations are among the most polluted ones in Europe. Recent years have seen grassroots initiatives, as well as very little governmental effort aiming to reduce pollution levels, with minor results. In 2021, the share of coal in the energy mix for the first time dropped below 70%, with a simultaneous 17% share of RES [20]. The current economic and political situation in Europe forces a radical reduction in fossil fuels; however, in the case of Poland, the achievement of the EC strategic goals is rather unrealistic [21,22]. However, an announcement [23] was aimed at bringing the Polish fuel mix closer to EU requirements. At the same time, local authorities of the most polluted conurbations are taking their own initiatives to reduce the level of the most troublesome pollutants, while researchers are looking for the adequate strategies to obtain this goal. For example, the authors of [24–27] identified the main sources responsible for standards violations and presented selected results focused on air pollution mitigation in Polish cities (Cracow and Warsaw). There are also numerous case studies dealing with air quality deterioration in major European cities, caused by both urban transport and other municipal sources. Most of the studies (e.g., [28–30]) focused on assessing the impact of urban transport on air quality in the city, where the authors of [30] additionally utilized the AERMOD modeling system to assess emission scenarios to reduce the high level of population exposure caused by the traffic-related NOx pollution in Trabzon, Turkey region. Furthermore, the authors of [31–33] analyzed hypothetical emission reduction strategies related to PM and NOx pollutions of urban areas.

This study was aimed at a quantitative assessment of the possible improvement of air quality in Warsaw, Poland, resulting from a reduction in pollution emissions mainly caused by road traffic and coal-fueled municipal installations. Projects recently launched by the Warsaw authorities are considered, regarding both residential heating and urban transportation, where the general trends in the transportation sector development and modernization were additionally taken into account. To fully assess environmental benefits coming from implementation of these corrective actions, the emission inventory for the year 2018/2019 (prior to the project initialization) was taken as the baseline dataset.

2. Methods

Modeling of atmospheric pollution dispersion in the city was carried out using the Gaussian modeling system CALPUFF, v.7.0 [34] with the CALMET meteorological preprocessor. The linear structure of the CAPUFF model enables independent analysis of the impact of the municipal and transportation sectors, while utilizing superposition of the results to estimate the final assessment of the decarbonization scenario. This linearity property was also used in the estimation of the final environmental effects (more information on CALPUFF system calculations and the model prediction accuracy and uncertainty can be found in [35–37]). Using this system, spatial maps of the annual average concentrations of the main air pollutants were obtained in order to determine areas where the permissible concentration levels of individual pollutants were exceeded and to identify the sources of emissions responsible for these exceedances.

The study domain considered in the computer simulation includes the Warsaw metropolitan domain (inside the capital administrative boundary), as well as the surrounding vicinity belt, about 30 km wide (Figure 1). The emission field combines a large number of sources that differ in technological parameters, emission characteristics, composition of emitted compounds, and assigned uncertainty [27,38]. In this study, the sources were divided into three basic categories, according to their emission parameters: (a) point sources (4073), (b) line sources of the transport networks (1806 + 4918), and (c) area sources of the municipal sector (1452 + 5819). All emitters for the pointwise sources were kept in one file (the number of sources in brackets). In the area and line sources, the emitters were divided into two subsets: sources located within the city domain and those in the outer belt (the respective components are shown in brackets) to help in assessing the contribution of the surrounding emission field to the total urban pollution. The external inflow of pollutants entering the study domain was also taken into account, calculated offline using the regional-scale CAMx model, with all categories of emission sources located in Poland, outside the analyzed region. The resulting concentration field, including primary and secondary pollutants (e.g., aerosols), was the pollution background to which the CALPUFF forecast was added.

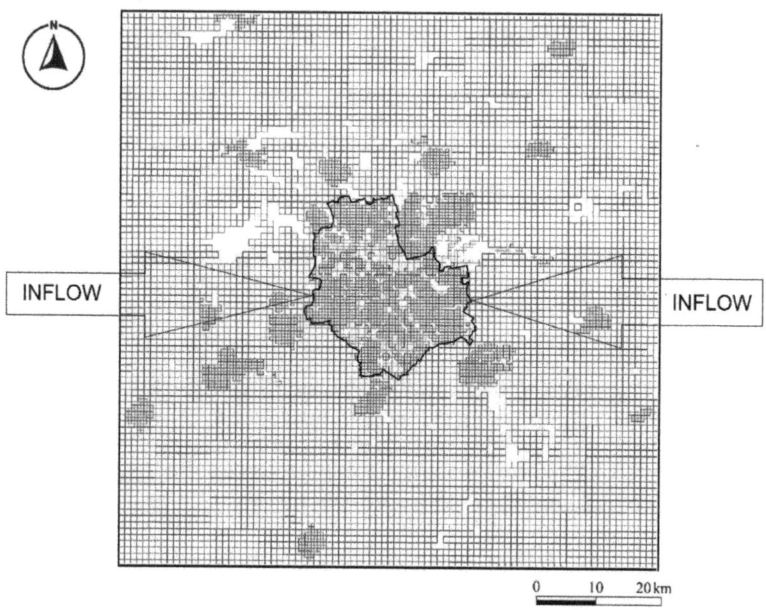

Figure 1. The study domain covering receptor area, inside Warsaw and in its vicinity.

For calculation of the pollutant concentrations, the Warsaw metropolitan area (about 520 km^2 within administrative boundaries) was digitized with a homogeneous grid 0.5 km × 0.5 km. The aggregated resolution of 1 km × 1 km was applied in the surrounding belt, where the satellite cities were also discretized using a fine-resolution grid. Using emission data given in these grids, the resulting concentrations were calculated in 2111 elementary mesh receptors inside Warsaw's administrative area (compare Figure 1). The calculations were performed for emission and meteorological data in 2018 [14,15]. According to the goal of this study, the following pollutants characterizing the urban atmosphere were analyzed: PM_{10}, $PM_{2.5}$, NOx, and BaP. The average annual concentration values of individual pollutants, averaged over the area of a unit grid element, were adopted as the final value for each receptor.

3. Air Pollution in the Baseline Year

Computer simulations provided annual mean concentration maps, showing distributions of the main pollutants that characterize the urban atmospheric environment. They also indicated which pollutants exceeded the WHO limit values and where these violations were the highest. The linear structure of the CALPUFF model allowed indication of the emission categories responsible for standards violation (source apportionment). Moreover, it was possible to quantify the percentage share of an individual emission source category in the total concentration at a given receptor point or in a district. This is very useful for elaborating an abatement strategy to improve the city air quality.

Figure 2 presents the resulting concentration maps of the key pollutants, NOx and $PM_{2.5}$, calculated for the baseline emission dataset in 2018. As compared with the previously obtained outputs [27,38,39], computation results differed in the spatial distribution of some pollutants, particularly PM and BaP, although not particularly in the overall health pollution impacts. This was mainly due to different meteorological conditions but also to some extent to conurbation development. The wind field in Warsaw, similarly to the entire territory of Poland, is usually dominated by western circulation. In 2018, however, the westerly and easterly winds were more balanced (compare Figure S1). Comparison of this and previous spatial distributions also indirectly indicated the high impact of pollutions inflowing from the surrounding belt on pollution concentrations in the city itself.

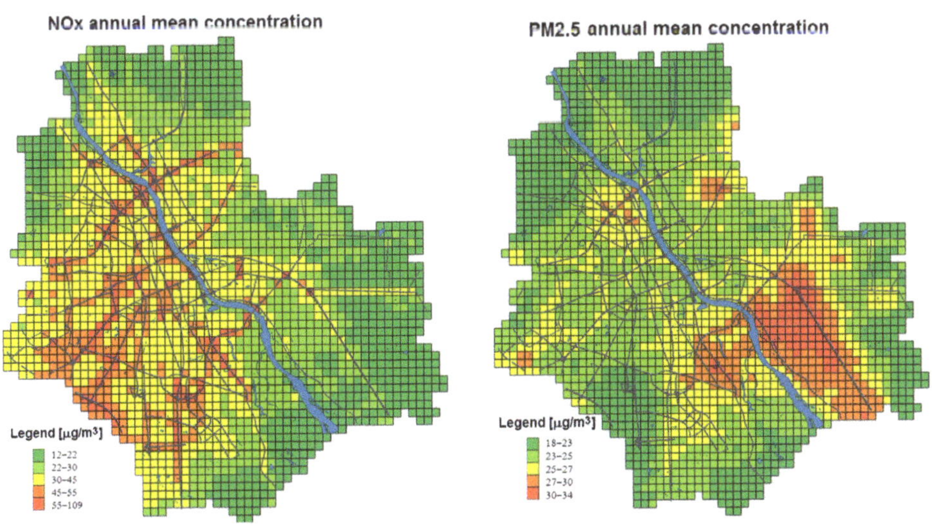

Figure 2. Concentrations for the baseline year dataset: NOx—**left**, $PM_{2.5}$—**right**. The geographical orientation and scales are the same as in Figure 1.

As shown in Figure 2, the air quality standards for the polluting compounds considered were exceeded in a large part of the city area. In particular, according to the current Polish regulations, as well as EU Air Quality Directives [40], the annual mean NOx concentration limit (30 μg/m^3) was exceeded in 50% of the receptor points, while the respective limit value for $PM_{2.5}$ (25 μg/m^3) was exceeded in about 62% of receptors. The calculated PM_{10} and BaP pollution concentrations also exceeded the respective limit values (40 μg/m^3) for PM_{10} in 33% of the receptors, as well as that (1 ng/m^3) for BaP in the entire urban area. This reflects a general situation related to the strong domination of coal in the fuel mix in Poland [4,10,26,38,41]. Furthermore, the modified and much more restrictive WHO air quality guidelines [6] for the pollutants discussed here were exceeded in the entire study domain.

The quoted above shares of receptor points where the limits were exceeded are smaller than those calculated for 2012 emission dataset [27,38], which were equal to 64% for NOx and 75% for $PM_{2.5}$. At the same time, the spatial distributions of pollutants for 2012 and 2018 differed quite significantly from each other. This was mainly seen for NOx, whose concentrations were strongly dispersed in the 2012 case study, with the maximum values in the central districts. On the other hand, in the year 2018, the car-induced pollutions were much less dispersed, but clearly concentrated along the main streets. This was caused, first of all, by strong administrative limits imposed in last few years on the car traffic in the city center. Moreover, the S8 expressway, launched in stages in 2013–2015, currently crosses the northern districts of the city (compare Figure 2) and handles very heavy traffic, including the intensive truck transit. This results in high pollution around the entire street, with the maximum NOx concentrations occurring just along this route. Spatial distributions of PM concentrations (including $PM_{2.5}$) for 2012 and 2018 case studies differed mainly due to different dominating wind directions (as stated above). However, the average BaP concentrations showed a significant increase (about 50%) in 2018 compared to 2012. This increase was, in similar proportions, due to three basic groups of the area sources: Warsaw's local emitters, sources located in the immediate vicinity of Warsaw, and the external inflow.

4. Projected Decarbonization Modernization

4.1. Official Plans and Scenarios Considered

The official document of the Ministry of Climate and Environment *On the state's energy policy until 2040* [23] presents activities aimed at bringing Polish fuel mix closer to the EU requirements, despite the existing limitations. According to this document, the share of coal in the structure of energy consumption in 2030 will be at most 56%, with a simultaneous increase in the share of RES to at least 23% in gross final energy consumption (at least 32% in the electricity sector). The increase in the share of renewable energy sources is expected to be mainly possible thanks to new onshore and offshore wind farms, photovoltaics, and heat pumps in the individual housing installations.

According to the program launched in 2020 by the Warsaw authorities [42], it is planned to substantially modernize municipal coal fueled installations within the next few years and substantially reduce $PM_{2.5}$ and BaP exposures. Currently, the impact of this program will be radically strengthened and extended to the entire Mazovian Voivodeship thanks to the initiative presented in Section 4.2 [42].

In the sequel, two emission abatement scenarios are discussed. They are connected with: (a) particulate matter pollutions (including PM_{10}, $PM_{2.5}$, and BaP), mainly depending on residential sector activity, and (b) NOx pollution where urban road traffic is the dominant contributing sector. However, the municipal emissions also partially contribute to NOx pollution, and the transport sector activity affects particulate matter concentrations.

The three basic scenarios considered in this study are as follows:

A. Baseline share of main emission categories (2018 emission dataset),
B. Reduction in the area emission from municipal sector, due to regulations on the limit on coal fired installations in Warsaw and vicinity, and the national level act on fuel mix decarbonization (horizon 2030)—Section 4.2,

C. On top of B, imposing the restrictive emission standard for the car fleet in Warsaw (to E4 norm at least) and taking into account the forecasted increase in the share of electric cars (horizon 2030)—Section 4.3.

4.2. Municipal Sector Modernization

Particulate matter, PM_{10} and $PM_{2.5}$, including carcinogenic BaP, constitutes a group of hazardous to health pollutants, whose dominant share comes from the area sources of the municipal sector (mainly due to coal combustion), located both in the city and in its vicinity In this case, the transboundary inflow (Figure 1) of pollutants also plays an important role, while the impact of road traffic is generally minor.

Regulations on the *Anti-Smog Resolution* [39] for Warsaw and its wide surroundings approved (26.04.2022) by the authorities of the Mazovian Voivodeship introduced the ban on burning coal in ordinary fireplaces in Warsaw from 2023 and throughout the Mazovian Voivodeship from 2028. The only approved furnace type is the low-emission class 5 boiler [43], which guarantees a radical reduction in particulate matter and organic compounds (BaP) emissions (compare Table S4). Thus, the above regulations mean a significant reduction in PM and BaP emissions from area sources, located both in the city and in its vicinity. The calculations performed for the baseline data show that this category of emission sources was responsible for about 50% of particulate matter pollutants and more than 70% of BaP.

Actually, efficiency of this modernization ranged from 20% to 60% and depended primarily on the parameters of the replaced boiler (Table S4). Since rather low-quality boilers (mostly of class 3) are expected in the case considered, the value 50% was assumed in the calculations as the average emission reduction factor. In the case of an action covering such a large area and a number of replaced sources, a certain number of installations usually cannot be modernized (for various reasons). Thus, it was assumed that the modernization would be implemented in at least 80% of the considered installations, which in turn would mean about a 40% reduction in the baseline emissions for PM_{10} and $PM_{2.5}$.

According to the official documents of the Ministry of Climate and Environment concerning decarbonization policy of the Polish economy [23], by the end of this decade, the share of coal in the country's energy needs is to drop to at least 56% (from the current 70%), with an equivalent increase in the use of renewable sources. This means about a 20% decrease in the volume of pollutants emitted nationwide, which translates into a corresponding reduction in the transboundary inflow.

A similar efficiency of BaP emission reduction ranges from 30% to 75% (Table S4) column OGC (organic gaseous carbon) that includes, in addition to BaP, other harmful PAHs (polycyclic aromatic hydrocarbons) and dioxins. Thus, on the basis of the same arguments, 50% can be taken as the average value of BaP emission abatement from the area sources. The overall BaP concentration is about 70% contributed from the area sources and about 30% contributed from the transboundary inflow (Figure 6 (A)). Application of class 5 boilers denotes in this case about a 60% reduction in BaP emission. Moreover, the planned fuel-mix decarbonization causes, as previously, a 20% reduction in the transboundary component. The resulting concentration at the end of 2030 is shown -in Section 5.

As shown in Figure 3(A), the area sources significantly contribute (about 20%) to the final NOx concentration for the baseline emission dataset. On the other hand, as follows from [43], low-emission class 5 boilers, meeting the *Ecodesign* Directive, also reduce NOx emissions by about 50–60%, as compared with the currently used, low-class installations (see also Table S4). Moreover, taking into account the approximately 16% reduction in the transboundary NOx inflow, due to the fuel mix modifications, we obtain an additional reduction of approximately 10% in the overall NOx exposure.

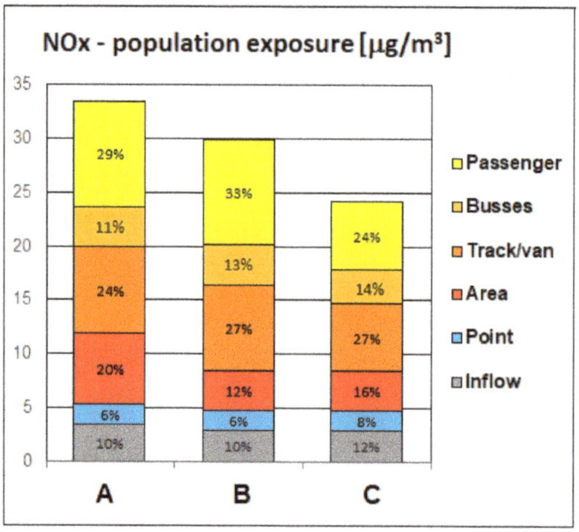

Breakdown of emision categories [μg/m3]

Scenario	A	B	C
Inflow	3.5	2.9	2.9
Point	1.9	1.9	1.9
Area	6.5	3.6	3.6
Track/van	8.0	8.0	6.3
Busses	3.8	3.8	3.2
Passenger	9.7	9.7	6.3
Sum	33.4	29.9	24.2

Figure 3. Modifications of NOx exposure.

4.3. Urban Transport Modernization

The scenario connected with the environmental influence of urban road transport takes into account several factors that, taken together, should visibly reduce the negative impact of this sector. First of all, on the basis of earlier results [27,28], effects of the passenger fleet modernization in terms of meeting EU emission standards should be considered. The analysis utilizes results of [28,29] where, using the national emission data, the NO_2 emission rates per fuel and Euro Norm for 30 European cities were collected. Furthermore, the Warsaw transportation system was fully specified there, including the fleet composition, fuel type used, yearly distance driven, and emission rates connected with the Euro norms depending on car production years. These data can be used to assess the emission reduction rate that may relate to the emission scenario being implemented. Below, we assume the respective emission rate reduction of cars from E0–E3 categories, which will be replaced by newer ones, meeting the Euro 4 standard, at least. The required emission data are shown in Table 1, which is an aggregated form of the detailed specification in [27,29].

Table 1. Reference NO_2 passenger car emissions in Warsaw per fuel and Euro norm.

	Euro Norm	Distance [km × 10^6]	Share [%]	Emiss. Rate [g/km]	Emission [kg × 10^3]	Share [%]
Gasoline	E0-E3	52,894	32%	0.32	17,000	19%
	E4-E6	13,684	9%	0.1	1368	2%
	distance	66,678	41%	emission	18,368	21%
Diesel	E0-E3	67,895	42%	0.77	51,190	59%
	E4-E6	26,316	16%	0.65	17,105	20%
	distance	94,211	59%	emission	68,295	79%
	TOTAL	160,789	100%		86,663	100%

Table 1 shows a dominant share of high emission (pre-E4) cars in Warsaw's car fleet. Using these data, it can be estimated that implementation of the assumed modernization scenario implies about a 22% emission reduction in the passenger car category.

The rapidly growing segment of electric cars will also contribute to reducing air pollution coming from road transport. This upward trend is additionally accelerated by appropriate financial incentives, as well as by the general EU policy in this field. Thus, the resulting cumulative number of battery electric vehicle (BEV) registrations in 2030 may exceed 955,000 cars [44,45]. There should be added to this number plug-in hybrids vehicles (PHEV), which in Poland in 2030 is forecasted at 653,000 cars. Thus, in 2030, up to 1.6 million electric cars (mainly passenger vehicles) may exist on Polish roads. According to estimates [45], about one-quarter of all electric vehicles have been registered in Warsaw, which means a cumulative number of about 400,000 electric cars in 2030, representing approximately 30% of all passenger cars in the city. Given the approximately 60% share of BEVs among electric vehicles, an additional reduction in emissions for this vehicle category can be estimated at about 15%.

Fulfilling the Euro 4 emission standard by the truck/van vehicles and taking into account a minor share of electric units will also result in a certain reduction in this category's emission. Taking into account appropriate unit norms for HDV cars (Table S3), the overall reduction for this vehicle category was estimated at about 22%.

At the end of 2021, 27% of the public transport buses were low-emission vehicles, i.e., LNG/CNG, along with rather few electric units. According to the plans of the Public Transport Authorities [46] in Warsaw, by the end of 2022, the share of this class of vehicles is to be 38%, while, in 2030, whole bus transport in the city will be performed by low-emission and zero-emission (electric) vehicles. The corresponding emission reduction resulting from this modernization has been estimated to be at around 16%.

The carbon monoxide (CO) concentration in the city mainly depends on the emissions of gasoline passenger cars. However, the calculations on the baseline dataset showed that the CO concentrations in the city are well below the relevant WHO limits [6].

Modernization of the car fleet also affects the total dust pollution, but rather insignificantly, as the contribution of this sector is 23% and 14% in the case of PM_{10} and $PM_{2.5}$ pollutants, respectively (compare Figures 4 and 5 (A)). In addition, in the case of particulate matter, the modernization process only reduces the emissions coming from the fuel combustion (primary), approximately 18% for PM_{10} and 40% for $PM_{2.5}$ of these emissions [13,27], while the secondary emissions (resuspended) dominate, especially in the case of PM_{10}.

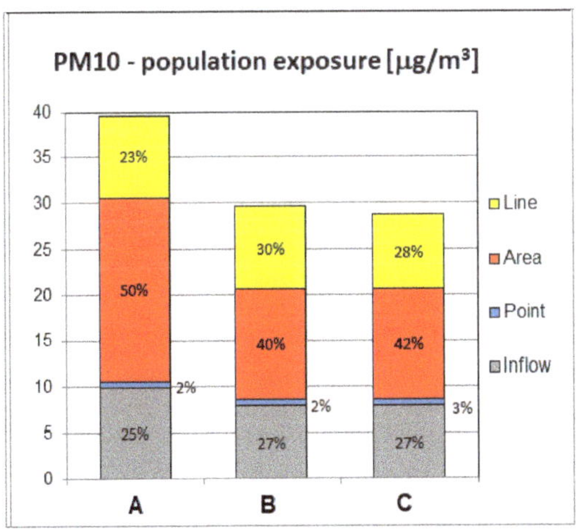

Breakdown of emision categories [µg/m3]

Scenario	A	B	C
Inflow	9.9	7.9	7.9
Point	0.7	0.7	0.7
Area	20.0	12.0	12.0
Line	9.0	9.0	8.1
Sum	39.6	29.6	28.7

Figure 4. Modifications of PM_{10} exposure.

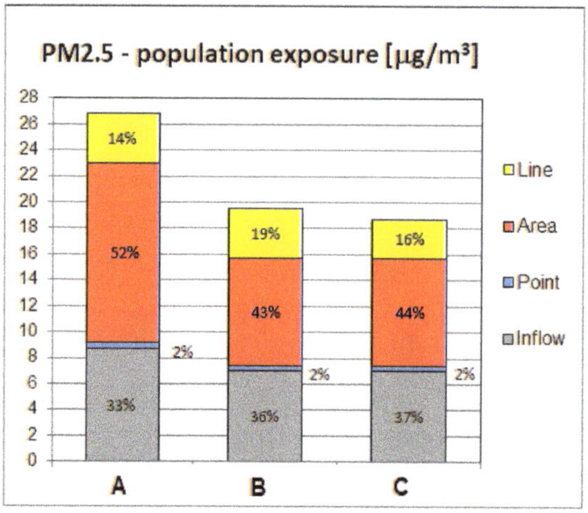

Breakdown of emission categories [µg/m³]

Scenario	A	B	C
Inflow	9.9	7.9	7.9
Point	0.7	0.7	0.7
Area	20.0	12.0	12.0
Line	9.0	9.0	8.1
Sum	39.6	29.6	28.7

Figure 5. Modifications of $PM_{2.5}$ exposure.

Table 1 shows that pre-E4 cars cover 75% of the total distance traveled by passenger vehicles. Extending this relation to other car vehicles and taking into account Euro Norm limits for dust (Tables S1–S3), it is possible to estimate the related reduction in PM emissions as 6% for PM_{10} and 11% for $PM_{2.5}$. The increase in the number of electric cars, mainly passenger ones, projected at the end of the decade, will also contribute to the reduction in dust emissions. As indicated earlier, the estimated share of BEV/HPEV cars is around 30% of all passenger cars. Taking into account the share of passenger cars in the entire vehicle fleet, the resulting emission reduction can be estimated at 3% and 5%, for the considered PM categories. In addition, a slight reduction in PM concentrations resulting from the modernization of truck/van vehicles should be taken into account (respectively 3% and 5%). Thus, the overall reduction in PM emission related to the contribution of the transportation sector can be estimated at 13% and 21% for PM_{10} and $PM_{2.5}$, respectively (see also Figures 4 and 5).

5. Comparative Simulation Results

The results presented below illustrate the environmental impact of the implementation of the emission reduction scenarios on the appropriate reduction of the negative influence of the four key pollutants discussed in the previous sections, NOx, PM_{10}, $PM_{2.5}$, and BaP. The final impact of each scenario defined in Section 4 is expressed as population exposure to the individual pollutant, calculated as the population-weighted averaged concentration within the Warsaw's receptor field (Figure 1, inside the administrative border).

For each pollutant, one of the scenarios is predominant, while the other has a complementary effect. In order to standardize the message, the same sequence of scenarios in all figures was adopted. Scenario B (municipal sector and national economy decarbonization) is a natural approach for particulate matter pollutants, while scenario C (car fleet modernization added to B scenario) gives obviously the most significant benefit for NOx, which can be seen when comparing B and C steps. However, the municipal sector contributes quite significantly to NOx pollution, via both the area and the transboundary inflow. Since low-emission boilers installed in municipal sector also reduce NOx emission, scenario B implementation implies approximately a 10% reduction in NOx pollution.

In the case of PMs pollutions, scenario B has a dominant impact on the emission reduction, which is visible in Figures 3–6 (B). The effect of modifying the transport sector, by reducing pollution from fuel combustion, is insignificant in this case. This is because resuspended particulate matter (secondary emissions) is predominant in PM_{10} emissions, while, in $PM_{2.5}$, where secondary emissions have a lower share, the total influence of linear sources is small. The total reduction attributable to the line sources is about 2%, which can be seen in the transition from scenarios B to C in Figures 4 and 5.

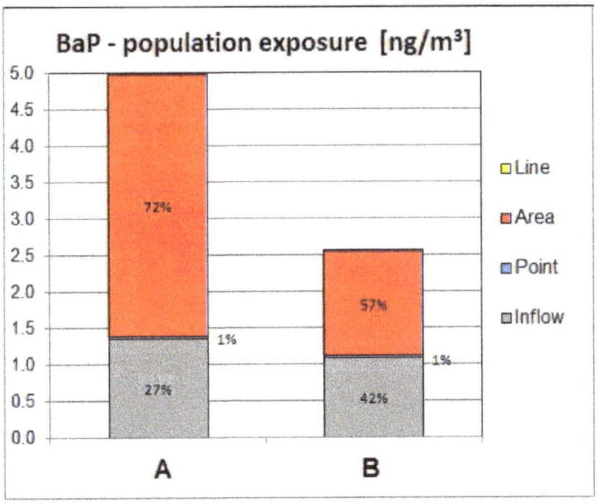

Breakdown of emission categories [μg/m³]

Scenario	A	B
Inflow	1.35	1.08
Point	0.03	0.03
Area	3.61	1.44
Line	0.01	0.01
Sum	5.00	2.56

Figure 6. Modifications of BaP exposure.

The concentration of BaP depends entirely on the emission of area sources, located both in the city and in its surroundings (the share of other types of sources is below 1%). As the low-emission boiler technology enables a significant reduction in BaP emission (Table S4), the total improvement in air quality in this case is close to 50%.

6. Discussion

According to many assessments [8,9,12], ambient air quality in Warsaw is one of the worst among large European cities. One, although not dominant reason, is the high level of car traffic and the associated emissions of nitrogen oxides, NOx. However, the main factor that determines bad air quality in Warsaw is the extremely high (for European conditions)concentration of particulate matter, especially $PM_{2.5}$ and the carcinogenic BaP. This is due to many years of dependence on coal (hard coal and lignite) of the Polish national energy and the municipal sectors. Therefore, high concentrations of PMs in Warsaw come from overlapping impacts of local sources, including to a large extent those located in the city's surroundings, as well as the transboundary inflow. As a result, the concentrations of PMs pollutants in a large area of the city exceed the WHO limit values, which were recently significantly tightened [6].

Emission abatement strategies discussed in the paper take into account technological progress and general trends observed, for example, in the transportation sector, as well as pro-ecological policies implemented by the EU regulations, including relevant legislative initiatives referring to both a national and a regional level. Using the 2018 emission inventory as the baseline dataset, two scenarios of emission reduction are considered. Scenario B, related mainly to PM pollution, follows the very restrictive legislative decision [47] of the Mazovian Voivodship concerning coal-fired installations in Warsaw and its vicinity (the latter emission abatement decisions are to be implemented by 2028), as well as the national-level legislation related to coal share reduction in the fuel mix with a simultaneous increase in the use of renewable sources. Scenario C, which mainly reflects the car fleet modernization (reduction of pre-E4 vehicles) and rapidly increasing share of electric cars (BEV/PHEV), is mainly responsible for NOx pollution.

A key factor of this strategy is the carbon footprint of electric cars, primarily related to battery manufacturing. As shown in the ICCT report [48], even when the battery-related footprint is added to the total life cycle of car production, the overall carbon footprint is definitely lower than that attributed to a comparable combustion car. The main problem in this case is the share of fossil fuels in the fuel mix of the electrical energy sector used for charging the car batteries, which is high in Poland. However, as shown in [49–51], even with such an unfavorable energy mix as for Poland in 2018, usage of electric cars lead to a reduction (albeit small one) in the carbon footprint. Therefore, taking into account the results presented above, the implementation of emission abatement scenarios discussed in this paper can result in a further reduction in the carbon footprint, associated with both the road traffic and the municipal sector.

In addition to the air pollution-related deaths, the climate policy will result in reduction in diet-related deaths and deaths due to physical inactivity [52]. The latter is related with active transportation [53], particularly cycling, which has been developed quite extensively in Warsaw. Although this change in transportation only marginally improves the air quality [27], the health impact connected with intensified physical activity is definitely positive in Warsaw air pollution conditions [54]. A negative factor of more intensive cycling and walking is caused by the connected increased rates of traffic incidents, although this is estimated to be minor in comparison with the health benefits gained by the physical activity [53].

An additional advantage of transportation changes is related to the reduction in noise. Driving technologically more advanced cars, particularly hybrid and electric ones, using electric scooters, cycling, or walking instead of driving significantly lowers the noise level coming from transportation, which is a main source of urban noise and reduces the connected health risk [39,53].

7. Conclusions

The paper presented an analysis of consequences of implementing the official scenario prepared for air quality improvement in Warsaw, Poland, particularly in terms of the population exposure and the associated health risk at the end of the present decade. The main tool used was a Gaussian system CALPUFF for modeling atmospheric pollution dispersion. Four pollutants, NOx, PM_{10}, $PM_{2.5}$, and BaP, all of which presently exceed environmental limits in Warsaw, were considered.

Results showed a reduction in population exposure attributed to the specific pollutants and scenarios being implemented. The final reduction in population exposure to NOx was about 28%, as depicted in Figure 3. This means that the base avoidable mortality assigned to this pollutant, 743 avoidable deaths [9], would be reduced by about 204 cases. The analogous result for $PM_{2.5}$ was more significant (Figure 5). Here again, the ~30% reduction in population exposure would reduce the avoidable yearly deaths by 607 cases, from the initial value of 2023 avoidable deaths. The reduction in exposure to PM_{10} (Figure 4) was similar, amounting to about 28%, while, for carcinogenic BaP (Figure 6), it was as much as approximately 50%. In the latter case, the concentration depended solely on the municipal sector emissions and the transboundary inflow, while the influence of other sources was negligible.

The considered undertakings would considerably improve air quality, but they are not sufficient. However, the processes of reducing the main polluting compounds, which are essential for air quality in the city and are presented in this paper, will be continued in the next decade. The basic EU regulations [16,18] will force the acceleration of the national economy decarbonization processes and a significant increase in the share of renewable energy sources in the fuel mix. European Union has an ambitious goal to become a climate-neutral continent by 2050 [40]. As a consequence, this will also force a further reduction in dust emissions, both from the local sources and from the transboundary inflow. In line with the above policy, electric vehicles (BEVs/PHEVs) are expected to have a dominant share in the car fleet, in both private and public transport [55,56]. To achieve this target, the European Commission announced a multitude of new legislative proposals in the coming years, where many of them target transport [57]. In particular, the European Parliament approved an effective EU ban on the sale of new petrol and diesel cars from 2035, to speed Europe's shift to electric vehicles [58].

Paradoxically, due to the country's significant backwardness compared to the EU average, Poland has potential opportunities to effectively implement recovery plans and to clearly improve air quality. This depends substantially on policies of the city, province, and country authorities. Implementation of these plans is subject to high uncertainty. Imagination of future conditions in even a couple of years may be far from reality. Recent events such as the COVID-19 pandemic, the energy resource crisis connected with the war in Ukraine, or economic turbulences caused by both these events, make this clear. During the pandemic, air quality tended to improve due to highly limited transportation because virtual meeting technologies reduced commuting and business travel. On the other hand, home delivery caused enhanced delivery truck and van traffic. Both these trends are seemingly continuing, despite being rather difficult to foresee when plans were made a few years ago. The energy resource crisis made the achievement of the presupposed goals questionable, also affected by the more difficult economic conditions. However, our considerations span to rather many years ahead; hence, we hope that these important environmental plans will not be substantially reduced or abandoned by these unforeseen perturbations.

Supplementary Materials: The following supporting information can be downloaded at: https://www.mdpi.com/article/10.3390/atmos13101613/s1, Table S1. Euro Norms for the gasoline cars; Table S2. Euro Norms for the diesel cars; Table S3. Euro Norms for HDV diesel cars; Table S4. Limit emission norms for low-emission boilers; Figure S1. Wind Rose for Warsaw 2018

Author Contributions: Conceptualization, P.H. and Z.N.; methodology, P.H.; software, A.K.; validation, Z.N., P.H. and A.K.; formal analysis, Z.N.; writing—original draft preparation, P.H.; writing—review and editing, Z.N.; visualization, A.K.; project administration, Z.N. All authors have read and agreed to the published version of the manuscript.

Funding: This research received no external funding.

Institutional Review Board Statement: Not applicable.

Informed Consent Statement: Not applicable.

Acknowledgments: Research conducted by Z. Nahorski was partially supported by the National Science Center, Poland under the Grant DEC-2018/30/Q/HS4/00764.

Conflicts of Interest: The authors declare no conflict of interest.

References

1. EEA. Europe's Air Quality Status 2022. Available online: https://www.eea.europa.eu/publications/status-of-air-quality-in-Europe-2022/europes-air-quality-status-2022 (accessed on 12 March 2021).
2. EC. Urban Air Pollution—What Are the Main Sources across the World? 2015. Available online: https://ec.europa.eu/jrc/en/news/what-are-main-sources-urban-air-pollution (accessed on 12 March 2021).
3. EC. Air Quality: Traffic Measures Could Effectively Reduce NO_2 Concentrations by 40% in Cities. 2019. Available online: https://ec.europa.eu/jrc/en/news/air-quality-traffic-measures-could-effectively-reduce-no2-concentrations-40-europe-s-cities (accessed on 12 March 2021).
4. EEA. Exceedance of Air Quality Standard in Europe. 2018. Available online: https://www.eea.europa.eu/ims/exceedance-of-air-quality-standards (accessed on 12 March 2021).
5. ETC/ATNI Report 2019/9. European Air Quality Maps for 2017. 2019. Available online: https://www.academia.edu/77543868/ETC_ATNI_Report_2019_9_European_air_quality_maps_for_2017_PM10_PM2_5_Ozone_NO2_and_NOx_spatial_estimates_and_their_uncertainties (accessed on 12 March 2021).
6. WHO. WHO Global Air Quality Guidelines. 2021. Available online: https://apps.who.int/iris/handle/10665/345329 (accessed on 12 March 2021).
7. WHO. Ambient (Outdoor) Air Pollution. 2018. Available online: https://www.who.int/news-room/fact-sheets/detail/ambient-(outdoor)-air-quality-and-health (accessed on 12 March 2021).
8. AIR8. Most Polluted Cities in Europe in 2021. Available online: https://air8.tech/most-polluted-cities-in-europe-in-2021/ (accessed on 12 March 2021).
9. ISGlobal—Ranking of Cities. 2021. Available online: https://isglobalranking.org/ranking/#air (accessed on 12 March 2021).
10. EEA. Report, No 10/2019. Air Quality in Europe—2019 Report. Available online: https://www.eea.europa.eu/publications/air-quality-in-europe-2019 (accessed on 20 July 2021).

11. EEA. Report 2020. Available online: https://www.eea.europa.eu/publications/air-quality-in-europe-2020-report (accessed on 15 September 2021).
12. EEA. Air Quality in Europe 2021. Available online: https://www.eea.europa.eu/publications/air-quality-in-europe-2021 (accessed on 20 September 2021).
13. GIOS. Yearly Assessment of Air Quality in the Mazovian Voivodship. Report for the Year 2018. 2019. Available online: https://powietrze.gios.gov.pl/pjp/rwms/publications/card/3 (accessed on 20 September 2021). (In Polish)
14. GIOS. Yearly Assessment of Air Quality in the Mazovian Voivodship. Report for the Year 2019. 2020. Available online: https://powietrze.gios.gov.pl/pjp/rwms/publications/card/1162 (accessed on 20 September 2021). (In Polish)
15. GIOS. Yearly Assessment of Air Quality in the Mazovian Voivodship. Report for the Year 2020. 2021. Available online: https://powietrze.gios.gov.pl/pjp/rwms/publications/card/1421 (accessed on 20 September 2021). (In Polish)
16. FIT55. EU Fit for 55 Package. 2021. Available online: https://europeanclimate.org/the-fit-for-55-package/ (accessed on 15 May 2022).
17. EC. EU Green Deal. 2019. Available online: https://ec.europa.eu/info/strategy/priorities-2019-2024/european-green-deal_en (accessed on 15 May 2022).
18. EC. 2021. Available online: https://ec.europa.eu/clima/eu-action/european-green-deal/2030-climate-target-plan_en (accessed on 15 May 2022).
19. EC. The 8th Environment Action Programme till 2030. Decision (EU) 2022/591 of the European Parliament. 2022. Available online: http://data.europa.eu/eli/dec/2022/591/oj (accessed on 15 May 2022).
20. Energy. Electricity Sources in Poland in 2020. 2021. Available online: https://wysokienapiecie.pl/35619-zrodla-energii-w-polsce-w-2020-mniej-wegla-wiecej-gazu-oze/ (accessed on 15 May 2022).
21. FIT55 2022. Will the Polish RES Target Be Consistent with the EU Climate Policy? Available online: https://instrat.pl/fit-for-55/ (accessed on 15 May 2022).
22. Energy 2022. Polityka Energetyczna Polski do 2040 r. Available online: https://bip.mos.gov.pl/strategie-plany-programy/polityka-energetyczna-polski-do-2040-r/ (accessed on 15 May 2022).
23. MP 2021. Monitor Polski. Announcement (2.03.2021) of the Minister of Climate and Environment on Poland's Energy Policy until 2040. Available online: https://www.gov.pl/web/climate/energy-policy-of-poland-until-2040-epp2040 (accessed on 15 May 2022).
24. Bogacki, M.; Oleniacz, R.; Rzeszutek, M.; Bździuch, P.; Szulecka, A.; Gorzelnik, T. Assessing the Impact of Road Traffic Reorganization on Air Quality: A Street Canyon Case Study. *Atmosphere* **2020**, *11*, 695. [CrossRef]
25. Juda-Rezler, K.; Reizer, M.; Maciejewska, K.; Błaszczak, B.; Klejnowski, K. Characterization of atmospheric PM2.5 sources at a Central European urban background site. *Sci. Total Environ.* **2020**, *713*, 136729. [CrossRef] [PubMed]
26. Holnicki, P.; Tainio, M.; Kałuszko, A.; Nahorski, Z. Burden of mortality and disease attributable to multiple air pollutants in Warsaw, Poland. *Int. J. Environ. Res. Public Health* **2017**, *14*, 1359. [CrossRef] [PubMed]
27. Holnicki, P.; Nahorski, Z.; Kałuszko, A. Impact of car fleet modernization on the traffic originated air pollution in an urban area—Case study. *Atmosphere* **2021**, *12*, 1581. [CrossRef]
28. Degraeuwe, B.; Thunis, P.; Clappier, A.; Weiss, M.; Lefebvre, W.; Janssen, S.; Vranckx, S. Impact of passenger car NOx emissions on urban NO_2 pollution—Scenario analysis for 8 European cities. *Atmos. Environ.* **2017**, *171*, 330–337. [CrossRef]
29. Degraeuwe, B.; Pisoni, E.; Peduzzi, E.; De Meij, A.; Monforti-Ferrario, F.; Bodis, K.; Mascherpa, A.; Astorga-Llorens, M.; Thunis, P.; Vignati, E. *Urban NO_2 Atlas (EUR 29943 EN)*; Publications Office of the European Union: Luxembourg, 2019.
30. Tezel-Oguz, M.N.; Sari, D.; Ozkurt, N.; Keskin, S.S. Application of reduction scenarios. on traffic-related NOx emissions in Trabzon, Turkey. *Atmos. Pollut. Res.* **2020**, *11*, 2379–2389. [CrossRef]
31. Karagulian, F.; Belis, C.A.; Carlos Francisco, C.; Dora, C.F.C.; Prüss-Ustün, A.M.; Bonjour, S.; Adair-Rohani, H.; Amann, M. Contributions to cities' ambient particulate matter (PM): A systematic review of local source contributions at global level. *Atmos. Environ.* **2015**, *120*, 475–483. [CrossRef]
32. Kiesewetter, G.; Borken-Kleefeld, J.; Schöpp, W.; Heyes, C.; Thunis, P.; Bessagnet, B.; Terrenoire, E.; Gsella, A.; Amann, M. Modelling NO_2 concentrations at the street level in the GAINS integrated assessment model: Projections under current legislation. *Atmos. Chem. Phys.* **2014**, *14*, 813–829. [CrossRef]
33. Mediavilla-Sahagún, A.; ApSimon, H.M. Urban scale integrated assessment for London: Which emission reduction strategies are more effective in attaining prescribed PM10 air quality standards by 2005? *Environ. Model. Softw.* **2006**, *21*, 501–513. [CrossRef]
34. Exponent. *CALPUFF Version 7—Users Guide Addendum*; Exponent, Inc.: Maynard, MA, USA, 2019; Doc. no. Z170308064614-0072.
35. Holnicki, P.; Kałuszko, A.; Nahorski, Z.; Stankiewicz, K.; Trapp, W. Air quality modeling for Warsaw agglomeration. *Arch. Environ. Prot.* **2017**, *43*, 48–64. [CrossRef]
36. Holnicki, P.; Kałuszko, A.; Trapp, W. An urban scale application and validation of the CALPUFF model. *Atmos. Pollut. Res.* **2015**, *7*, 393–402. [CrossRef]
37. Holnicki, P.; Nahorski, Z. Emission data uncertainty in urban air quality modeling—Case Study. *Environ. Model. Assess.* **2015**, *20*, 583–597. [CrossRef]
38. Holnicki, P.; Kałuszko, A.; Nahorski, Z.; Tainio, M. Intra-urban variability of the intake fraction from multiple emission sources. *Atmos. Pollut. Res.* **2018**, *9*, 1184–1193. [CrossRef] [PubMed]
39. Tainio, M. Burden of disease caused by local transport in Warsaw, Poland. *J. Transp. Health* **2015**, *2*, 423–433. [CrossRef] [PubMed]

40. EU. Air Quality Standards. 2021. Available online: https://www.eea.europa.eu/themes/air/air-quality-concentrations/air-quality-standards (accessed on 15 May 2022).
41. EEA. Air Quality in Europe—2018 Report; EEA Report, No 12/2018. 2018. Available online: https://www.eea.europa.eu/publications/air-quality-in-europe-2018 (accessed on 15 May 2022).
42. Anti-smog. 2022. Available online: https://www.powietrze.mazovia.pl/uchwala-antysmogowa/informacje-ogolne-na-temat-uchwaly-antysmogowej (accessed on 15 May 2022).
43. LE-Boil. 2021. Available online: https://czysteogrzewanie.pl/podstawy/normy-emisji-i-sprawnosci-dla-domowych-kotlow (accessed on 15 May 2022).
44. PSPA. 2021. Available online: https://www.teraz-srodowisko.pl/aktualnosci/elektromobilnosc-raport-PSPA-9889.html (accessed on 15 May 2022).
45. EV Outlook. 2021. Available online: https://pspa.com.pl/2021/raport/najnowsza-edycja-jedynej-na-rynku-kompleksowej-analizy-sektora-e-mobility-w-polsce-polish-ev-outlook-2021-jest-juz-dostępna/ (accessed on 15 May 2022).
46. BUS. 2021. Available online: https://www.transport-publiczny.pl/wiadomosci/mza-warszawa-do-konca-2022-r-38-autobusow-zero (accessed on 15 May 2022).
47. Smoglab. 2022. Available online: https://smoglab.pl/zakaz-spalania-wegla-od-pazdziernika-2023/ (accessed on 15 May 2022).
48. Messagie, M.; Boureima, F.-S.; Coosemans, T.C.; Macharis, C.; Van Mierlo, J.A. Range-Based Vehicle Life Cycle. Assessment Incorporating Variability in the Environmental Assessment of Different Vehicle Technologies and Fuels. *Energies* **2014**, *7*, 1467–1482. [CrossRef]
49. Hamilton, I.; Kennard, H.; McGushin, A.; Höglund-Isaksson, L.; Kiesewetter, G.; Lott, M.; Milner, J.; Purohit, P.; Rafaj, P.; Sharma, R.; et al. The public health implications of the Paris Agreement: A modelling study. *Lancet Planet. Health* **2021**, *5*, e74–e83. [CrossRef]
50. Mueller, N.; Rojas-Rueda, D.; Cole-Hunter, T.; de Nazelle, A.; Dons, E.; Gerike, R.; Götschi, T.; Int Panisi, L.; Kahlmeier, S.; Nieuwenhuijsen, M. Health impact assessment of active transportation: A systematic review. *Prev. Med.* **2015**, *76*, 103–114. [CrossRef] [PubMed]
51. Tainio, M.; de Nazelle, A.J.; Götschi, T.; Kahlmeier, S.; Rojas-Rueda, D.; Nieuwenhuijsen, M.J.; de Sá, T.H.; Kelly, P.; Woodcock, J. Can air pollution negate the health benefits of cycling and walking? *Prev. Med.* **2016**, *87*, 233–236. [CrossRef] [PubMed]
52. Mehling, D.; Woodward, H.; Oxley, T.; Holland, M.; ApSimon, H. Electrification of Road transport and the Impacts on Air Quality and Health in the UK. Centre for Environmental Policy, Imperial College London. *Atmosphere* **2021**, *12*, 1491. [CrossRef]
53. UN Environment Programme. Available online: https://www.unep.org/resources/report/used-vehicles-and-environment-progress-and-updates-2021 (accessed on 15 May 2022).
54. VIRTA. 2022. Available online: https://www.virta.global/blog/this-is-how-eu-regulation-accelerates-the-electric-vehicle-revolution (accessed on 15 May 2022).
55. EU. 2022. Available online: https://www.france24.com/en/europe/20220609-eu-parliament-approves-ban-on-new-fossil-fueled-cars-b (accessed on 15 May 2022).
56. ICCT. A Global Comparison of the Life-Cycle Greenhouse Gas Emissions of Combustion Engine and Electric Passenger Cars. 2021. Available online: https://theicct.org/publication/a-global-comparison-of-the-life-cycle-greenhouse-gas-emissions-of-combustion-engine-and-electric-passenger-cars/ (accessed on 15 May 2022).
57. Messagie, M. Life Cycle Analysis of the Climate Impact of Electric Vehicles (2017) Semantic Scholar. Available online: https://www.semanticscholar.org/paper/Life-Cycle-Analysis-Of-The-Climate-Impact-Of-Messagie/1eb5b7c002404326dca2391b2004daa56dc18028 (accessed on 15 May 2022).
58. EV F-PRINT. 2021. Available online: https://ev.auto-swiat.pl/wiadomosci/auta-elektryczne-juz-dzis-pozostawiaja-mniejszy-slad-weglowy-niz-spalinowe/m1ct39h (accessed on 15 May 2022).

atmosphere

Article

Characteristics of Ozone Pollution in Tai'an and Topographic Effects of Mount Tai

Yanzhen Ge [1], Chunqiang Chen [2], Yaping Jiang [3], Tongsuo Yang [2], Hongyang Kang [4], Jingzhong Li [1], Xinran Zhao [1], Yibo Zhang [3], Mengying Li [3], Tangyan Hou [3], Jianmin Chen [5] and Shaocai Yu [3,*]

1 Tai'an Ecological Environment Protection and Control Center, Tai'an Ecological Environment Bureau, Tai'an 271000, China
2 Shandong Academy of Environmental Sciences Co., Ltd., Jinan 250000, China
3 Key Laboratory of Environmental Remediation and Ecological Health, Ministry of Education, Research Center for Air Pollution and Health, College of Environmental and Resource Sciences, Zhejiang University, Hangzhou 310058, China
4 Beijing Insights Value Technology Co., Ltd., Beijing 100000, China
5 Research Center of Analysis and Measurement, Department of Environmental Science & Engineering, Fudan University, Shanghai 200433, China
* Correspondence: shaocaiyu@zju.edu.cn

Abstract: Through the analyses of the observed concentrations of ozone (O_3) in Tai'an from 2016 to 2021, the results show that O_3 pollution was relatively serious, with the mean concentrations of the 90th percentile of daily maximum 8-h O_3 (O_3-8h-90per) above 180 μg/m^3. O_3 pollution in Tai'an mainly occurred from May to September, accounting for 69%~100% of the total O_3 pollution days, of which the most serious pollution occurred in June. Combined with the analyses of temperature, humidity, and wind speeds, the probability of O_3 exceedances in Tai'an increased significantly under the conditions with the temperatures higher than 30 °C, the relative humidity of 20%~40%, and the wind speeds of 1~3 m/s. The dominant wind directions on O_3 pollution days in Tai'an were southerly winds, based on the analyses of wind directions and their clustering trajectories. Based on the results at three monitoring stations at different distances from Mount Tai from May to September in 2021, the average O_3 concentrations at the Renkou School station near the mountain was about 13~15 μg/m^3 higher than those at the other two stations (Dianli College and Shandong First Medical University) which are far away from the mountain, indicating that the Renkou School site was more affected by the obstruction of the mountain. In addition, the WRF-CMAQ model was used to simulate ten O_3 pollution events in 2021, showing that the average O_3 concentrations in Tai'an were reduced by 1.7~7.5 μg/m^3 after changing the topographic height of Mount Tai.

Keywords: ozone; pollution characteristics; terrain; mount tai; WRF-CMAQ

Citation: Ge, Y.; Chen, C.; Jiang, Y.; Yang, T.; Kang, H.; Li, J.; Zhao, X.; Zhang, Y.; Li, M.; Hou, T.; et al. Characteristics of Ozone Pollution in Tai'an and Topographic Effects of Mount Tai. *Atmosphere* **2022**, *13*, 1299. https://doi.org/10.3390/atmos13081299

Academic Editors: Elena Hristova, Manousos Ioannis Manousakas, Anikó Angyal and Maria Gini

Received: 6 July 2022
Accepted: 12 August 2022
Published: 16 August 2022

1. Introduction

In recent years, ozone (O_3) has gradually become the primary pollutant affecting air quality in China, and it has shown a rapid growth and deterioration trend in vast areas of China [1–5]. Ground-level O_3 is generated due to complex photochemical reactions of volatile organic compounds (VOCs) and nitrogen oxides (NOx) under sunlight [6–9]. The surface O_3 can stimulate the respiratory tract and cause changes in lung function and respiratory inflammation and asthma, seriously threatening human health [10–13]. In order to improve the regional air quality and reduce the serious harm of O_3 pollution to the human body, effectively reducing the concentrations of surface O_3 has become an arduous task for local environmental protection departments.

Since 2016, O_3 pollution in Tai'an has been serious. In 2016, the 90% percentile of annual O_3 concentrations ranked 168th among 168 key cities specially selected by the Ministry of Ecology and Environment for the national urban air quality ranking in China.

In 2021, it ranked 166th among 168 cities, showing that O_3 has become the primary pollutant affecting air quality in Tai'an [14,15]. Tai'an city has a special geographical environment, with the terrain high in the north and low in the south. It is located at the foot of Mount Tai, with the Taishan Mountains in the north and the Culai Mountain in the east, being favorable for pollutants to accumulate at the foot of the mountain and the downtown area [15]. VOCs and NOx are transported northward by the southerly winds, providing precursors for the photochemical reactions of O_3. In addition, Mount Tai is rich in vegetation, thus VOCs emitted by vegetation can provide certain precursors to O_3 generations in Tai'an. This study analyzed the characteristics of O_3 pollution in Tai'an from 2016 to 2021, and explored the effects of Mount Tai on O_3 concentrations. It may provide scientific basis and technical support for the causes of O_3 pollution and air pollution prevention and control measures in Tai'an.

2. Materials and Methods

2.1. Observation Data

The observation data of O_3, CO, SO_2, NO_2, $PM_{2.5}$, and PM_{10} at six air quality stations of Tai'an, as shown in Figure 1, from 2016 to 2021 were obtained from the Tai'an ambient air quality automatic monitoring system. The daily data were obtained from the average of observations at these six air quality observation stations, and the monthly and annual average O_3 concentrations were taken as the 90th percentile of the maximum 8 h average concentration of O_3 (O_3-8h-90per). The O_3 observation data of other cities were from China National Environmental Monitoring Centre. The observational data of temperatures, relative humidity, and wind speeds in Tai'an were provided by the Tai'an ambient air quality automatic monitoring system, and the wind direction data were obtained from the weather post report website (http://tianqihoubao.com/weather/top/taian.html) (accessed on 1 January 2022).

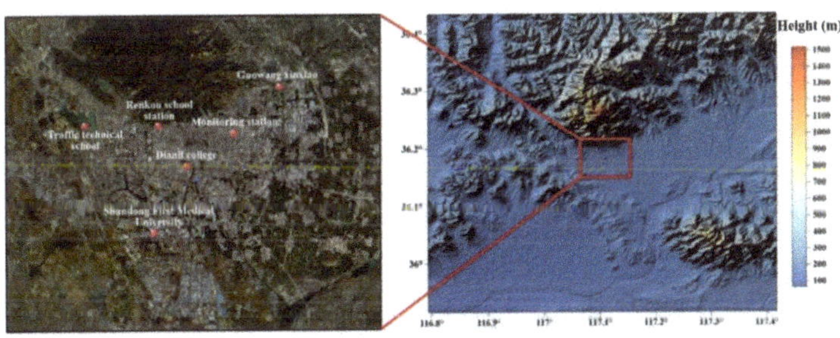

Figure 1. Map of Tai'an and distributions of six air quality monitoring stations in Tai'an.Map data © 2022, Gaode Map.

2.2. Analytical Method

According to the relevant regulations of the "Ambient Air Quality Standards" in China (GB 3095-2012), this study defined an O_3 pollution day when the maximum 8 h average concentration of O_3 (O_3-8h) exceeded 160 $\mu g/m^3$, and then conducted monthly and annual compliance evaluations based on the "Ambient Air Quality Assessment Technical Specifications (Trial)" (HJ 663-2013). According to the relevant requirements of the "Ambient Air Quality Index (AQI) Technical Regulations (Trial)" (HJ 633-2012), the air quality was divided into five levels depending on the concentrations of O_3-8h: excellent (0–100 $\mu g/m^3$), good (101–160 $\mu g/m^3$), mild (161–215 $\mu g/m^3$), moderate (216–265 $\mu g/m^3$), and heavy pollutions (266–800 $\mu g/m^3$).

2.3. Model Settings

The Community Multiscale Air Quality model (CMAQ) was applied to simulate the O_3 concentrations in Tai'an (https://www.epa.gov/cmaq) (accessed on 9 August 2022). The model meteorological field was provided by the Weather Research and Forecasting Model (WRF) (https://www.mmm.ucar.edu/weather-research-and-forecasting-model) (accessed on 9 August 2022). The input data for the WRF model were obtained from the Meteorological Reanalysis Data (FNL) of the US National Center of Environmental Prediction (NCEP). The CMAQ model adopted the CB05 carbon bond chemistry mechanism and the AERO6 aerosol chemistry mechanism [16,17]. The model had a horizontal resolution of 36 km and covered central and eastern China, as shown in Figure 2. Figure 2b,c show the terrain heights in the baseline simulation and the sensitivity simulation scenario, respectively, and the difference in the height of Mount Tai, as can be seen in Figure 2d. The anthropogenic emission sources were provided by the Multi-resolution Emission Inventory for China (MEIC) developed by Tsinghua University, and the biological sources were calculated by the online Biogenic Emission Inventory System version 3.14 (BEISv3.14).

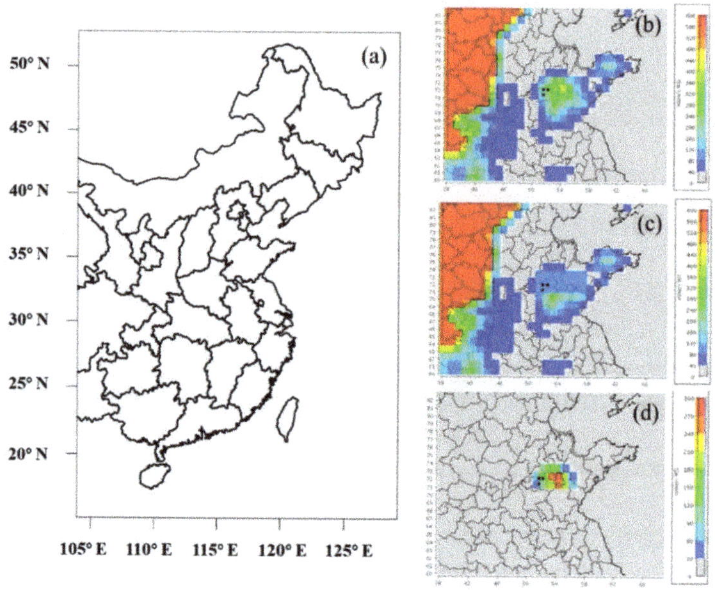

Figure 2. Map of (**a**) the model domain; (**b**) default terrain height (m); (**c**) modified terrain height (m); and (**d**) the terrain height difference between (**b**) and (**c**) in the model.

3. Results and Discussion

3.1. Overview of Ozone Pollutions in Tai'an

The annual average concentrations of O_3-8h in Tai'an from 2016 to 2021 were selected to analyze the interannual variation characteristics of O_3 pollution. As shown in Table 1, the annual average concentrations of O_3-8h-90per in Tai'an from 2016 to 2021 were above 180 μg/m^3 (except in 2018), suggesting that the O_3 pollution in Tai'an has been very serious in recent years. Among them, the highest annual average concentration of O_3-8h occurred in 2019, reaching 195 μg/m^3, while the lowest concentration was in 2018. From 2019 to 2020, the O_3 concentrations in Tai'an showed a decreasing trend with the reduction of 12 μg/m^3, while O_3 concentrations in 2021 were similar to those in 2020, indicating that the O_3 pollution situation in Tai'an was still relatively severe.

Table 1. Annual average concentrations of pollutants from 2016 to 2021 and O_3 pollution days for different wind directions in Tai'an during May–September from 2016 to 2021. The quantities in parenthesis represent the proportion of polluted days in a certain wind direction to the total polluted days.

Year	Pollutants (μg/m³)						O_3 Pollution Days (%) with Different Wind Directions		
	O_3-8h-90per	CO-95per	SO_2	NO_2	$PM_{2.5}$	PM_{10}	Southerly	Northerly	Easterly
2016	180	2.0	34	35	61	108	46 (74)	15 (24)	1 (2)
2017	193	1.6	23	36	53	97	58 (82)	12 (17)	1 (1)
2018	178	1.6	17	33	48	96	37 (74)	9 (18)	4 (8)
2019	195	1.5	15	34	53	98	62 (79)	16 (20)	1 (1)
2020	183	1.5	14	30	47	83	46 (78)	13 (22)	0 (0)
2021	182	1.3	12	28	42	76	52 (80)	12 (18)	1 (2)

Ground-level O_3 is mainly produced due to complex photochemical reactions of volatile organic compounds (VOCs) and nitrogen oxides (NOx) under the conditions of strong light, high temperature, and low humidity [18–21]. In view of the strong solar radiation, high temperature, and low humidity from May to September in Tai'an (Figure S1), these meteorological factors are favorable for generation of near-surface O_3. Therefore, this study selected the period of May–September to conduct a statistical analysis of the O_3 pollution in Tai'an from 2016 to 2021 (Figure 3a).

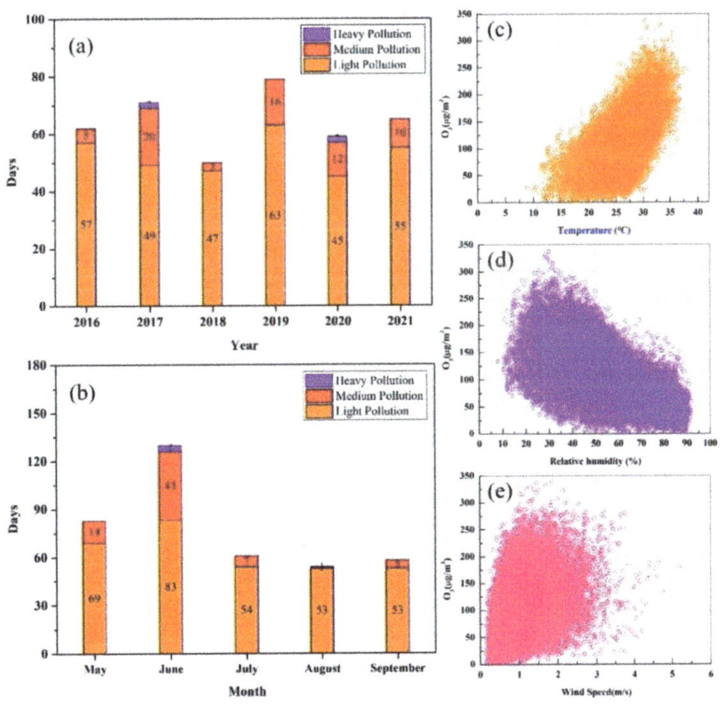

Figure 3. The number of days of O_3 pollution in different levels in Tai'an for (**a**) the annual values from 2016 to 2021 and (**b**) monthly values during May–September from 2016 to 2021, scatter plots between hourly O_3 concentrations and temperature (**c**); relative humidity (**d**); and wind speeds (**e**) in Tai'an during May–September from 2016 to 2021.

As seen from Figure 3a, the number of O_3 pollution days in Tai'an from May to September was in the range of 50~79 days from 2016 to 2021 for the annual values, accounting for 32.7% to 51.6% of the total days from May to September every year. Among them, the number of pollution days in 2019 was the most (79 days), and in 2021 it was 65 days, indicating that the O_3 pollution in Tai'an was still relatively severe. From the perspective of pollution extent, O_3 pollution in Tai'an was mainly mild pollution, accounting for 69.0%~94.0% of the total pollution days from 2016 to 2021. The number of days with moderate and above O_3 pollution gradually decreased from 16 days in 2019 to 14 days in 2020, and then to 10 days in 2021 (Table 2). The O_3-8h-90per concentrations also decreased year by year in the recent three years. This shows that Tai'an achieved certain improvements in O_3 controls in recent years.

Table 2. Monthly mean concentrations of maximum 8 h O_3 and NO_2 ($\mu g/m^3$) in Tai'an during May–September from 2016 to 2021.

Year	May		June		July		August		September		Average	
	O_3	NO_2	O_3	NO_2	O_3	NO_2	O_3	NO_2	O_3	NO_2	O_3	NO_2
2016	185	31	208	24	183	19	167	23	203	35	193	26
2017	240	30	236	31	177	22	190	21	182	28	225	26
2018	186	26	212	22	182	16	189	15	148	32	192	22
2019	186	27	225	22	211	22	169	22	211	31	216	25
2020	211	23	244	22	186	19	183	19	183	31	214	23
2021	206	19	239	20	188	14	190	17	188	17	200	17

Table 2 shows the statistical results of O_3-8h-90per in Tai'an during May–September from 2016 to 2021. The concentrations of O_3-8h-90per in May from 2020 to 2021 were both above 200 $\mu g/m^3$, indicating that the occurrence times of O_3 pollution in Tai'an had a tendency to advance. June was the most polluted month, with the most days of pollution exceedance, the concentrations of O_3-8h-90per ranging from 208–244 $\mu g/m^3$ and a total of 130 pollution days from 2016 to 2021, accounting for 34.7% of the total pollution days (Figure 3b). In addition, the moderate and above pollution in Tai'an occurred most frequently in June, with 43 days of O_3 moderate pollution, accounting for 61.4% of the total days of O_3 moderate pollution from 2016 to 2021. Notably, the four heavy pollution processes from 2016 to 2021 all occurred in June, which may be attributed to the relatively strong solar radiation and high temperature in June, resulting in strong photochemical reactions of O_3 in the atmosphere and high concentrations of ground-level O_3 [17].

3.2. *Analyses of Ozone Precursors and Meteorology*

Based on the analyses of NO_2 concentrations in Tai'an from 2016 to 2021, it was found that the overall NO_2 concentrations showed a downward trend, as shown in Table 2 and Figure S2. The average NO_2 concentration during May–September in 2021 was 17 $\mu g/m^3$, while the NO_2 concentration in 2018 was relatively low, which might be caused by meteorological conditions such as typhoons. The NO_2 concentrations were relatively high from May to June in 2017, being consistent with the high levels of O_3 concentrations. From 2018 to 2020, the NO_2 concentrations were the highest in September, followed by May, while the highest NO_2 concentration occurred in June (20 $\mu g/m^3$) and May (19 $\mu g/m^3$) in 2021, being consistent with the months with high O_3 levels.

According to the analysis of temperature, humidity, and wind speed during the high-O_3 period (May–September) in Tai'an from 2016 to 2021, the probability of O_3 exceedances rose sharply under the conditions with the temperature higher than 30 °C, relative humidity of 20–40%, and wind speeds of 1–3 m/s, as shown in Figure 3c–e.

According to the statistical analysis of the wind directions during the O_3 pollution periods in Tai'an from 2016 to 2021, the dominant wind directions leading to O_3 pollution days were southerly (southwest, south, and southeast), accounting for 65%~100% of the total number of pollution days each year (see Table 1). The pollution days dominated by the northerly winds accounted for 17%~24%, and those dominated by easterly winds accounted for 0%~8%. In 2021, number of the O_3 pollution days in Tai'an dominated by southerly winds were 52 days, accounting for 80% of the total pollution days from May to September.

In order to study the effects of air flow trajectory differences on O_3 concentrations in Tai'an, cluster analysis was performed on the hourly backward trajectory during May–September from 2020 to 2021. Results show that the summer winds were dominated by southwesterly, northeasterly, and southeasterly winds, and the proportion of trajectory sources varied slightly between different years (see Table 3). Moreover, combining the hourly O_3 observation values in Tai'an with the backward trajectories, the trajectory pollution characteristics were explored. Referring to the secondary standard of O_3-8h (160 $\mu g/m^3$) in "Ambient Air Quality Standards" (GB 3095-2012), the backward trajectories were divided into clean and polluted trajectories: the polluted trajectory was defined as the one with the corresponding O_3 concentrations greater than 160 $\mu g/m^3$, otherwise as the clean trajectory [22]. The statistical results of various trajectories are shown in Figure 4 and Table 3. Results show that there were significant differences in the O_3 concentrations corresponding to different airflow trajectories and the proportions of pollution trajectories. In summer, the pollution trajectories in the southwesterly wind direction accounted for the highest proportion. The above analysis shows that due to the effects of southern transport and local topography, the proportion of O_3 pollution in Tai'an increased significantly under the southerly winds. The transports of ozone and its precursors were considered in the all simulation cases. Since Tai'an is located in the south of Mount Tai, O_3 concentrations in the other eight cases decreased after reducing the height of Mount Tai, indicating that the existence of Mount Tai led to the accumulation of pollutants blown from the south and east in Tai'an. A typical O_3 pollution event on 21 June 2021 was selected to compare the O_3 concentrations in Tai'an, Jinan, and Jining, and analyze the O_3 and NO_2 data in Tai'an. On 21 June 2021, the maximum concentration of O_3-8h in Tai'an was 254 $\mu g/m^3$ (corresponding to AQI of 189), which reached a moderate pollution level, significantly higher than that those in Jining (219 $\mu g/m^3$) and Jinan (230 $\mu g/m^3$) (Figure 5a). For the hourly variations of O_3 concentrations in Tai'an (Figure 5b), the O_3 concentrations gradually decreased from 0:00 to 7:00 (stage 1) due to the titration of nitrogen oxides (NOx) on O_3 [23,24]. From 7:00 to 13:00, the O_3 concentrations gradually increased and reached a peak at 13:00, and O_3 and NO_2 showed a significant negative correlation during the hours of 7:00–15:00 (stage 2), indicating that the near-surface O_3 during this period was mainly generated from the photochemical reactions [22]. From 13:00 to 20:00, the NO_2 concentrations gradually increased, which may be affected by the upwind transports and local effects. From 13:00 to 18:00, the O_3 concentrations remained above 250 $\mu g/m^3$ for 6 consecutive hours, significantly higher than that those in Jining and Jinan. Considering that the dominant wind direction in Tai'an was southerly with the wind power of 2~3 during this period, the persistent high O_3 in Tai'an was caused by the mountain blocking effects, and then the O_3 concentrations gradually decreased due to the titration of NOx.

Table 3. Statistics of occurrence frequencies of O_3 trajectories and their corresponding mean O_3 concentrations in Tai'an during May–September in 2020 and 2021.

Year	Type	Through Regions	All Trajectories		Pollution Trajectories		
			Percentage (%)	O_3 (μg/m^3)	O_3 (μg/m^3)	The Class Percentage (%)	The Year Percentage (%)
2021	1	Bohai Sea, Dongying, Binzhou, Zibo, Jinan, Tai'an	11.4	97	183	9.8	6.8
	2	Rizhao, Linyi, Tai'an	19.7	111	199	16.3	19.7
	3	Huaibei, Xuzhou, Jining, Taian	29.9	121	187	23.1	42.4
	4	Yellow Sea, Rizhao, Linyi, Jinan, Tai'an	12.9	93	198	6.8	5.3
	5	Inner Mongolia, North of Shanxi, Center of Hebei, Dezhou, Jinan, Tai'an	4.9	104	175	8.3	2.50
	6	Jinan, Tai'an	21.1	107	192	17.9	23.2
2020	1	Cangzhou, Dezhou, Jinan, Tai'an	18.0	94	196	9.1	10.6
	2	Bohai Sea, Dongying, Zibo, Jinan, Tai'an	13.2	102	196	13.8	11.8
	3	Heze, Jining, Tai'an	19.9	111	201	20.9	26.9
	4	Linyi, Tai'an	23.8	99	191	14.3	22.1
	5	Yellow Sea, Rizhao, Linyi, Jinan, Tai'an	9.5	102	194	10.9	6.7
	6	Lianyungang, Linyi, Jining, Tai'an	15.6	123	196	21.7	21.9

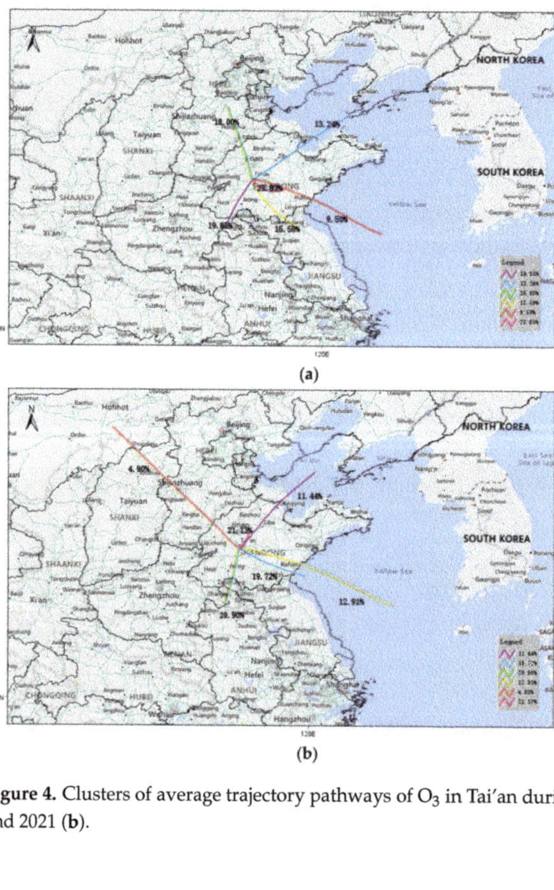

Figure 4. Clusters of average trajectory pathways of O_3 in Tai'an during May–September in 2020 (**a**) and 2021 (**b**).

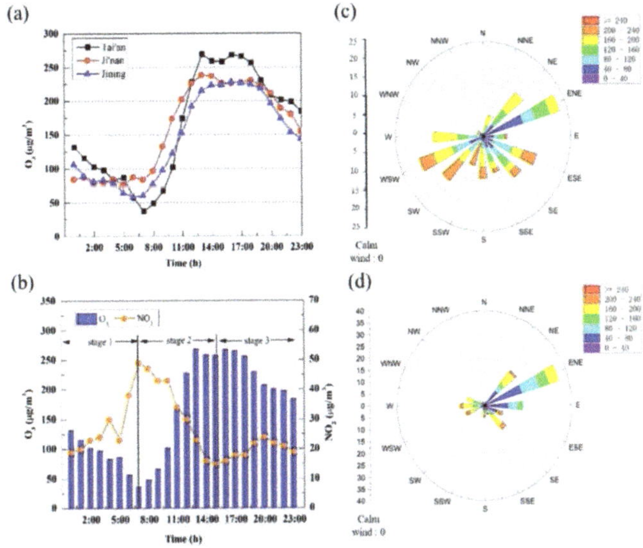

Figure 5. Comparisons of (**a**) diurnal variations of O_3 among Tai'an and surrounding cities and (**b**) diurnal variations of hourly O_3 and NO_2 concentrations in Tai'an on 21 June 2021; wind rise of observed O_3 concentrations in Tai'an during (**c**) 3–9 July and (**d**) 7–12 August 2021.

3.3. The Effects of Mount Tai on O_3

Three air quality stations in Tai'an, namely, Renkou School, Dianli College, and Shandong First Medical University, were selected to analyze the effects of the mountain on O_3 (see Table 4). The Renkou School, Dianli College, and the Shandong First Medical University stations are about 0.5, 3.5, and 8 km away from the mountain, respectively. The specific distributions of the three stations are shown in Figure 1. Based on the comparisons of monthly average concentrations of O_3 at the three stations from May to October 2021, the average O_3 concentration at the Renkou School station closer to the mountain was 118 μg/m^3, significantly higher than that at the Dianli College and Shandong First Medical University stations, which were relatively far away from the mountain. For different months from May to October 2021, the O_3 concentrations at the Renkou School station were 8~19 μg/m^3 higher than those at the Dianli College station, and 10~20 μg/m^3 higher than those at the Shandong First Medical University station. In summary, the Renkou School station was relatively more affected by the blocking of the mountain due to being closer to the mountain, leading to the significantly higher O_3 concentrations relative to those of other stations.

Table 4. Comparison of mean O_3-8h-90per concentrations (μg/m^3) at three monitoring stations in Tai'an from May to September 2021.

Stations	May	June	July	August	September	Average
Renkou School	120	144	104	120	102	118
Dianli College	112	125	91	105	84	103
Medical University	109	130	94	100	90	105

By setting the baseline simulations and sensitivity simulation scenarios of reducing the Mount Tai terrain heights to 150 m (Tai'an terrain height) (see Figure 2), the WRF-CMAQ air quality model was applied to simulate the changes of O_3 concentrations in Tai'an through the adjustment of topographic heights of Mount Tai. Note that biogenic VOCs in Mount Tai were not removed in the sensitivity simulations. A total of 10 O_3 pollution events that occurred in Tai'an from May to September 2021 were selected to compare the changes of O_3 concentrations between the baseline and sensitivity simulations. Results show that after changing the terrain height of Mount Tai in the 10 selected pollution events, the simulated O_3 concentrations in eight pollution events decreased by 1.7~7.5 μg/m^3 relative to the baseline simulations, suggesting the positive contributions of the topography of Mount Tai to the occurrences of ozone pollutions in Tai'an (see Table 5). The simulated O_3 concentrations after reducing the height of Mount Tai in another two pollution events increased relative to the base simulations. Among the three monitoring stations, the reductions of O_3 concentrations caused by reducing terrain height were 2.4~8.8, 2.4~8.8, and 0.3~6.3 μg/m^3 for the Renkou School, Dianli College, and Shandong First Medical University stations, respectively. In the simulations, Renkou School and Dianli College were on the same grid, so the reductions of O_3 concentrations were the same. Through the analyses of the wind directions in the two pollution events in which the O_3 concentrations increased (3–9 July 2021 and 7–12 August 2021), it was found that the prevailing wind directions in Tai'an were northeasterly during the two pollution events (see Figure 5c,d). This might affect the transports of ozone and precursors from the upwind areas and result in an increase in O_3 concentrations after reducing the terrain heights of Mount Tai.

Table 5. Comparisons of O_3 concentrations between the baseline and scenario simulations by reducing the Mount Tai terrain heights for ten O_3 pollution events in 2021 in Tai'an (μg/m^3).

Pollution Events	Average			$O_3 \geq 160$ (μg/m^3)			$O_3 \geq 200$ (μg/m^3)		
	Observation	Base	Scenario	Observation	Base	Scenario	Observation	Base	Scenario
2021.5.9–5.12	119.1	122.7	115.3	189.5	167.6	163.6	213.9	197.9	193.9
2021.5.18–5.21	125.8	130.9	129.2	169.6	180.2	177.4	/	/	/
2021.5.29–6.2	155.2	146.1	143.0	198.9	180.6	176.0	220.2	199.2	193.1
2021.6.4–6.13	151.6	138.1	136.1	212.3	180.7	176.6	233.8	195.2	190.9
2021.6.18–7.1	137.4	138.5	136.6	205.4	189.9	185.5	235.8	209.3	197.9
2021.7.3–7.9	124.2	132.3	142.0	188.0	185.0	195.0	207.2	209.9	214.6
2021.8.2–8.4	131.3	110.0	102.6	180.5	161.8	154.9	206.9	197.4	183.7
2021.8.7–8.12	119.6	128.9	124.8	187.5	168.5	172.2	218.4	190.5	186.4
2021.8.15–8.21	122.8	111.8	105.3	183.1	175.6	173.1	211.5	199.7	199.7
2021.9.3–9.6	84.3	69.1	63.8	/	/	/	/	/	/

4. Conclusions

In recent years, Tai'an has suffered from serious O_3 pollution, with the annual average concentrations of O_3-8h-90per above 180 μg/m^3 from 2016 to 2021. In this study, we analyzed the O_3 pollution situation in Tai'an from May to September, and found that the pollution during this period was mainly light pollution, accounting for 69% to 100% of the total polluted days, with June being the most polluted month. High temperature (>30 °C), low relative humidity (20%~40%), and low wind speed (1~3 m/s) provided favorable meteorological conditions for the generation of O_3, especially with prevailing southerly wind, resulting in light and above pollution in Tai'an. The results of backward trajectory analysis showed that in the summer pollution trajectories, the pollution trajectories in the southwest direction accounted for the highest proportion of all kinds of trajectories. When the southerly winds dominated, the accumulated O_3 concentrations were significantly higher than those of the surrounding cities due to the blocking of the northern mountains. Comparing the three monitoring stations in Tai'an, the average O_3 concentrations from May to September 2021 at the Renkou School station closer to the mountain were 13~15 μg/m^3 higher than those at the Dianli College and the Shandong First Medical University stations, which are relatively far away from the mountains, indicating that the Renkou School station was more affected by the obstructions of the mountains. The WRF-CMAQ was applied in this study to quantify the blocking effect of mountains, and results indicated that the average O_3 concentrations in Tai'an decreased about 1.7~7.5 μg/m^3 after reducing the terrain height of Mount Tai in the model. Our results are of great significance for guiding

Tai'an to control O_3 in the future, and measures can be formulated in advance based on our conclusions.

Supplementary Materials: The following supporting information can be downloaded at: https://www.mdpi.com/article/10.3390/atmos13081299/s1. Figure S1: The variations of average monthly concentrations of temperature, relative humidity, precipitation, and wind speed in Tai'an from 2016 to 2021; Figure S2: The monthly mean concentrations of maximum 8-h O_3 (a) and NO_2 (b) in Tai'an during May September from 2016 to 2021.

Author Contributions: S.Y. and Y.G. designed this research; C.C., T.Y., H.K., J.L. and X.Z. performed data analyses; Y.Z., M.L., T.H. and Y.J. performed model simulations; J.C. made contributions to discussion; S.Y., Y.G., C.C., Y.Z., Y.J. and M.L. wrote the paper. All authors have read and agreed to the published version of the manuscript.

Funding: This research was funded by the National Natural Science Foundation of China (Nos. 42175084, 21577126, and 41561144004), Department of Science and Technology of China (Nos. 2018YFC0213506 and 2018YFC0213503), and National Research Program for Key Issues in Air Pollution Control in China (No. DQGG0107).

Institutional Review Board Statement: Not applicable.

Informed Consent Statement: Not applicable.

Data Availability Statement: The observation data and model results are available upon request.

Acknowledgments: The authors would like to thank Qiang Zhang of Tsinghua University for providing the MEIC emission inventory in East Asia for this study. This study was guided and assisted by the tracking team of "One City, One Policy".

Conflicts of Interest: The authors declare no conflict of interest.

References

1. Cheng, L.J.; Wang, S.; Gong, Z.Y.; Yang, Q.; Wang, Y.Y. Pollution trends of ozone in ozone monitoring pilot cities from 2008 to 2016. *Environ. Monit. China* **2017**, *33*, 26–32. [CrossRef]
2. Yu, Y.J.; Meng, X.Y.; Wang, Z.; Zhou, W.; Yu, H.X. Driving Factors of the Significant Increase in Surface Ozone in the Beijing-Tianjin-Hebei Region, China, During 2013–2018. *Environ. Sci.* **2020**, *41*, 106–109. [CrossRef]
3. Zhang, M.; Ding, C.; Li, Y.; Wang, G.X.; Lin, J.J.; Meng, H. Spatial and Temporal Distribution of Ozone and Influencing Factors in Shandong Province. *Environ. Sci.* **2021**, *12*, 5723–5735. [CrossRef]
4. Zhao, W.; Gao, B.; Lu, Q.; Zhong, Z.Q.; Liang, X.M.; Liu, M.; Ma, S.X.; Sun, J.R.; Chen, L.G.; Fan, S.J. Ozone Pollution Trend in the Pearl River Delta Region During 2006–2019. *Environ. Sci.* **2021**, *42*, 92–101. [CrossRef]
5. Zhou, M.W.; Kang, P.; Wang, K.K.; Zhang, X.L.; Hu, C.Y. The spatio-temporal aggregation pattern of ozone concentration in China from 2016 to 2018. *China Environ. Sci.* **2020**, *40*, 110–121. [CrossRef]
6. Pu, Q.; Li, Z.L.; Zhang, Y.; Gao, Y.H.; Lv, P.J.; Zhang, W.D.; Zhai, C.Z. Characterization of atmospheric circulation and transmission in Chongqing City during ozone polluted days. *China Environ. Sci.* **2021**, *41*, 18–27. [CrossRef]
7. Sun, Y.C.; Yan, X.Y.; Gou, X.H.; Liu, Y.; Wang, X.; Zuo, H.J.; Liu, Y.L. Characteristics and Correlation of Ozone and Its Precursors in Typical Cities in China. *Res. Environ. Sci.* **2020**, *33*, 50–59. [CrossRef]
8. Wang, X.D.; Yin, S.S.; Yang, J.; Yuan, M.H.; Zhang, R.Q.; Li, Y.S.; Lu, X. Characteristics, Meteorological Influences, and Transport Source of Ozone Pollution in Zhengzhou City. *Environ. Sci.* **2021**, *42*, 604–615. [CrossRef]
9. Yan, R.N.; Ye, H.; Lin, X.; He, X.; Chen, C.; Shen, J.D.; Xu, K.E.; Zheng, X.J.; Wang, L.J. Characteristics and influence factors of ozone pollution in Hangzhou. *Acta Sci. Circumst.* **2018**, *38*, 1128–1136. [CrossRef]
10. Cui, M.R.; Bai, L.Y.; Feng, J.Z.; Lin, X.S.; Li, H.N.; Gao, W.W.; Li, Z.W. Analysis of temporal and spatial variations of ozone coupling with dynamics of meteorological factors in the Beijing-Tianjin-Tangshan region. *Acta Sci. Circumst.* **2021**, *41*, 373–385. [CrossRef]
11. Huang, J.P.; Zhou, C.H.; Lee, X.H.; Bao, Y.; Zheng, Y. The effects of rapid urbanization on the levels in tropospheric nitrogen dioxide and ozone over East China. *Atmos. Environ.* **2013**, *77*, 558–567. [CrossRef]
12. Jiang, M.Q.; Lu, K.D.; Su, R.; Tan, Z.F.; Zhang, Y.H. Ozone formation and key VOCs in typical Chinese city clusters. *Chin. Sci. Bull.* **2018**, *63*, 1130–1141. [CrossRef]
13. Li, L.; Liu, Y.X.; Xiao, S.R.; Wang, S.Y.; Mi, J.L. Temporal and Spatial Variation of O_3 Concentration Near the Surface of Shandong Peninsula and Analysis of Potential Source Areas. *Environ. Sci.* **2022**, *43*, 1256–1267. [CrossRef]
14. Li, K.; Liu, M.; Mei, R.B. Pollution characteristics and sensitivity analysis of atmospheric ozone in Taian city. *Environ. Sci.* **2020**, *41*, 3539–3541. [CrossRef]
15. Wang, Y. Study on characteristics of ozone pollution in Tai'an City. *Environ. Dev.* **2018**, *10*, 42–43. [CrossRef]

16. Eder, B.; Yu, S. A performance evaluation of the 2004 release of Models-3 CMAQ. *Atmos. Environ.* **2006**, *40*, 4811–4824. [CrossRef]
17. Zhang, Y.; Chen, X.; Yu, S.; Wang, L.; Li, Z.; Li, M.; Liu, W.; Li, P.; Rosenfeld, D.; Seinfeld, J.H. City-level air quality improvement in the Beijing-Tianjin-Hebei region from 2016/17 to 2017/18 heating seasons: Attributions and process analysis. *Environ. Pollut.* **2021**, *274*, 116523. [CrossRef]
18. Cheng, N.L.; Li, Y.T.; Zhang, D.W.; Chen, T.; Wang, X.; Huan, N.; Chen, C.; Meng, F. Characteristics of ozone over standard and its relationships with meteorological conditions in Beijing city in 2014. *Environ. Sci.* **2016**, *37*, 2041–2051. [CrossRef]
19. Huang, X.G.; Shao, T.J.; Zhao, J.B.; Cao, J.J.; Yue, D.P. Impact of meteorological factors and precursors on spatial distribution of ozone concentration in eastern China. *China Environ. Sci.* **2019**, *39*, 2273–2282. [CrossRef]
20. Qi, Y.J.; Yu, S.J.; Yang, J.; Yin, S.S.; Cheng, J.H.; Zhang, R.Q. Analysis of characteristics and meteorological influence factors of ozone pollution in Henan province. *Environ. Sci.* **2020**, *41*, 587–590. [CrossRef]
21. Yang, J.; Yin, S.S.; Yu, S.J.; Wang, L.J.; Zhang, R.Q.; Hao, J.L.; Su, F.C.; Qi, Y.J. Characteristic of surface ozone and meteorological parameters analysis in Anyang city. *Environ. Sci.* **2020**, *41*, 115–119. [CrossRef]
22. Ren, C.B.; Wu, L.X.; Zhang, Y.Y.; Li, J.L.; Chai, M.; Xiang, C.C. Analyze to the seasonal differences of transport pathways and potential source-zones of Beijing Urban $PM_{2.5}$. *China Environ. Sci.* **2016**, *36*, 2591–2598. [CrossRef]
23. He, B.Q.; He, C.; Yang, L. Study on the spatiotemporal variation of surface ozone and the main driving factors in North China. *China Environ. Sci.* **2022**, *42*, 1562–1574. [CrossRef]
24. Zhang, Y.J.; Zhao, T.L.; Yin, C.Z.; Wang, Z.F.; Ge, B.Z.; Liu, R.Y.; Du, X.X. Seasonal variation of the relationship between surface $PM_{2.5}$ and O_3 concentrations in Xuzhou. *China Environ. Sci.* **2019**, *5*, 2267–2272. [CrossRef]

atmosphere

Article

Characteristics of Resuspended Road Dust with Traffic and Atmospheric Environment in South Korea

Sungjin Hong, Hojun Yoo, Jeongyeon Cho, Gyumin Yeon and Intai Kim *

Department of Transportation Engineering, Myongji University, Youngin 17058, Korea; ahsj9927@mju.ac.kr (S.H.); dbghwns0@naver.com (H.Y.); wjddus942@naver.com (J.C.); bx1002@naver.com (G.Y.)
* Correspondence: kit1998@mju.ac.kr; Tel.: +82-10-9767-9017

Abstract: Characterizing the influencing factors of resuspended dust on paved roads according to the atmospheric environment and traffic conditions is important to provide a basis for road atmospheric pollution control measures suitable for various road environments in the future. This study attempts to identify factors in the concentration of resuspended dust according to the level of road dust loading and PM_{10} emission characteristics according to atmospheric weather environment and traffic conditions using real-time vehicle-based resuspended PM_{10} concentration measuring equipment. This study mainly focuses on the following main topics: (1) the increased level of resuspended dust according to vehicle speed and silt loading (sL) level; (2) difference between atmospheric pollution at adjacent monitoring station concentration and background concentration levels on roads due to atmospheric weather changes; (3) the correlation between traffic and weather factors with resuspended dust levels; (4) the evaluation of resuspended dust levels by road section. Based on the results, the necessity of research to more appropriately set the focus of analysis in order to characterize the resuspended dust according to changes in the traffic and weather environment in urban areas is presented.

Keywords: silt loading; resuspended dust and lane traffic level; vehicle-based real-time PM measured system; yellow dust; PM_{10}

Citation: Hong, S.; Yoo, H.; Cho, J.; Yeon, G.; Kim, I. Characteristics of Resuspended Road Dust with Traffic and Atmospheric Environment in South Korea. *Atmosphere* **2022**, *13*, 1215. https://doi.org/10.3390/atmos13081215

Academic Editors: Elena Hristova, Manousos Ioannis Manousakas, Anikó Angyal and Maria Gini

Received: 8 July 2022
Accepted: 28 July 2022
Published: 1 August 2022

1. Introduction

Particulate matter (PM) with diameters of less than 10 μm can cause various damages to the human body, such as respiratory failure and mental illness [1]. Additionally, their concentrations are often elevated close to roadways [2]. PM is caused through the contribution of effective road management, so the degree of contamination must be determined by including a quantitative analysis of contamination to determine if it is important enough to implement control measures. Road dust (RD) is generated and diffused by traffic, and it is one of the main causes of road pollution in cities. It is categorized as an exhaust emission (EE). For example, exhaust emissions include emissions from internal combustion locomotives and automobiles. In addition, nonexhaust emissions (NEEs) are generated by vehicles driving on roads [3]. In recent decades, it has been known that direct emissions from tailpipes have been reduced significantly due to the strengthening of exhaust emission standards applied to vehicle air masses and the extensive implementation of catalytic converters and diesel particulate filters [4]. However, it has been suggested that the main cause of failure to derive an effective reduction in PM_{10} in large cities is the undervaluation and neglect of NEEs, which consist of road dust, tire wear, road wear, and brake wear particles caused by tire–road friction [5]. It also emphasizes that emissions from road traffic contribute to PM concentrations to at least the same extent as exhaust emissions [3,4,6–8].

According to the 2017 Clean Air Policy Support System (CAPSS), which provides the extent of the emissions of air pollutants in the Republic of Korea, resuspended dust from roadways makes up 45% of the total concentration of dust. Therefore, as an alternative to the main management of suspended dust in urban components, road cleaning programs by local governments are used extensively. It is important to properly design and implement road cleaning in consideration of road dust characteristics and climate and weather characteristics because of the economic uncertainty caused by short-term effects without considering cleaning methods, regional characteristics, and weather characteristics [9]. However, the reason for the difficulty of establishing the characteristics of dust that is resuspended from roadways is that the study to estimate the concentration of resuspended dust through the establishment of related processes and modeling is still at the basic level [10,11]. Furthermore, due to the large fluidity characteristics of very small substances, i.e., smaller than PM_{10}, research to determine the chemical profile of dust on roads composed of meteorological environments, transportation activities, and construction requires significant resources and time.

In this study, the concentration of silt loading (sL), resuspended dust of PM_{10}, and floating dust PM_{10} on paved roads separated by space was measured, and the results were used to analyze the concentration of resuspended dust on paved roads in urban areas. The PM_{10} concentration measured during operation quantified the characteristics of PM_{10} based on the analysis of PM_{10} information of the air pollution monitoring station, weather information of the competent meteorological station, and traffic information measured through the use of a vehicle detection system (VDS). Previous studies have shown the physical and chemical properties of fine dust on Mars by testing re-entrained aerosol kinetic emissions from road (TRAKER) equipment for real-time monitoring. The equipment has the advantage of being able to immediately investigate dust emissions on many roads with long sections [12–14]. This clearly states that the fine dust concentration measured by the system can be used to quantify the relationship between major factors for increasing road particle emissions by utilizing repeated measurements of resuspended dust to supplement the realistic limitations of AP-42. Various studies in these fields were mainly for identifying the emission characteristics or chemical profiles of the resuspended dust from roads [15–21] and a few studies considered the spatial and time variations in the major parameters [22].

The main purpose of this study was to provide an experimental basis for future road environmental changes by analyzing and characterizing the main emission characteristics of the PM_{10} concentrations measured on paved roads that are caused by sL, air pollution, meteorological variables, and traffic conditions. In this study, we focused on this and tried to establish a process to secure data reliability by constructing a resuspended PM_{10} measurements system and by observing the concentration of resuspended dust in separate sections of a road, i.e., sections less than 500 m in length.

2. Materials and Methods

2.1. Study Area

To present a reasonable classification of the characteristics of PM_{10} emissions on roads, locations were selected that were representative of vehicle-operating conditions and environmental characteristics. Measurements were held during periods without traffic impact observed in four sections from September to October 2021. During the measurement period, there were no weather events such as rainfall (0.01 inches of rain) that could have had a major effect on the fine dust concentration [23,24]. In addition, another measurement was determined on the concentration of resuspended dust on paved roads where traffic operated (Figure 1). Additionally, background PM_{10} data (Bg) and weather variables and traffic condition data were collected through the urban atmosphere and traffic monitoring system at the same time. The traffic data were collected in six directions in three adjacent areas where traffic volume VDS was installed (Table 1).

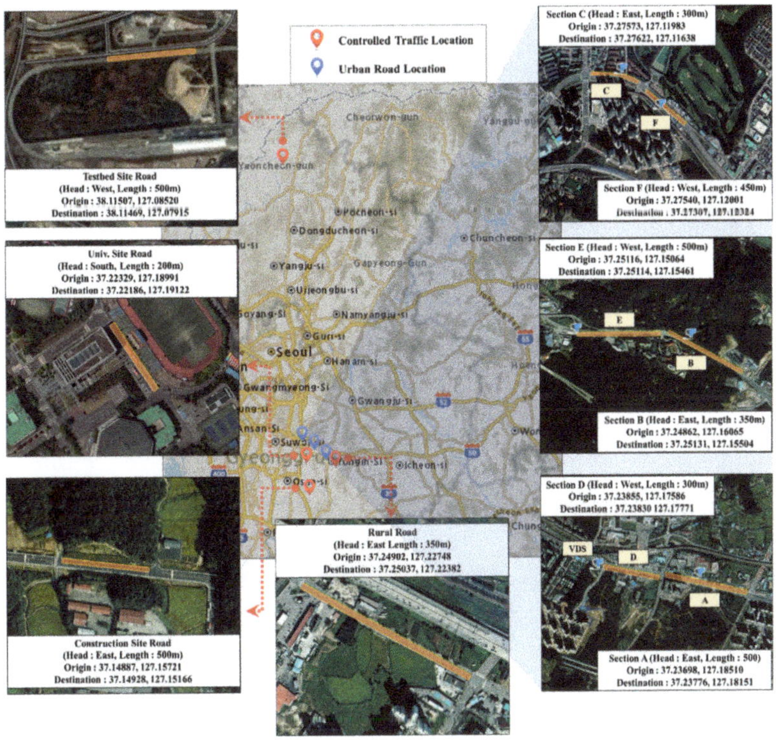

Figure 1. Study area in South Korea (Controlled traffic location and urban road location).

Table 1. Road conditions of the measurement section and distance from the observation.

Category	Site	Length (m)	Road Width/Condition	Distance of the Observation	
				City Bg PM_{10}	Meteorological Variables
Controlled Traffic	Testbed	500	One-lane road	With a radius of 2.5 km	With a radius of 24 km
	Construction Site	500	Two-lane road	With a radius of 3.5 km	With a radius of 20 km
	Univ.	200	One-lane road	With a radius of 1.6 km	With a radius of 14.5 km
	Rural	350	Two-lane road	With a radius of 2.8 km	With a radius of 17.5 km
Operating road	A	500	Four-lane road/ 905 (Veh/time/day)	With a radius of 8 km	With a radius of 14.5 km
	B	350	Three-lane road/ 633 (Veh/time/day)	With a radius of 5 km	With a radius of 11.5 km
	C	300	Four-lane road/ 1006 (Veh/time/day)	With a radius of 1 km	With a radius of 8.2 km
	D	300	Four-lane road/ 850 (Veh/time/day)	With a radius of 8 km	With a radius of 13.2 km
	E	500	Three-lane road/ 431 (Veh/time/day)	With a radius of 5 km	With a radius of 11 km
	F	450	Four-lane road/ 697 (Veh/time/day)	With a radius of 1 km	With a radius of 8.5 km

2.2. *Resuspended PM_{10} Measuring System*

Road dust generated by vehicles driving on urban roads is known to have the greatest impact on the concentration of PM depending on the size of the sL collected on the surface of the road and the load and speed of the vehicles being driven on the road [11,14]. However, the method of measuring fine dust by performing and comparing direct dust collection activities through human activities on roads has safety and efficiency problems, and there are practical limitations in observing real-time concentrations. Mobile resuspended dust measurement systems were developed and implemented in the early 2000s. Scamper, TRAKER, and others generally use a method to calculate the difference between background concentration and tire measurements [12].

By establishing a real-time mobile dust measurement system based on the Desert Research Institute (DRI)'s TRAKER method, the temporal and spatial distribution characteristics of the resuspended dust on urban pavement can be checked and the impact of road dust sources can be confirmed by traffic and environmental factors [25].

The resuspended PM_{10} measurement system (Figure 2) used an optical sensor that inhaled PM every second to determine the PM_{10} concentration on the road through the front (mounted on the hood) inlet of the vehicle and the rear (right front wheel) inlet of the tire [25]. In quantifying the optical and physical characteristics of resuspended dust, the measurement of real-time dust concentration used scattered light to measure the weight of dust passing through. The measurement method clearly states that the concentration of resuspended dust can be measured [26].

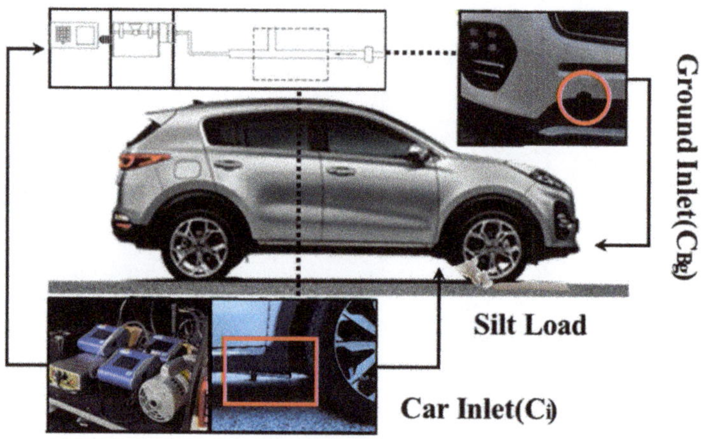

Figure 2. Configuration of the resuspended PM_{10} measurement system.

To summarize, the equipment contains six main components, i.e.,:

1. Vehicle specifications: 2016 SUV Sportage 2.02 WD, Tire Model 225/60R17.
2. PM_{10} Sensor: The 8530 Dust-Trak Optical PM Sensor Model (TSI Company, Shoreview, MN, USA) with PM_{10} inlet is a portable device that can measure the concentrations of various sizes of dust particles in real time, such as PM_{10} and $PM_{2.5}$.
3. Inlet: The right tire line inlet is 175 mm above the ground and 50 mm behind the tire. The PM_{10} concentration (C_i) obtained from the back of the tire during movement indicates that the air velocity measured along the center of the tire was not affected significantly by the surrounding wind direction at a distance less than 100 mm from the tire [27].
4. Uniform Velocity Flow Sampling Inhalation: A vacuum pump is controlled by a customized suction port to adjust the pressure of the suction port appropriately, thereby producing a pressure-free suction port, so the air flow that is being sampled has the same fluid flow as the surrounding air flow.
5. Data Collection System: Laptop computers were used to collect and transmit Dust-Trak data at one-second intervals.
6. Monitoring Camera: A camera capable of monitoring the inside and outside of the vehicle can be installed to visually cross-verify the measurement data in the driving section and monitor the speed observed on the vehicle's speedometer.

2.3. AP-42 Silt Loading Collection

In order to collect fine dust on the surface of a paved road, a section of the road was sampled using an 800 W vacuum cleaner, 3 m^2 frame of silt loading sampling, a fine dust brush, and a 3 kW gasoline generator [23].

Dust was immediately collected in the chamber, and it was eventually collected from the quartz fiber filters. At least three samples with different aerodynamic diameters were taken from a surface area of 3 m^2 for field sampling to reduce errors [28]. The samples were weighed using 200 mesh, standard Taylor screens, and a sieve filter machine.

2.4. Collecting City PM_{10}, Meteorological Variables, and Traffic Data

The data were collected as external impact indicators to analyze the effects of changes in the meteorological variables and traffic conditions on the PM_{10} concentration levels observed through the resuspended PM_{10} measurements system [29,30]. The collected elements were PM_{10}, the temperature of the ground (Gr temp), relative humidity (RH), wind speed (WS), sea level pressure, visibility, and time of after rainfall (rainfall time). The traffic data observed with adjacent VDS devices were divided into sections during the measurement period.

3. Results and Discussion

3.1. Verification of the Real-Time Measuring System

As a method of quantifying the concentration of road dust by sections using a mobile, real-time, PM_{10} monitoring system, the concentration data measured with the inlet built into the front bumper of the vehicle were set to the background PM_{10} concentration (C_{Bg}). This value was obtained by subtracting the background concentration from the resuspended PM_{10} concentration (C_i) measured using the dust resuspended on the road along with the background concentration on the road through friction between the tires and the surface of the road. This value was quantified [31]. C_{res} was calculated according to the following equation:

$$C_{res} = \frac{1}{n} \cdot \sum_{i=1}^{n} C_i - C_{Bg} \quad (1)$$

where C_{res} is the concentration of resuspended dust from the road surface ($\mu g/m^3$), C_i is the PM_{10} concentration measured in inches on the back of the tire resuspended dust concentration ($\mu g/m^3$), C_{Bg} is the PM_{10} concentration measured at the front bumper inlet of the vehicle background concentration ($\mu g/m^3$).

Monitoring was performed to analyze the change in the concentration of resuspended dust due to sL in the space and time, where the concentration measured during the interval transit time varied by the inlet. The large changes meant that the C_{Bg} measurement was more than 100 $\mu g/m^3$ within a short measurement period, and C_i measured on the back of the tire was more than 1000 $\mu g/m^3$. Unlike the sL effects, this was proven to be an effect of the road dust as well as the external effects, such as the exhaust gas of other vehicles and the addition and deceleration of vehicles. Considering the average PM_{10} concentration was 50 $\mu g/m^3$ and the 24 h average PM_{10} concentration was 100 $\mu g/m^3$ level, it was excluded as an outlier. This was based on previous studies that indicated that the change in the C_{Bg} level when the vehicle was not moving was not affected significantly by the exhaust of other vehicles, and the concentration of C_i measured on the back of the tire could be used as the concentration of resuspended dust [13].

The process of verifying that the monitoring system can perform data collection functions for quantifying the concentration of road dust was conducted under three experimental conditions that could determine whether the measured concentration of the dust was significant when the vehicle moved.

(1) We ensured that the measured concentration did not vary significantly when the vehicle stopped, and tried to calculate the range of the measured speed by comparing

the concentration measured by increasing the speed as it passed the section at a constant speed.

(2) If a certain measurement speed was maintained, the average C_{res} concentration measured in a short section and the amount of dust accumulated on the surface of the road and the concentration of the resuspended dust due to vehicle travel were quantified by the sL collected at three different locations in the section.

(3) The correlation was analyzed by comparing the concentration of C_{res} measured by the size of sL in different sections (regions) selected by classification according to the major causes of the inflow of the expected material.

In the absence of peripheral vehicles, the monitoring system maintained a constant stop period and a constant speed, and the C_{Bg} value measured during the driving period maintained a constant slight change of approximately 2 to 6 μg/m^3 per second. When the sL level collected in the section was 0.43 g/m^2 for the mean and 0.16 g/m^2 for the SD, the concentration was measured by maintaining 30, 40, 50, and 60 km/h, respectively, to analyze the effect on changes in the C_{res} concentration as the speed increased during the section (Table 2).

Table 2. C_{res} concentration mean and standard deviation value measured with car speed in testbed.

Car Speed (km/h)	C_{res} Mean (μg/m^3)	C_{res} SD (μg/m^3)
30	23.0	3.4
40	54.6	14.8
50	69.1	18.9
60	93.0	21.3

The expression for increasing C_{res} concentration by velocity was the same as in Equation (2) (Figure 3).

$$y = 7.2971e^{0.0442x} \tag{2}$$

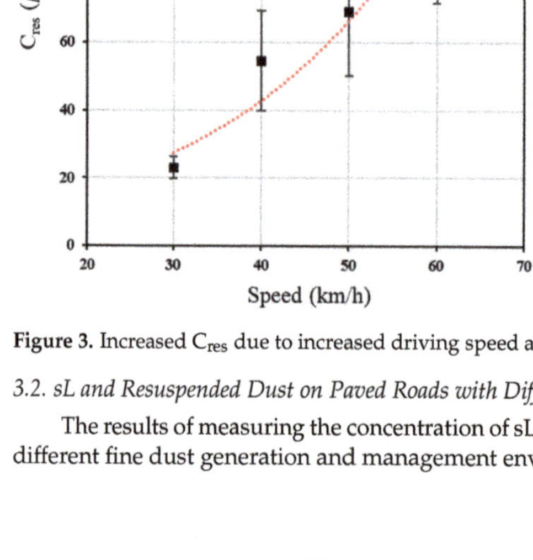

Figure 3. Increased C_{res} due to increased driving speed at sL (0.043 g/m^2).

3.2. sL and Resuspended Dust on Paved Roads with Different Inflow Conditions

The results of measuring the concentration of sL and resuspended dust in sections with different fine dust generation and management environments while maintaining a speed of 50 km/h, which usually is used as the speed limit on roads operating in the center of the city, are presented in Table 3 and Figure 4. The sL collected at the entrance and exit of the construction vehicle next to the cement plant, where the traffic of medium-sized vehicles mainly occurred, was the highest at 0.093 g/m^2, and the concentration of resuspended dust was observed to be 225 μg/m^3. For four consecutive days on the controlled testbed and low-traffic, university roads, the sL levels measured at the same time were found to continue to increase in deposition in the absence of specific events for rainfall, and the weather conditions observed were at a 11 μg/m^3 C_{res} concentration (Table 3). As an influencing factor of resuspended dust generated on paved roads, it is known that the concentration level deposited on the section varies greatly depending on the heavy vehicle configuration and characteristics of traffic [32]. It has been shown that the resuspended dust from the road due to vehicle travel can be deposited, removed, moved into the air [23], or affected by weather conditions, geometric conditions, building geometry, etc. [29,33,34], or by regular road cleaning [9].

Table 3. Concentration of sL, resuspended PM_{10} on paved roads without any traffic.

Section		sL 1 (g/m^2)	sL 2 (g/m^2)	sL 3 (g/m^2)	Mean sL (g/m^2)	Bg PM_{10} (μg/m^3)	Mean C_i PM_{10} (μg/m^3)	Mean C_{Bg} PM_{10} (μg/m^3)	Mean C_{res} PM_{10} (μg/m^3)
Construction Site Road		0.115	0.089	0.074	0.093	30	242	17	225
Testbed Road	Day 1	0.024	0.038	0.039	0.034	13	51	4	47
	Day 2	0.043	0.044	0.042	0.043	18	77	4	73
	Day 3	0.062	0.068	0.063	0.064	26	93	11	81
	Day 4	0.069	0.08	0.075	0.075	21	91	10	81
Univ. Site Road	Day 1	0.002	0.002	0.002	0.002	10	40	11	29
	Day 2	0.003	0.003	0.003	0.003	16	41	17	24
	Day 3	0.005	0.005	0.007	0.006	23	50	22	28
	Day 4	0.009	0.01	0.008	0.009	36	58	37	21
Rural Road		0.007	0.006	0.003	0.005	58	77	66	11

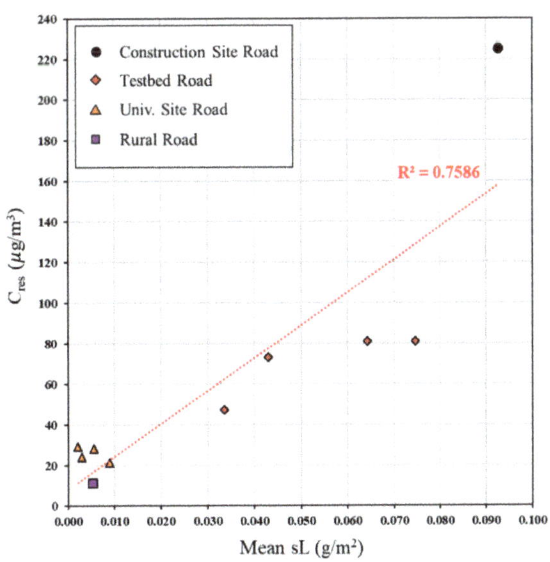

Figure 4. Trend of C_{res} growth due to sL level in other measured sites.

An analysis of the correlation between the C_{res} concentration and the collected sL measured when no ambient traffic occurred during the measurement and the space maintained was constant at 50 km/h speed on other roads showed a strong correlation, i.e., R^2 = 0.7586 (Figure 4). Depending on the increase in the sL level, the amount of C_{res} increases depended on the site being measured, which could vary depending on the source of the dust, but roughly represented a pattern in which the amount of resuspended dust along with the sL increased with the size of the sL.

3.3. Relationship between Section Background Concentration of Dust and Its Resuspended PM_{10}

Due to the relationship between C_{Bg} measured through the monitoring system and dust C_i resuspended due to driving, the impact of the sL size varied from site to site (Figure 5). If the observed concentration level was closer to the Y-axis, the tendency to increase the concentration of C_i compared to C_{Bg} was stronger, and the size of sL was less than 0.002 to 0.009 g/m^2 compared to 0.034 to 0.093 g/m^2, and the inclination of the C_{Bg} and C_i concentrations were closer to one. When the size of the sL gradually became higher than 0.034 g/m^2, it was observed that the concentration level of C_i increased linearly to show a difference of more than 47 to 225 μg/m^3. In the 0.002 to 0.009 g/m^2 range, the difference of the C_i level of C_{Bg} was 11 to 29 μg/m^3.

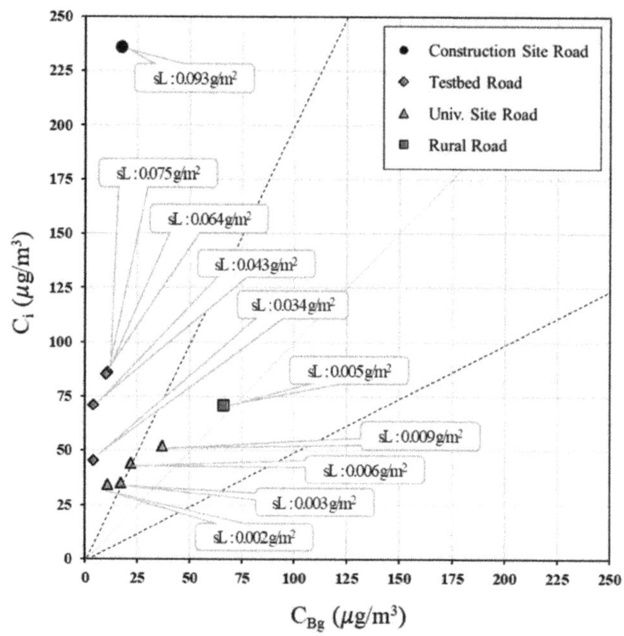

Figure 5. C_{Bg} vs. C_i concentration with sL in other measured sites.

Such a level of sL below 0.01 g/m^2 was a level that could be collected on roads where traffic was continuous, such as general highways and highways where particulate matter was not special [35] and low resuspended dust was observed. C_{Bg} measured in a surrounding vehicle-free environment was found to increase with the level of air pollution (Bg) measured at adjacent stations at each measurement site, but the correlations were analyzed and found to have a strong correlation of R^2 = 0.8484 (Figure 6).

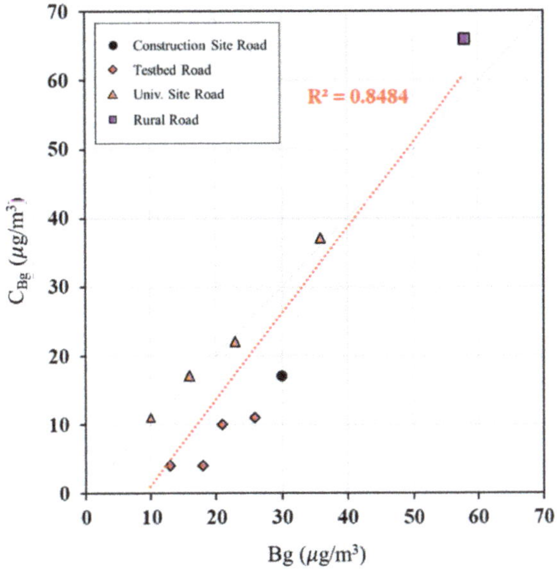

Figure 6. Compared concentration of Bg and C_{Bg} on paved roads (without any traffic).

The measured C_{Bg} was 4 to 14 μg/m^3, in Bg 10 to 20 μg/m^3, 10 to 22 μg/m^3, in Bg 20 to 30 μg/m^3, 17 to 37 μg/m^3, in Bg 30 to 40 μg/m^3, and 50 to 60 μg/m^3 in Bg 50 to 60 μg/m^3. However, the increase may have varied depending on the spatial difference between the observed sections. This level was considered to be a significant result when comparing the report [36,37] that the mortality rates of all diseases, cardiopulmonary disease, and lung cancer increased significantly by 4%, 6%, and 8%, respectively, for every 10 μg/m^3 increase in the concentration of particulate matter.

The relationship between Bg and C_{Bg}, measured during the running of the monitoring system on urban roads, was divided into two main patterns, i.e., (1) the yellow dust storm effect and (2) the no yellow dust storm effect. The yellow dust storm effect is a kind of dust fog, which is a phenomenon in which a large amount of yellow dust, which is called the continental yellow dust, drifts over a given area [38], indicating that the level of Bg and C_{Bg} measured at adjacent stations can vary greatly. The correlation between Bg and C_{Bg} measured on roads during traffic operation was relatively low compared to that of Bg and C_{Bg} measured on roads with limited traffic, and $R^2 = 0.4697$ was shown when the yellow dust phenomenon was excluded (Figure 7). It was found that differences in C_{Bg} also existed between sections in the same Bg and weather conditions observed by time zones at adjacent measuring stations. The difference was 20.45 μg/m^3, with different levels of C_{Bg} on roads where the main sources of fine dust and inflow were judged to be similar. In addition, the occurrence of yellow dust was characterized by the opposite characteristics of the correlation $R^2 = 0.1033$ as the Bg level of the adjacent station became very high. The Regional Bg could be expected to have a quantitative impact on the overall pollution on roads, but when yellow dust occurred, the difference between C_{Bg} and Bg on roads could cause different concentration levels of C_{res} in the process of the continuous floating of road resuspended dust.

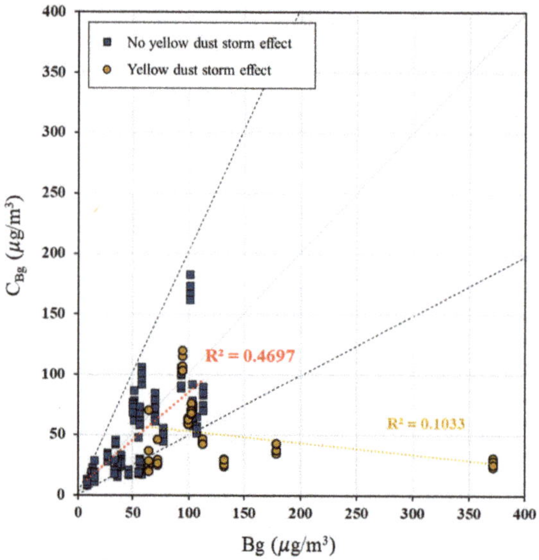

Figure 7. Concentration of Bg vs. C_{Bg} on paved roads during traffic operation (classification due to the effects of yellow dust).

The correlation between C_{Bg} and C_i concentrations measured on roads during traffic operation showed a strong correlation of R^2 = 9253 without the influence of yellow dust, a weak correlation of R^2 = 0.4256 with yellow dust, and a correlation of R^2 = 0.8031 in respect to the overall concentration (Figure 8). The average concentration at which C_i was derived was 226 μg/m^3, and the C_i concentration measured at the rear of the tire was the highest compared to C_{Bg}.

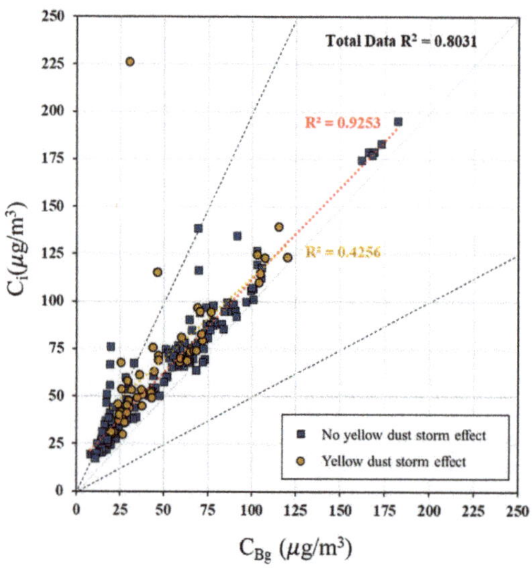

Figure 8. Concentration of C_{Bg} vs. C_i on paved roads during traffic operation (classification due to yellow dust impact).

3.4. Analysis of PM_{10} Concentration Characteristics of Resuspended Dust by Influencing Factors

For the correlation analysis between the concentration of resuspended dust measured on the road, atmosphere, meteorological variables, and traffic factors, dust measured data were collected (excluding yellow dust days data from the entire data) in six sections.

To analyze the effects of the average C_{res} concentration observed by road section on atmospheric and meteorological environments and traffic conditions, the C_{res} data samples were divided into five grades within the statistical analysis. (Lower quartile (25th): 7.7 μg/m^3; median: 12.7 μg/m^3; upper quartile (75th): 15.6 μg/m^3; high C_{res} (sL level: over 0.009 g/m^2): 29 μg/m^3 were used.)

- Low-level type: under 7.7 μg/m^3.
- Low- to middle-level type: 7.7 to 12.7 μg/m^3.
- Middle- to high-level type: 12.7 to 15.6 μg/m^3.
- High-level type: 15.6 to 29 μg/m^3.
- Bad-level type: over 29 μg/m^3 and the number of measurements by C_{res} grade was the same as Figure 9.

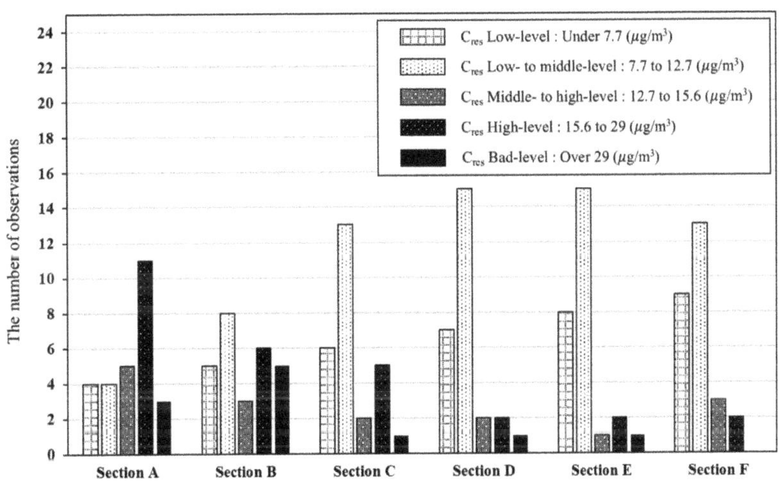

Figure 9. Comparison of the number of C_{res} level type observations.

Among them, the results of C_{res} measured as the bad level by section were observed in a wide range of Bg up to 15 to 113 μg/m^3 (Table 4). In different measured sections, C_{res} with the bad level was observed even at 15 μg/m^3 in the level of concentration of Bg, and did not increase linearly. It is known that the Bg concentration and weather effects have a major influence on road atmosphere pollution, but it is difficult to represent them thoroughly, and it can be expected that external factors that are applied differently for each section are measured in the same time.

Since PM_{10} on roads is known to be generated and diffused through various traffic elements, including traffic volume, traffic speed, road environment, and virtual elements [33], it is important to quantify the effect on the PM_{10} concentration by reflecting the traffic characteristics of each section. The order of average C_{res} concentration and traffic by liver level was Location 2—Section B (18.0 μg/m^3, 296 veh/time/lane) > Location 1—Section A (17.1 μg/m^3, 324 veh/time/lane) > Location 3—Section C (12.7 μg/m^3, 341 veh/time/lane) > Location 1—Section D (10.0 μg/m^3, 318 veh/time/lane) > Location 2—Section E (9.4 μg/m^3, 212 veh/time/lane) > Location 3—Section F (8.5 μg/m^3, 245 veh/time/lane) (Figure 10).

Table 4. Comparison of observed C_{res} bad level concentrations by section and Bg concentration of the same time zone.

Section	C_{res} ($\mu g/m^3$)	Bg ($\mu g/m^3$)
A	33.1	103
	34.4	39
	47.6	55
B	31.4	15
	33.5	55
	36.7	46
	46.7	113
	69.4	104
C	56.7	58
D	32.6	55
E	29.5	58

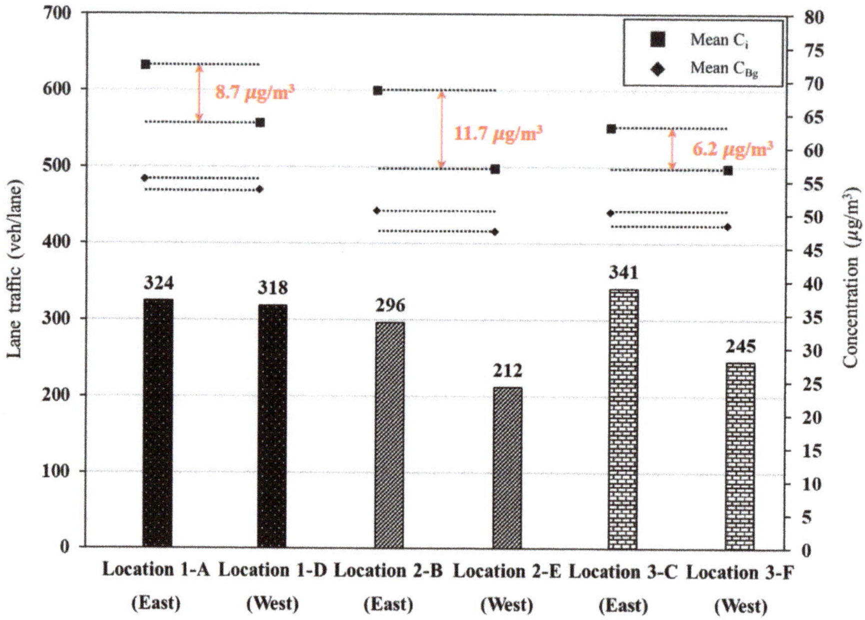

Figure 10. Comparison of lane traffic level and average values of C_i and C_{Bg} in the measurement section (the location of each measurement section was compared with Locations 1, 2, and 3, and the direction of the vehicle was the opposite side).

C_i and C_{Bg} showed a relatively high consistent tendency when traffic was high by location, and C_{res} results when traveling in the eastern direction were higher than C_{res} when traveling in the western direction, but the C_{res} concentration did not appear in order depending on the traffic difference. In order to confirm the distribution of the lane traffic volume data sample, the lower quartile (25th) value or less were classified into lane traffic low level, the lower quartile (25th) value to median (50th) value were divided into lane traffic low to middle level, the median(50th) value to upper quartile (75th) value were divided into middle to high level, and the upper quartile (75th) value or higher value was divided into high level type (Figure 11).

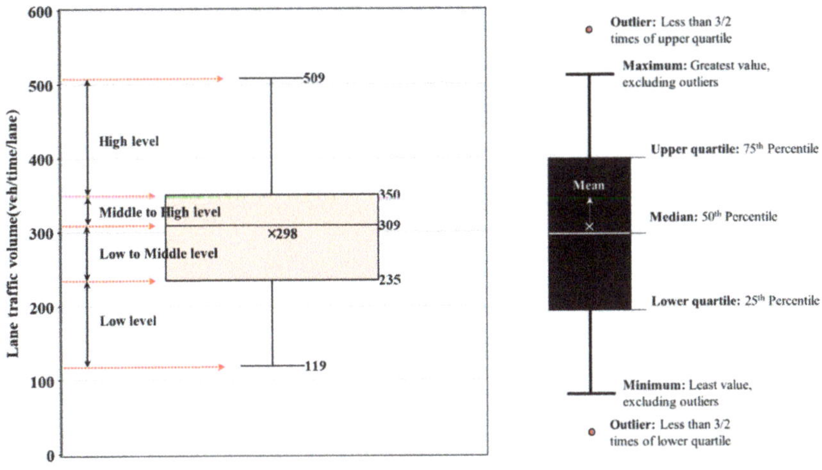

Figure 11. (**a**) Lane traffic level type classification considering lane traffic volume data distribution; (**b**) definition of boxplot used in this study.

Figure 12 shows the graph of the level type and road pollution levels C_i and C_{Bg} of lane traffic generated for each section. Relatively high C_{res} values were observed in sections with a relatively high frequency of occurrence of high-level lane traffic by location, and the results were as follows: Location 1—Section A (11 times, C_{res}: 17.1 μg/m^3) vs. Section D (7 times, C_{res}: 10 μg/m^3); Location 2—Section B (5 times, C_{res}: 18 μg/m^3) vs. Section E (1 time, C_{res}: 9.4 μg/m^3); Location 3—Section C (16 times, C_{res}: 12.8 μg/m^3) vs. Section F (0 time, C_{res}: 8.5 μg/m^3).

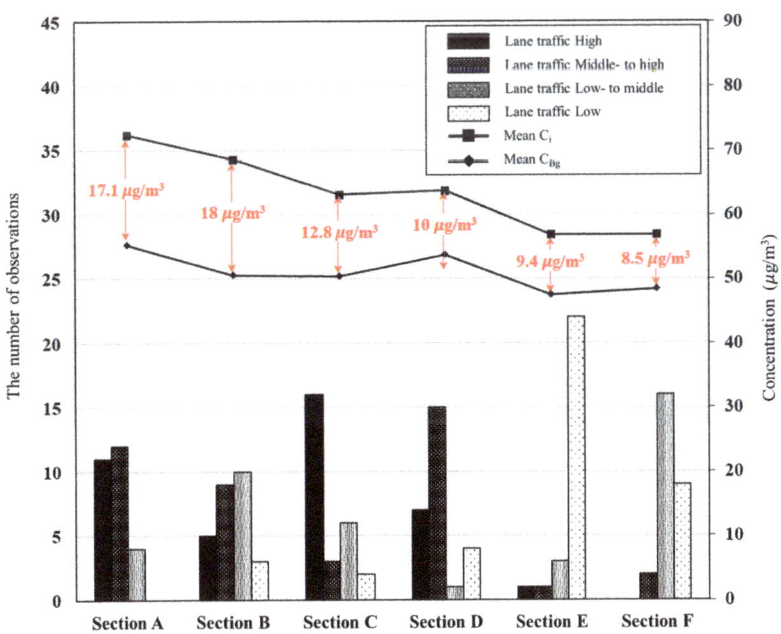

Figure 12. Comparison of the number of lane traffic level type observations and C_{res} concentrations.

In addition, the lowest average C_{Bg} (47.5 μg/m^3) and average C_i (56.9 μg/m^3) concentrations were measured in Section E, where low-level lane traffic appeared 22 times.

3.5. Characteristics of C_i, C_{Bg} Resuspended Dust by Lane Traffic Level by Measurement Section through Statistical Probability Comparison

As Table 5, the weather factor of Bg and the correlation coefficient R ≥ 0.4 were derived as the rainfall time, and a positive (+) correlation of R: 0.41 was observed. Additionally, as the visibility decreased, a negative (−) correlation of R: 0.35 was observed, C_{Bg} for each observed section also derived a positive (+) correlation between the rainfall time and R: 0.35 to 0.50, and a negative (−) correlation between visibility and R: 0.33 to 0.60, and rainfall time and visibility, were shown to affect Bg and C_{Bg} with similar properties.

Table 5. Bg, C_{Bg}, C_i by interval and correlation analysis of major factors.

Independent		Correlation R-Value								
		C_i	Bg	Lane Traffic	Gr Temp.	RH	WS	Sea Level Pressure	Visibility	Rainfall Time
Bg		0.66		0.04	−0.12	−0.14	−0.13	0.19	−0.35	0.41
Section A	C_{Bg}	0.95	0.62	−0.21	−0.24	0.36	−0.30	0.35	−0.60	0.44
	C_i		0.71	−0.14	−0.12	0.24	−0.27	0.25	−0.52	0.34
Section B	C_{Bg}	0.93	0.67	0.05	−0.28	0.29	−0.29	0.47	−0.59	0.48
	C_i		0.77	0.00	−0.11	0.18	−0.21	0.35	−0.46	0.43
Section C	C_{Bg}	0.96	0.70	−0.09	−0.29	0.16	−0.32	0.38	−0.50	0.44
	C_i		0.74	−0.17	−0.27	0.14	−0.30	0.34	−0.41	0.38
Section D	C_{Bg}	0.99	0.49	0.29	−0.14	0.08	−0.36	0.47	−0.40	0.35
	C_i		0.52	0.27	−0.14	0.04	−0.32	0.45	−0.35	0.34
Section E	C_{Bg}	0.98	0.52	0.53	−0.19	0.26	−0.33	0.35	−0.36	0.49
	Ci		0.56	0.55	−0.22	0.24	−0.29	0.34	−0.28	0.43
Section F	C_{Bg}	0.99	0.69	0.35	−0.31	0.23	−0.30	0.48	−0.56	0.47
	C_i		0.70	0.34	−0.30	0.24	−0.27	0.47	−0.53	0.46

This was similar to the results of a study that showed that the fine dust on the road increased over time after rainfall [39]. Additionally, the correlation coefficient between C_{Bg} and other weather factors was R: −0.13 to −0.31 in Gr temp., R: −0.16 to −0.36 in WS, R: 0.27 to 0.48 in sea level pressure, and the correlation coefficient between Bg and factors was higher than the Gr temp. (R: −0.12), WS (R: −0.13) and sea level pressure (R: 0.19).

As a result of analyzing the correlation between C_{Bg} and C_i concentrations and meteorological factors by section, the correlation R-value with C_i was consistently lower than the C_{Bg} correlation R-value and major weather factors (sea level pressure, visibility, and rainfall time) with R ≥ 0.4. However, a consistent trend was not observed in the correlation R-value with C_i and C_{Bg} concentrations according to the change of lane traffic by section.

We tried to analyze the characteristics of resuspended PM_{10} concentration according to the lane traffic level type by classifying lane traffic into level types and comparing the observed atmospheric weather environment level and resuspended PM_{10} concentration.

Table 6 shows the average, 75th, median, and 25th values classified to compare C_{res} and Bg and the distribution of weather factor data according to the lane traffic level type.

Table 6. Distribution of 25th, median, mean, and 75th percentile data by lane traffic level factors.

Lane Traffic Level	Percentile of Data	C_{res} (μg/m³)	Bg (μg/m³)	Gr Temp (°C)	RH (%)	WS (m/s)	Sea Level Pressure (hPa)	Visibility (m)	Rainfall Time (Days)
High (over 350 veh/time/lane)	75th	14.0	93.0	19.6	54.0	4.0	1025.9	1996.3	7.0
	Mean	11.9	53.2	15.9	48.1	3.0	1022.1	1685.2	5.4
	Median	9.6	50.0	13.3	49.5	3.1	1020.6	1932.0	4.0
	25th	6.1	23.5	11.5	36.8	1.6	1019.5	1635.0	3.0
Middle to high (309 to 350 veh/time/lane)	75th	16.5	93.0	26.0	68.0	3.7	1021.6	1995.0	7.0
	Mean	12.6	57.8	21.0	47.5	2.6	1019.6	1703.5	5.9
	Median	11.1	56.0	21.4	44.0	2.5	1019.5	1937.0	6.0
	25th	7.2	34.0	13.3	31.0	1.5	1018.0	1527	4.0
Low to middle (235 to 309 veh/time/lane)	75th	20.1	70.0	24.8	53.0	3.8	1025.6	1995.0	6.0
	Mean	15.3	54.0	19.0	46.1	2.8	1021.1	1736.0	4.7
	Median	10.7	51.0	19.6	46.0	2.8	1020.5	1932.0	4.0
	25th	8.0	34.0	13.3	31.0	1.6	1018.8	1685.0	4.0
Low (under 235 veh/time/lane)	75th	11.6	61.0	25.5	68.3	3.8	1024.8	1986.8	6.0
	Mean	10.1	54.4	17.0	53.0	2.6	1021.0	1588.0	4.8
	Median	8.9	51.0	18.7	52.0	2.5	1020.6	1895.5	5.0
	25th	8.3	35.5	9.8	35.0	1.6	1018.0	913	3.8

Among the data samples classified by the lane traffic level type, the order in which the C_{res} mean observed was found to be large was lane traffic low to middle level (15.3 μg/m³) > lane traffic middle to high level (12.6 μg/m³) > lane traffic high level (11.9 μg/m³) > lane traffic low level (10.1 μg/m³). Results measured at the lane traffic high-level Bg were mean: 53.2 μg/m³; median: 50.0 μg/m³; 25th to 75th value: 23.5 μg/m³ to 93.0 μg/m³; then, the Bg of the good level (under the results measured at the broadest range of Bg levels, from 31 μg/m³) to the bad level (81 to 151 μg/m³). Relatively high WS, sea level pressure, rainfall time 75th value (7.0 days), and high visibility 25th value (1635 m) were measured.

Results measured at the lane traffic middle to high level Bg were: mean: 57.8 μg/m³; median: 56.0 μg/m³; 25th to 75th value: 34.0 μg/m³ to 93.0 μg/m³; highest Bg concentration and Gr temp; rainfall time values were measured.

Results measured at the lane traffic low to middle level Bg were: mean: 54.0 μg/m³; median: 51.0 μg/m³; 25th to 75th value: 34.0 μg/m³ to 70.0 μg/m³; relatively high WS (1.6 to 3.8 m/s); sea level pressure (1018.8 to 1025.6 hPa); visibility (1685 to 1995 m) in a wide range.

Results measured at the lane traffic low level at the Bg level in the narrowest range were: mean: 54.4 μg/m³; median: 51.0 μg/m³; 25th to 75th values: 35.5 μg/m³ to 61.0 μg/m³. RH had a relatively high 75th value (68.3%); the visibility 25th value (913 m) was the lowest.

Figure 13 shows a box plot that compares the C_{Bg} and C_i mean values and the 25th, median, and 75th values calculated by the lane traffic level type, comparing the distribution of data samples, and analyzing the characteristics of road resuspended dust concentrations measured by traffic class. The maximum 25th value of C_{Bg} was 19.7 μg/m³ (lane traffic high level), and the minimum value was 18.1 μg/m³ (lane traffic low to middle level). The maximum median value of C_{Bg} was 47.8 μg/m³ (lane traffic low level), and the minimum value was 32.7 μg/m³ (lane traffic low to middle level). The maximum 75th value of C_{Bg} was 75.3 μg/m³ (lane traffic high Level), and the minimum value was 67.1 μg/m³ (lane traffic low Level). The maximum 25th% value of C_i was 36.6 μg/m³ (lane traffic low to Middle level), and the minimum value was 30.4 μg/m³ (lane traffic low level), The maximum median value of C_i was 63.9 μg/m³ (lane traffic low Level), and the minimum value was 54.5 μg/m³ (lane traffic low to middle Level). The maximum ^{75th} value was 85.1 μg/m³ (lane traffic high level), and the minimum value was 77.7 μg/m³ (lane traffic low level).

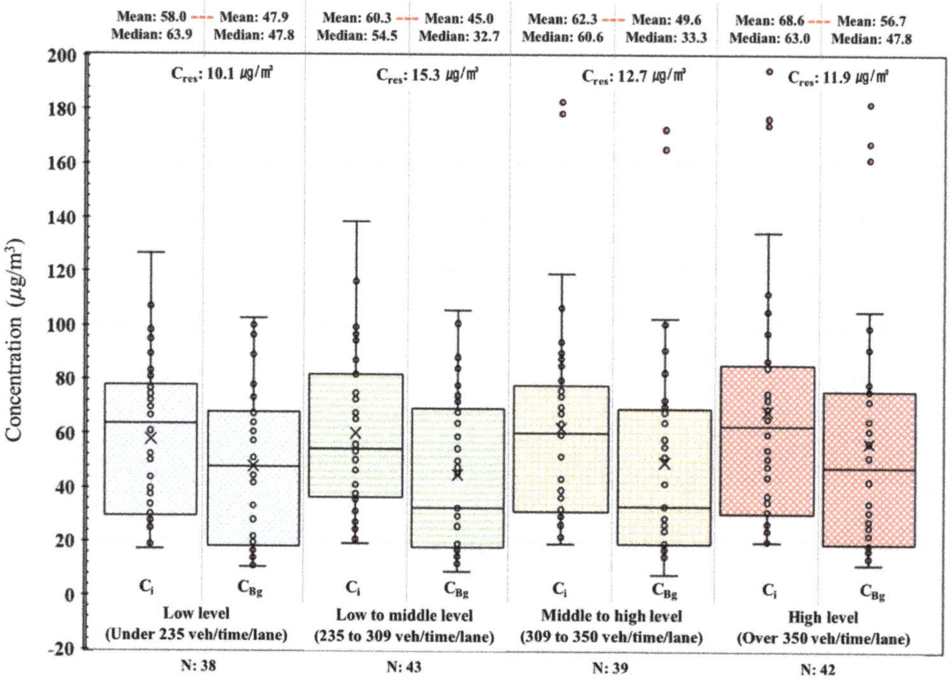

Figure 13. Comparison of Box plot by lane traffic level types.

The average level of lane traffic level for C_{Bg} was observed to be relatively high in the order of lane traffic low to middle level (45.0 $\mu g/m^3$) < lane traffic low level (47.9 $\mu g/m^3$) < lane traffic middle to high level (49.6 $\mu g/m^3$) < lane traffic high level (56.7 $\mu g/m^3$). The average level of C_i was observed in the order of lane traffic low level (58.0 $\mu g/m^3$) < lane traffic low to middle level (60.3 $\mu g/m^3$) < lane traffic middle to high level (62.3 $\mu g/m^3$) < lane traffic high level (68.6 $\mu g/m^3$). Compared to over the middle-level lane traffic, under middle-level lane traffic produced an average small difference of Bg and observed high C_{Bg} and C_i levels, resulting in low C_i and C_{Bg} for relatively low volumes of traffic and high C_i and C_{Bg} for high volumes of traffic. The overall pollution level of the road may have increased as the lane traffic level increased. However, when the traffic volume increased, the relative road atmospheric pollution also increased, so the C_{res} level did not differ significantly.

4. Conclusions

The main goal in this paper was to quantify the characteristics of weather and traffic conditions on PM_{10} concentrations measured using the resuspended PM_{10} measurements system and to identify the main characteristics of PM_{10} concentrations. Technical statistics, a correlation analysis, and a comparison and analysis were carried out according to changes in the traffic levels.

The four roads, which varied by region with limited traffic, were easy sections to observe the concentration of resuspended dust (C_{res}) increased by speed and dust load (sL) on the road. In addition, the six roads where traffic was observed were composed of auxiliary highways in urban areas, where it was easy to collect traffic information at the same time as conducting air pollution monitoring of adjacent stations and have conditions to grasp the effect of PM_{10} concentration. The following are the results identifying the characteristics of sL measured on roads with limited traffic, the PM_{10} concentration of resuspended dust, and PM_{10} measured in six sections of the urban area during traffic operation.

C_i increased with increasing speed from 30 km/h to 60 km/h, and the highest increase was when the speed increased from 30 km/h to 40 km/h. At this time, C_{Bg} showed little change in the same measured environment. Therefore, the dispersion of the resuspended dust measured was expected to have an effect on the sL in the section due to the wind generated by the speed of driving.

On different measured roads, higher sL was observed compared with urban roads operating in areas with a clear external dust inflow, such as roads in construction complexes. In addition to the empty site, the sL collected on the same road was observed to increase over time without an adjustment of the external environment and traffic intervention.

The sL measured on paved roads in the testbed (restricted access to traffic) and on the construction site was observed to be 0.034 g/m^2 to 0.093 g/m^2. In the case of 0.034 g/m^2 with sL, the concentration of resuspended dust was found to be in the range of 47 μg/m^3~225 μg/m^3. Compared to this value, an sL distribution below 0.009 g/m^2 could produce from 11 μg/m^3 up to 29 μg/m^3 of resuspended dust, which was judged to be lower than 0.034 g/m^2 of sL. Therefore, with a higher level of sL, one can generate a high level of road resuspended dust.

Even if sL could produce effective road resuspended dust, the background concentration (C_{Bg}) during driving was not affected by the surrounding area. Although it had almost the same level as the concentration of adjacent stations (Bg) (R^2 = 0.84), it had a low correlation with the concentration of fine dust floating due to it being suspended continuously on roads where traffic operated. In the event of yellow dust, the difference between the background concentration level and the air pollution level on the road was large, indicating a negative correlation (R^2 = 0.10).

Using the resuspended dust PM_{10} monitoring results for six sections of roads under traffic operation during the same period, the correlation analysis results between traffic and atmospheric weather environmental factors were as follows:

- A significant correlation coefficient between Bg and R > 0.3 was derived from time after rain (R = 0.41) and visibility (R = −0.35), C_{Bg} by section was time after rain (R = 0.35 to 0.50) and visibility (R = −0.60 to −0.33), and time after rain and visibility were found to have similar characteristics to Bg and C_{Bg}.

As a result of analyzing the correlation between C_{Bg} and C_i concentrations and meteorological factors by section, the correlation R-value with C_i was consistently lower than the C_{Bg} correlation R-value and major weather factors (sea level pressure, visibility, and rainfall time) with $R \geq 0.4$. However, a consistent trend was not observed in the correlation R-value with C_i and C_{Bg} concentrations according to the change of lane traffic by section.

The results of resuspended dust PM_{10} monitoring of six road sections operating during the same period were compared and analyzed by location, direction, and lane traffic level types in the measurement areas as follows:

- A lane traffic level type for whole data was compared by dividing the upper distribution of the lower quartile (25th) value into a high-level type, median to 75th value distribution grade into middle to high-level type, and the 25th to median value range distribution grade into low-level type.
- Higher C_{res} values were observed consistently in sections with a relatively high frequency of occurrence of high levels of lane traffic among sections traveling in the same direction on roads divided by location.
- The lowest average C_{Bg} (45.0 μg/m^3) and average C_i (56.7 μg/m^3) concentrations were measured in Section E, where low levels of lane traffic were observed the most, i.e., 22 times.
- Among the data samples classified by lane traffic level type, the order in which the C_{res} mean was observed and found to be large was lane traffic low to middle level (15.3 μg/m^3) > lane traffic middle to high level (12.7 μg/m^3) > lane traffic high level (11.9 μg/m^3) > lane traffic low level (10.1 μg/m^3).

- The average level of lane traffic level C_{Bg} was observed to be relatively high in the order of lane traffic low to middle level (45.0 μg/m^3) < lane traffic low level (47.9 μg/m^3) < lane traffic middle to high level (49.6 μg/m^3) < lane traffic high level (56.7 μg/m^3). The average level of C_i was observed in the order of lane traffic low level (47.9 μg/m^3) < lane traffic low to middle level (49.6 μg/m^3) < lane traffic middle to high level (62.3 μg/m^3) < lane traffic high level (68.6 μg/m^3).
- The overall pollution level of the road may have increased as lane traffic level increased. However, when the traffic volume increased, the relative road atmospheric pollution also increased, so the C_{res} level did not differ significantly.

The characteristics of the difference between the concentration of resuspended dust (C_i) and the background concentration of roads (C_{Bg}) and the background of city atmosphere (Bg) concentration measured were compared with the effects of traffic and weather conditions. In the case of Korea, PM reduction measures are being implemented according to the occurrence of high concentrations of PM_{10} and $PM_{2.5}$ provided by the city Bg observations. However, results of this study suggested the need for an efficient alternative considering the effect of yellow dust over time, because due to the occurrence of yellow dust, Bg can differ to the resuspended dust concentration. Additionally, the concentration of resuspended dust on roads may differ significantly from that of the adjacent Bg observation caused by vehicles driving on roads. Therefore, it is suggested that a more frequent occurrence of high levels of resuspended dust may occur, and research should be continued to quantify the influence of dust collected through mobile measurements and to provide accurate forecasts.

Author Contributions: Conceptualization, I.K. and S.H.; methodology, H.Y. and G.Y.; investigation, J.C. and H.Y.; original draft preparation, S.H.; writing—review and edition, I.K. and G.Y. All authors have read and agreed to the published version of the manuscript.

Funding: This work was supported by the Korea Agency for Infrastructure Technology Advancement (KAIA) grant funded by the Ministry of Land, Infrastructure, and Transport (grant number 21POQWB152342-03) and the National Research Foundation of Korea (grant number NRF-2019R1A2C1007224).

Institutional Review Board Statement: Not applicable.

Informed Consent Statement: Not applicable.

Data Availability Statement: The urban weather observation Automated Synoptic Observing System(ASOS) data provided by Korea Meteorological Administration (KMA). Available online: https://data.kma.go.kr, accessed on 27 July 2022; The road traffic observation Vehicle Detect System(VDS) data provided by Yong-In city in Korea. Available online: http://its.yongin.go.kr/trafficStats/road.do, accessed on 27 July 2022.

Acknowledgments: The authors would like to thank the members of the research team, KAIA, and MOLIT for their guidance and support throughout the project.

Conflicts of Interest: The authors declare no conflict of interest.

References

1. Pope, C.A.; Thun, M.J.; Namboodiri, M.M.; Dockery, D.W.; Evans, J.S.; Speizer, F.E.; Heath, C.W. Particulate air pollution as a predictor of mortality in a prospective study of US adults. *Am. J. Respir. Crit. Care Med.* **1995**, *1513*, 669–674. [CrossRef] [PubMed]
2. Askariyeh, M.H.; Venugopal, M.; Khreis, H.; Birt, A.; Zietsman, J. Near-road traffic-related air pollution: Resuspended $PM_{2.5}$ from highways and arterials. *Int. J. Environ. Res. Public Health* **2020**, *17*, 2851. [CrossRef] [PubMed]
3. Lewis, A.; Moller, S.J.; Carslaw, D. Non-Exhaust Emissions from Road Traffic. 2019. Available online: https://eprints.whit-erose.ac.uk/156628/ (accessed on 27 July 2022).
4. Rienda, I.C.; Alves, C.A. Road dust resuspension: A review. *Atmos. Res.* **2021**, *261*, 105740. [CrossRef]
5. Penkała, M.; Ogrodnik, P.; Rogula-Kozłowska, W. Particulate matter from the road surface abrasion as a problem of non-exhaust emission control. *Environments* **2018**, *5*, 9. [CrossRef]

6. Amato, F.; Alastuey, A.; De La Rosa, J.; Gonzalez Castanedo, Y.; Sánchez de la Campa, A.M.; Pandolfi, M.; Lozano, A.; Contreras González, J.; Querol, X. Trends of road dust emissions contributions on ambient air particulate levels at rural, urban and industrial sites in southern Spain. *Atmos. Chem. Phys.* **2014**, *14*, 3533–3544. [CrossRef]
7. Amato, F.; Cassee, F.R.; van der Gon, H.A.; Gehrig, R.; Gustafsson, M.; Hafner, W.; Harrison, R.M.; Jozwicka, M.; Kelly, F.J.; Moreno, T.; et al. Urban air quality: The challenge of traffic non-exhaust emissions. *J. Hazard. Mater.* **2014**, *275*, 31–36. [CrossRef]
8. Matthias, V.; Arndt, J.A.; Aulinger, A.; Bieser, J.; Denier van der Gon, H.; Kranenburg, R.; Kuenen, J.; Neumann, D.; Pouliot, G.; Quante, M. Modeling emissions for three-dumensional atmospheric chemistry transport models. *J. Air Waste Manag. Assoc.* **2018**, *68*, 763–800. [CrossRef] [PubMed]
9. Choi, Y.; Kim, Y.; Lee, H. A study on road cleaning to reduce resuspension of road dust. *Seoul Inst. Policy Res.* **2018**, 1–135. Available online: http://lib.seoul.go.kr/oak/handle/201302/20733 (accessed on 27 July 2022).
10. Kakosimos, K.E.; Hertel, O.; Ketzel, M.; Berkowicz, R. Operational Street Pollution Model (OSPM)—A review of performed application and validation studies, and future prospects. *Environ. Chem.* **2010**, *7*, 485–503. [CrossRef]
11. Padoan, E.; Ajmone-Marsan, F.; Querol, X.; Amato, F. An empirical model to predict road dust emissions based on pavement and traffic characteristics. *Environ. Pollut.* **2018**, *237*, 713–720. [CrossRef]
12. Kuhns, H.; Etyemezian, V.; Landwehr, D.; MacDougall, C.; Pitchford, M.; Green, M. Testing re-entrained aerosol kinetic emissions from roads: A new approach to infer silt loading on roadways. *Atmos. Environ.* **2001**, *35*, 2815–2825. [CrossRef]
13. Han, S.; Youn, J.S.; Jung, Y.W. Characterization of PM_{10} and $PM_{2.5}$ source profiles for resuspended road dust collected using mobile sampling methodology. *Atmos. Environ.* **2011**, *45*, 3343–3351. [CrossRef]
14. Li, D.; Chen, J.; Zhang, Y.; Gao, Z.; Ying, N.; Gao, J.; Zhang, K.; Zhu, S. Dust emissions from urban roads using the AP-42 and TRAKER methods: A case study. *Atmos. Pollut. Res.* **2021**, *12*, 101051. [CrossRef]
15. Amato, F.; Pandolfi, M.; Viana, M.; Querol, X.; Alastuey, A.; Moreno, T. Spatial and chemical patterns of PM_{10} in road dust deposited in urban environment. *Atmos. Environ.* **2009**, *43*, 1650–1659. [CrossRef]
16. Amato, F.; Pandolfi, M.; Escrig, A.; Querol, X.; Alastuey, A.; Pey, J.; Pérez, N.; Hopke, P.K. Quantifying road dust resuspension in urban environment by multilinear engine: A comparison with PMF2. *Atmos. Environ.* **2009**, *43*, 2770–2780. [CrossRef]
17. Hussein, T.; Johansson, C.; Karlsson, H.; Hansson, H.C. Factors affecting non-tailpipe aerosol particle emissions from paved roads: On-road measurements in Stockholm, Sweden. *Atmos. Environ.* **2008**, *42*, 688–702. [CrossRef]
18. Fan, S.; Tian, G.; Li, G.; Huang, Y.; Qin, J.; Cheng, S. Road fugitive dust emission characteristics in Beijing during Olympics Game 2008 in Beijing, China. *Atmos. Environ.* **2009**, *43*, 6003–6010.
19. Zhu, D.; Kuhns, H.D.; Brown, S.; Gillies, J.A., Etyemezian, V., Gertler, A.W. Fugitive dust emissions from paved road travel in the Lake Tahoe basin. *J. Air Waste Manag. Assoc.* **2009**, *59*, 1219–1229. [CrossRef]
20. Kauhaniemi, M.; Stojiljkovic, A.; Pirjola, L.; Karppinen, A.; Harkonen, J.; Kupiainen, K.; Kangas, L.; Aarnio, M.A.; Omstedt, G.; Denby, B.R.; et al. Comparison of the predictions of two road dust emission models with the measurements of a mobile van. *Atmos. Chem. Phys.* **2014**, *14*, 9155–9169. [CrossRef]
21. Kwak, J.; Lee, S.; Lee, S. On-road and laboratory investigations on non-exhaust ultrafine particles from the interaction between the tire and road pavement under braking conditions. *Atmos. Environ.* **2014**, *97*, 195–205. [CrossRef]
22. Zhang, W.; Ji, Y.; Zhang, S.; Zhang, L., Wang, S. Determination of silt loading distribution characteristics using a rapid silt loading testing system in Tianjin, China. *Aerosol Air Qual. Res.* **2017**, *17*, 2129–2138. [CrossRef]
23. U.S. Environmental Protection Agency. Emission Factor Documentation for AP-42, Section 13.2.1 Paved Roads. EPA Contact No. 68-D0-0123, Work Assignment No.44 MRI Project No. 9712-44(1993). Available online: https://www3.epa.gov/ttn/chief/ap42/ch13/bgdocs/b13s0201.pdf (accessed on 27 July 2022).
24. Askariyeh, M.H.; Zietsman, J.; Autenrieth, R. Traffic contribution to $PM_{2.5}$ increment in the near-road environment. *Atmos. Environ.* **2020**, *224*, 117113. [CrossRef]
25. Jung, Y.W.; Han, S.H.; Won, K.H.; Jang, K.W.; Hong, J.H. Present status of emission estimation methods of resuspended dusts from paved roads. *J. Korean Soc. Environ. Eng.* **2006**, *28*, 1126–1132.
26. Perera, I.E.; Litton, C.D. Quantification of optical and physical properties of combustion-generated carbonaceous aerosols (<PM2.5) using analytical and microscopic techniques. *Fire Technol.* **2015**, *51*, 247–269. [CrossRef] [PubMed]
27. Zhu, D.; Kuhns, H.D.; Gillies, J.A.; Etyemezian, V.; Gertler, A.W.; Brown, S. Inferring deposition velocities from changes in aerosol size distributions downwind of a roadway. *Atmos. Environ.* **2011**, *45*, 957–966. [CrossRef]
28. U.S. Environmental Protection Agency. *Environmental Protection Agency Compilation of Air Pollution Emission Factors, AP-42*, 5th ed.; U.S. Environmental Protection Agency: Washington, DC, USA, 2003; Volume I: Stationary Point and Area Sources, Section 13.2.1 Paved Roads.
29. Amato, F.; Schaap, M.; van der Gon, H.A.D.; Pandolfi, M.; Alastuey, A.; Keuken, M.; Querol, X. Effect of rain events on the mobility of road dust load in two Dutch and Spanish roads. *Atmos. Environ.* **2012**, *62*, 352–358. [CrossRef]
30. Lundberg, J. Non-exhaust PM10 and road dust. *Dr. Diss. KTH R. Inst. Technol.* **2018**. Available online: https://www.diva-portal.org/smash/get/diva2:1179589/FULLTEXT02 (accessed on 27 July 2022).
31. Kuhns, H.; Gillies, J.; Etyemezian, V.; Dubois, D.; Ahonen, S.; Nikolic, D.; Durham, C. Spatial Variability of Unpaved Road Dust PM_{10} Emission Factors near El Paso, Texas. *J. Air Waste Manag. Assoc.* **2005**, *55*, 3–12. [CrossRef] [PubMed]
32. Lin, S.; Liu, Y.; Chen, H.; Wu, S.; Michalaki, V.; Proctor, P.; Rowley, G. Impact of change in traffic flow on vehicle non-exhaust $PM_{2.5}$ and PM_{10} emissions: A case study of M25 motorway, UK. *Chemosphere* **2022**, *303*, 135069. [CrossRef]

33. Sahanavin, N.; Prueksasit, T.; Tantrakarnapa, K. Relationship between PM_{10} and $PM_{2.5}$ levels in high-traffic area determined using path analysis and linear regression. *J. Environ. Sci.* **2018**, *69*, 105–114. [CrossRef]
34. Denby, B.R.; Sundvor, I.; Johansson, C.; Pirjola, L.; Ketzel, M.; Norman, M.; Kupiainen, K.; Gustafsson, M.; Blomqvist, G.; Omstedt, G. A coupled road dust and surface moisture model to predict non-exhaust road traffic induced particle emissions (NORTRIP). Part 1: Road dust loading and suspension modelling. *Atmos. Environ.* **2013**, *77*, 283–300. [CrossRef]
35. Han, S.; Won, K.; Jang, K.; Son, Y.; Kim, J.; Hong, J.; Jung, Y. Development and Application of Real-time Measurement System of Silt Loading for Estimating the Emission Factor of Resuspended Dust from Paved Road. *J. Korean Soc. Atmos. Environ.* **2007**, *23*, 596–611. [CrossRef]
36. Pope, C.A.; Burnett, R.T.; Thun, M.J.; Calle, E.E.; Krewski, D.; Ito, K.; Thurston, G.D. Lung cancer, cardiopulmonary mortality, and long-term exposure to fine particulate air pollution. *JAMA* **2002**, *287*, 1132–1141. [CrossRef] [PubMed]
37. Yang, W. Changes in Air Pollutant Concentrations Due to Climate Change and the Health Effect of Exposure to Particulate Matter. *J. Popul. Health Stud.* **2019**, *269*, 20–31.
38. Lee, G.H. Understanding and Contrasting Yellow Dust. *J. Korean Soc. Soc. Secur.* **2009**, *2*, 18–27.
39. Thorpe, A.J.; Harrison, R.M.; Boulter, P.G.; McCrae, I.S. Estimation of particle resuspension source strength on a major London Road. *Atmos. Environ.* **2007**, *41*, 8007–8020. [CrossRef]

atmosphere

Article

Impact of Land Use on Atmospheric Particulate Matter Concentrations: A Case Study of the Beijing–Tianjin–Hebei Region, China

Haoran Zhai [1,2], Jiaqi Yao [2,3,*], Guanghui Wang [2,4] and Xinming Tang [2,3]

1 College of Resource Environment and Tourism, Capital Normal University, Beijing 100048, China; 2@by2studio.cn
2 Land Satellite Remote Sensing Application Center, Ministry of Natural Resources, Beijing 100048, China; wanggh@lasac.cn (G.W.); txm@lasac.cn (X.T.)
3 College of Geomatics, Shandong University of Science and Technology, Qingdao 266590, China
4 School of Environment and Spatial Informatics, China University of Mining and Technology, Xuzhou 221116, China
* Correspondence: 201981020027@sdust.edu.cn

Abstract: The increasing frequency of human activities has accelerated changes in land use types and consequently affected the atmospheric environment. In this manuscript, we analyze the relationships between the particulate matter concentration and land use changes in the Beijing–Tianjin–Hebei (BTH) region, China, from 2015 to 2018. The experimental results indicate that (1) an improved sine function model can suitably fit the periodic changes in the particulate matter concentration, with the average R^2 value increasing to 0.65 from the traditional model value of 0.49, while each model coefficient effectively estimates the change characteristics of each stage. (2) Among all land use types, the particulate matter concentrations in construction land and farmland are high, with a large annual difference between high and low values. The concentration decreases slowly in spring and summer but increases rapidly in autumn and winter. The concentrations in forestland and grassland are the lowest; the difference between high and low values is small for these land use types, and the concentration fluctuation pattern is relatively uniform. Natural sources greatly influence the concentration fluctuations, among which frequent dusty weather conditions in spring impose a greater influence on forestland and grassland than on the other land use types. (3) The landscape pattern of land use exerts a significant influence on the particulate matter concentration. Generally, the lower the aggregation degree of patches is, the higher the fragmentation degree is, the more complex the shape is, the higher the landscape abundance is, and the lower the particulate matter concentration is. The higher the construction land concentration is, the more easily emission sources can be aggregated to increase the particulate matter concentration. However, when forestland areas are suitably connected, this land use type can play a notable role in inhibiting particulate matter concentration aggravation. This conclusion is of great relevance to urban land use planning and sustainable development.

Keywords: land use and land cover; $PM_{2.5}$; PM_{10}; spatiotemporal characteristics; air pollution; remote sensing; China

Citation: Zhai, H.; Yao, J.; Wang, G.; Tang, X. Impact of Land Use on Atmospheric Particulate Matter Concentrations: A Case Study of the Beijing–Tianjin–Hebei Region, China. *Atmosphere* **2022**, *13*, 391. https://doi.org/10.3390/atmos13030391

Academic Editors: Elena Hristova, Manousos Ioannis Manousakas, Anikó Angyal, Maria Gini and Rajasekhar Balasubramanian

Received: 9 December 2021
Accepted: 21 February 2022
Published: 26 February 2022

1. Introduction

Land not only provides an important basis of human survival but is also an essential resource for human development [1]. Land use and land cover are two different but closely related concepts. Land use refers to the dynamic and purposeful utilization of natural land resources by humans, while land cover constitutes the synthesis of natural and human-made buildings covering the Earth's surface. Both of these concepts are important land system attributes [2,3]. All human activities are inseparable from land. These activities have affected the distribution patterns and change mechanisms of the original land system and exerted far-reaching influences on resources, the environment, geographical processes, and biodiversity [4,5]. The urban environment is the result of the combined influence of industrial production and urban construction. Land use types have increasingly changed from natural types, such as farmland, forestland, and grassland, to artificial types, which is one of the important characteristics of urbanization. The increasing frequency of human activities has accelerated land use type changes, altered the original urban landform and regional energy balance, damaged the natural ecological functions of the surface system, and driven air flow exchange variations between surface particles and the atmosphere, affecting the generation, diffusion, dilution, and collection of particles and thus affecting the atmospheric environment [6,7]. The compositional structure, content, and properties of the atmosphere can change due to changes in land use/land cover, and the chemical properties and processes of the atmosphere can also be affected, which can impact the gas-generation mechanism, thus influencing the circulation of atmospheric materials within a given region. In the long term, these effects may alter the local climate and even affect the global climate [8]. Landscape patterns describe the spatial distribution of landscape patches of different shapes and sizes, which is an important manifestation of landscape heterogeneity and can reflect the effects of various ecological processes. The study of landscape patterns can elucidate the inherent spatial distribution of seemingly disorderly patches and quantitatively describe their changes over time [9]. Analyzing changes in land use/land cover types and the evolution of landscape patterns can provide insights into the extent, directions, and characteristics of changes in particulate matter [10]. Environmental protection, environmental improvement, and the optimization of land use patterns in development to improve urban air quality are all major issues related to the effects of land use/land cover on atmospheric particulate matter.

In recent years, the contribution of human activity-related emissions to particulate matter has gradually increased. Vehicle exhaust, road dust, industrial emissions, domestic emissions, fuel combustion-related emissions and building dust continuously drive particulate matter emissions into the atmosphere, and these activities mainly occur in construction land areas. Buildings in cities also have a significant impact on particulate matter dispersion [11,12]. Many research results revealed that the higher the construction land proportion is, the higher the air pollution degree. Xu et al. [8] found that the construction land and road area proportion in the Changsha–Zhuzhou–Xiangtan urban agglomeration was significantly positively correlated with the concentrations of NO_2 and particles with aerodynamic diameters of 2.5 μm or smaller ($PM_{2.5}$), and the influence on NO_2 was greater than that on $PM_{2.5}$. In contrast, the influence on particles with an aerodynamic diameter of 10 μm or smaller (PM_{10}) was unstable, which may have been because vehicle exhaust emissions comprised the main NO_2 source and were closely related to construction land and road areas. Peng et al. [13] determined that forestland in Chengdu exerted an obvious influence on NO_2 concentration variations; urban land was highly correlated with the highest total suspended particulate matter concentration; and the spatial distributions of NO_2 and total suspended particulate matter concentrations were highly consistent with those of urban land, industrial and mining land, and transportation land areas. Font et al. [14] reported that PM_{10} greatly increased during and after road expansion in London, but the impact on $PM_{2.5}$ was relatively limited, indicating that construction activities contributed more notably to PM_{10}. Zhai et al. [15] observed a highly positive correlation between the NO_2 column concentration in the troposphere and the coverage rates of impervious surfaces in various cities and municipal districts, and the spatial distribution patterns in these two hot-spot areas were consistent. The mean NO_2 column concentration in the troposphere above open, compact, intensive, and highly intensive municipal districts increased sequentially, and the standardized concentration index of the NO_2 column in the troposphere above areas exhibiting impervious surface expansion exhibited an upward trend. Tang et al. [10] found that hazy days in Beijing showed obvious positive correlations with construction land, residential areas, industrial and mining land, and transportation land. There was a significant positive correlation between the building area and $PM_{1.0}$ concentration in the 0.5- and 1-km buffer zones. Wei et al. [16] determined that an obvious negative correlation existed between urban land use and pollution-free weather; however, positive correlations with other weather conditions were found, and these correlations gradually increased. Mo et al. [17] reported that the concentrations of $PM_{2.5}$, PM_{10}, and other particulate matter types in Beijing were positively correlated with the resident population density, regional gross domestic product (GDP), and other factors that significantly reflected the urbanization level.

Forestland and grassland, as typical vegetation types, are generally considered to reduce particulate matter. First, human activities in forestland and grassland are limited; few anthropogenic emission sources occur; and natural emissions are the main source of particulate matter. Second, vegetation cover can effectively separate the surface and atmosphere, thus preventing floating dust near the surface from rising into the atmosphere. Third, vegetation with a complex canopy structure can reduce the wind speed and prevent particles from entering local areas. In addition, particles fall onto vegetation surfaces due to dry sedimentation and are adsorbed or captured by leaves, stems, and other organs, which can play notable roles in dust retention. Xu et al. [8] found that forestland, green land, and cultivated land areas were negatively correlated with NO_2 and $PM_{2.5}$ concentrations. Wei et al. [16] reported that forestland was positively correlated with nonpolluted and slightly polluted weather conditions but negatively correlated with highly polluted weather conditions. Shi et al. [11] determined that a decrease in cultivated land and forestland areas and an increase in construction land areas contributed to a decrease in the number of foggy days but aggravated the haze occurrence degree and frequency. Mo et al. [17] observed that the trend of the concentrations of $PM_{2.5}$, PM_{10}, and other particulate matter types in Beijing was the opposite of the development trend of forest coverage, exhibiting a significant negative correlation. Tang et al. [10] found that an obvious negative correlation existed between ecological land and cultivated land areas and the number of haze days in Beijing, and an obvious negative correlation between the green space area proportion and $PM_{1.0}$ concentration in 0.5- and 1-km buffer zones was identified. Sun [18] reported that the PM_{10} concentrations associated with various land use types in the Pearl River Delta region followed the ascending order of forestland, grassland, farmland, construction land, and desert wasteland. Generally, the influence of farmland on particulate matter is bidirectional. During the crop growing season, vegetation cover causes farmland to exert a certain dust-retention effect, but the corresponding efficiency may be lower than that of forestland and grassland [19,20]. After crop harvesting, dust emitted during straw burning represents an important source of particulate matter [21,22]. Moreover, in winter, without vegetation coverage, a large amount of dust becomes exposed on the surface and is easily lifted into the atmosphere [19,20]. The influence of water on particles is complicated. On the one hand, when particles fall on the water surface due to sedimentation, these particles are not resuspended. On the other hand, the presence of water causes the temperature to decrease and the humidity to increase, which may affect the accumulation and diffusion of particulate matter [20,23]. Due to the lack of vegetation coverage, a large amount of floating dust is exposed near the surface of unused land and is easily lifted into the atmosphere under the action of wind. Generally, floating dust readily becomes the main source of particles, and the corresponding contribution to particles with a large aerodynamic equivalent diameter is relatively high.

The temporal and spatial variability of particulate matter concentrations is a complex system that is driven by a combination of influencing factors. The effects of meteorological factors, human activities, and topographic conditions on particulate matter have been studied by several scholars [24–27]. In this manuscript, we focus on the intrinsic patterns of land use, and it is essential to explore its impact on the atmospheric environment to guide land use planning scientifically. In this paper, land use distribution and changes are analyzed using land use classification data in the Beijing–Tianjin–Hebei (BTH) region in 2015–2018. The differences in different land use types on the temporal changes in atmospheric particulate matter concentrations are explored with the help of an improved sine function model. The correlation between landscape patterns and atmospheric particulate matter is analyzed using landscape ecology and statistical methods. The results of this paper are important for the prevention and control of atmospheric pollution and the sustainable development of resources and the environment and can also provide a scientific reference for the optimal allocation of land resources.

2. Methods

2.1. Optimized Particle Concentration Fitting Model

Systematic mathematical models can be used to fit concentration change patterns. Comparison of the parameters in the fitted results among different regions is a powerful tool for analyzing the changes in pollutant concentrations. The concentration exhibits a significant periodicity in annual cycles, with a peak and a trough in each cycle. To analyze the characteristics of the variation in particulate matter concentration over time quantitatively, many scholars fitted periodic quantitative mathematical function models, among which the sinusoidal function model is the most widely adopted [28–32]. The traditional sine model is based on the least-squares method and comprises linear and sine functions. The general form of the sine function model is shown in Figure 1a. The corresponding equation is defined as follows:

$$y = \mathrm{A_o}x + \mathrm{B_o} + \mathrm{C_o} \cdot \sin\left(\frac{2\pi}{\mathrm{D_o}}(x + \mathrm{E_o})\right) \quad (1)$$

where y is the particulate matter concentration, and the first two terms on the right side of the equal sign constitute a linear function, i.e., $\mathrm{A_o}x + \mathrm{B_o}$, representing the interannual variation trend of the particulate matter concentration. The third term is a sine function, i.e., $\mathrm{C_o} \cdot \sin\left(\frac{2\pi}{\mathrm{D_o}}(x + \mathrm{E_o})\right)$, which is applied to describe the periodic variation characteristics of the particulate matter concentration on a monthly scale. Scholars evaluated and applied this model in monthly analyses of mean particulate matter concentrations at the country, region, and city spatial scales and confirmed that it can describe the general temporal variation patterns of particulate matter concentrations. However, due to its relatively simple function shape, the model is limited in its ability to provide detailed information on the variation in particulate matter. To capture the actual trends in changes in particulate matter concentrations more closely, describe temporal changes in monthly mean concentrations more accurately, and determine the inherent mechanisms, this paper improves the traditional sinusoidal function model in the following aspects [33]:

(1) Regarding general concentration change trends, a quadratic function is employed instead of a linear function, and this function can suitably fit various situations, such as initially rising and then falling, initially falling and then rising, accelerated rising or falling, decelerated rising or falling, and steadily rising or falling concentrations.
(2) In particulate matter concentration fluctuations, when the overall concentration increases, the fluctuation range also increases. When the overall concentration decreases, the fluctuation amplitude also decreases. Therefore, the first half of a quadratic function, i.e., $\mathrm{A_t}x^2 + \mathrm{B_t}x + \mathrm{C_t}$, is added to the coefficient of the original sine function. This improvement ensures that the amplitude no longer remains fixed over time but varies with the overall concentration change trend.
(3) At the monthly scale, changes in the particulate matter concentration are often not symmetrical periodic fluctuations, and often, the durations of periods with low values are long, while the durations of periods with high values are short. Notably, narrower peaks and wider valleys or the opposite, i.e., wider peaks and narrower valleys, may occur. Considering these concentration change characteristics, this paper replaces the original sine function $\sin(\frac{2\pi}{\mathrm{D_t}}(x + \mathrm{E_t}))$ with $2 \cdot (\frac{1}{2} \cdot (1 + \sin(\frac{2\pi}{\mathrm{D_t}}(x + \mathrm{E_t}))))^{\mathrm{F_t}}$ and adds a deformation factor $\mathrm{F_t}$ to control the shape of the sine function based on its magnitude.

(4) The particulate matter concentration variation during a given cycle is often not symmetrical or uniform, and the concentration increase and decrease rates in this case are not equal. In particular, the time from one wave peak to the next wave trough is not the same as the time from one wave trough to the next wave peak. According to this feature, an oscillation term with the same period and a $\pi/2$-phase difference is added to x in the sine function; x is replaced with $x + H_t \cdot \sin(\frac{2\pi}{D_t}(x + E_t))$; and the left/right shifting of the peak or trough is controlled by the magnitude of the offset factor H_t.

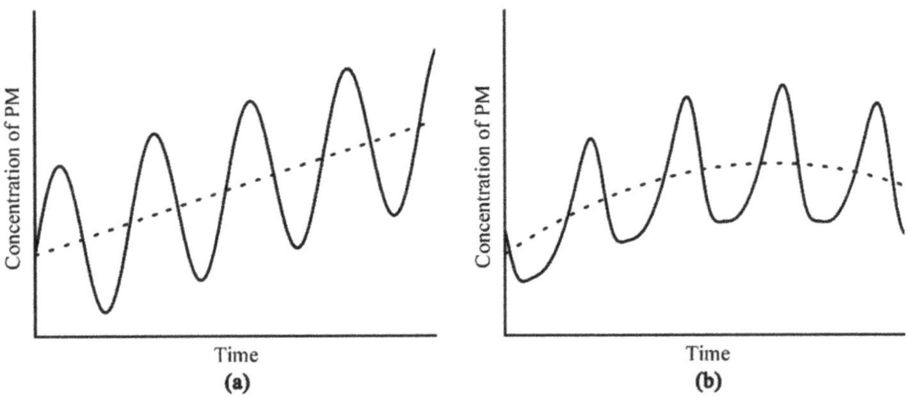

Figure 1. Schematic diagram of the traditional (**a**) and improved (**b**) sine function models.

After the above 4 improvements, the general form of the improved sine function model is schematically shown in Figure 1b. The improved function is reorganized, and an improved sine function model equation is obtained as follows:

$$y = \left(A_t x^2 + B_t x + C_t\right) \cdot \left(1 + D_t\left(\frac{1}{2}\left(1 + \sin\left(\frac{2\pi}{E_t}\left(x + H_t \cdot \sin\left(\frac{2\pi}{E_t}(x + G_t)\right) + G_t\right)\right)\right)\right)^{F_t}\right) \quad (2)$$

where y is the particulate matter concentration, and the terms on the right side of the equal sign comprise a quadratic function and the improved sine function that captures uneven fluctuations. There are a total of 8 parameters from A_t to H_t in the formula, and the meaning and description of each parameter are shown in Table 1.

Table 1. Parameters and their descriptions in the improved sine function model.

Parameters	Definition	Range	Description
A_t	Quadratic coefficient	$(-\infty, \infty)$	A_t determines the opening direction and magnitude of the parabolic curve, namely: (1) for $A_t > 0$, the parabolic curve opens upwards; (2) for $A_t < 0$, the parabolic curve opens downwards. Additionally, the larger the absolute value of A_t is, the smaller the opening of the parabolic curve, while the smaller the absolute value of A_t is, the larger the opening of the parabolic curve.
B_t	Primary term coefficient	$(-\infty, \infty)$	B_t and A_t jointly determine the position of the axis of symmetry.
$B_t/2A_t$	Axis of symmetry		(1) When the axis of symmetry occurs within the studied time range, for $A_t > 0$, the concentration first decreases and then increases. Conversely, for $A_t < 0$, the concentration first increases and then decreases. (2) When the axis of symmetry occurs on the left side of the studied time range, for $A_t > 0$, the concentration continuously increases, and the rate of increase rises. Conversely, for $A_t < 0$, the concentration continuously decreases, but the rate of decrease rapidly declines.

Table 1. *Cont.*

Parameters	Definition	Range	Description
			(3) When the axis of symmetry is located on the right side of the studied time range, for $A_t > 0$, the concentration continuously decreases, but the rate of decrease declines. Conversely, for $A_t < 0$, the concentration continuously increases, but the rate of increase declines. In addition, when the axis of symmetry is located far from the studied time range, the concentration can be considered to increase or decrease at an approximately constant rate.
C_t	Constant	$(-\infty, \infty)$	C_t determines the intersection of the parabolic curve and Y-axis and represents the overall particulate matter concentration at the starting time of the analysis.
D_t	Amplitude	$[0, \infty)$	D_t indicates the differences between peak and valley values and determines the oscillation amplitude. The larger its value, the stronger the fluctuations are.
E_t	Cycle	12	E_t represents the fluctuation duration. Because particulate matter fluctuation over a year is considered in this paper, a fixed E_t value of 12 is adopted.
F_t	Deformation factor	$(0, \infty)$	F_t indicates the duration of periods with high or low values during a given cycle: (1) for $F_t = 1$, the curve of the sine function uniformly oscillates; (2) for $F_t > 1$, the function curve exhibits narrow peaks and wide valleys, and the larger the F_t value is, the stronger the deformation; (3) for $0 < F_t < 1$, the function curve exhibits wide peaks and narrow valleys, and the closer the Ft value is to 0, the stronger the deformation.
G_t	Phase	$[0, 12]$	G_t determines the occurrence positions of peaks. During a given period, the larger the value, the later peaks and valleys occur.
H_t	Offset factor	$(-\infty, \infty)$	H_t indicates the relative positions of peaks and valleys, namely: (1) for $H_t = 0$, the left and right positions of valleys are located at the centers of two adjacent peaks, i.e., the time when one peak reaches the next valley is equal to the time when one valley rises to reach the next peak; (2) for $H_t > 0$, the trough moves towards the right, i.e., the time required for one peak to reach the next trough increases, and the time required for one trough to reach the next peak decreases; (3) for $H_t < 0$, the trough moves towards the left, i.e., the time required for one peak to reach the next trough decreases, and the time required for one trough to reach the next peak increases. The larger the absolute value of H_t is, the larger the offset.

2.2. Landscape Pattern Index

In this paper, the landscape pattern changes reflected by common landscape pattern indices at the landscape and patch levels were evaluated. It should be noted that certain indices may yield different meanings at different levels and should be examined separately [34,35].

2.2.1. Landscape Level

At the landscape level, the calculation methods and specific meanings of the patch density (PD), largest patch index (LPI), edge density (ED), contagion index (CONTAG), splitting index (SPLIT), Shannon's diversity index (SHDI), Simpson's diversity index (SIDI), aggregation index (AI), and eight other landscape pattern indices are explained below.

PD is equal to the number of patches in the landscape divided by the total landscape area (m^2), which is then multiplied by a scaling factor of 1,000,000 (conversion into 100 hectares), with a value range of (0, +∞). Moreover, PD is expressed as the number of patches per 100 hectares. When each unit involves an individual patch, the PD value is the largest, which indicates that based on the number of patches per unit area, landscapes of different sizes can be compared. PD can be calculated as follows:

$$\text{PD} = \frac{\text{N}}{\text{A}} \cdot (1{,}000{,}000) \tag{3}$$

where N is the number of patches in the landscape, and A is the total landscape area (m^2).

LPI is expressed as a percentage and is equal to the percentage coverage of the largest patch in the area, with a value range of (0, 100). When the largest patch in the landscape decreases in coverage, the LPI gradually approaches 0. When the whole landscape comprises only one patch, i.e., the largest patch accounts for 100% of the landscape area, LPI = 100. This index is a simple way to measure dominance. The LPI equation is

$$\mathrm{LPI} = \frac{\max(a_{ij})}{A} \cdot (100) \tag{4}$$

where a_{ij} is the area of patch ij (m^2), and A is the total landscape area (m^2).

ED is equal to the sum of the lengths (m) of all edge segments related to the corresponding patch types divided by the total landscape area (m^2), which is then multiplied by 10,000 (conversion into hectares), with a value range of (0, +∞) in meters per hectare. When no graded edges occur in the landscape, i.e., when the whole landscape comprises only one patch, ED = 0. This index can be adopted to compare landscapes of different sizes. ED can be calculated as

$$\mathrm{ED} = \frac{\sum_{k=1}^{m} e_{ik}}{A} \cdot (10{,}000) \tag{5}$$

where e_{ik} is the total length (m) of the edge segments in the landscape of patch type i, and A is the total landscape area (m^2).

CONTAG represents the highest possible contagion degree observed for a given number of patch types, with a value range of (0, 100) expressed as a percentage. When the patch types are decomposed and dispersed to the greatest extent, CONTAG approaches 0. CONTAG = 100 when all patch types exhibit the maximum aggregation degree. CONTAG reflects the aggregation or expansion degrees of different patches in the landscape. A high value indicates a good relationship between certain major patch types, while a low value indicates that the landscape encompasses a multielement intensive pattern with a high degree of landscape fragmentation. The CONTAG calculation equation is

$$\mathrm{CONTAG} = \left[1 + \frac{\sum_{i=1}^{m}\sum_{k=1}^{m}\left(P_i \cdot \frac{g_{ik}}{\sum_{k=1}^{m} g_{ik}}\right) \cdot \left(\ln\left(P_i \cdot \frac{g_{ik}}{\sum_{k=1}^{m} g_{ik}}\right)\right)}{2\ln(m)}\right] \cdot (100) \tag{6}$$

where P_i is the proportion of the landscape occupied by patch type I; g_{ik} is the number of connections between pixels of patch types i and k based on the double-counting method; and m is the number of patch types in the landscape.

SPLIT is equal to the square of the total landscape area (m^2) divided by the sum of the squared patch areas (m^2), and its value range is [1, n^2], where n is the total number of pixels in the area (dimensionless). When the landscape comprises a single patch, SPLIT = 1. SPLIT reaches its maximum value when the landscape is subdivided to the greatest extent. SPLIT is based on the regional distribution of accumulated patches and can characterize the degrees of separation between various patch types. SPLIT can be calculated as follows:

$$\mathrm{SPLIT} = \frac{A^2}{\sum_{i=1}^{n}\sum_{j=1}^{n} a_{ij}^2} \tag{7}$$

where a_{ij} is the area of patch ij (m^2), and A is the total landscape area (m^2).

SHDI denotes the proportional abundance of each patch type among all negative patch types multiplied by the sum of the proportions, and its value range is [0, +∞). When the landscape contains only one patch type, SHDI = 0. As the proportional distribution of areas among various patch types becomes increasingly uniform, SHDI increases. SHDI can reflect landscape heterogeneity and is particularly sensitive to an unbalanced distribution of various patch types in the landscape. If the land use diversity and fragmentation degree are high, the information content in the uncertainty and the SHDI value will increase. The SHDI equation is

$$\text{SHDI} = -\sum_{i=1}^{m}(P_i \cdot \ln P_i) \tag{8}$$

where P_i is the proportion of the landscape occupied by patch type i.

SIDI is equal to 1 minus the sum of the proportional abundance of the square of each patch type among all patch types, with a unitless range of [0, 1]. When the landscape contains only one patch type, SIDI = 0. With an increasing number of different patch types (i.e., increasing patch abundance), SIDI approaches 1. SIDI is another popular diversity index borrowed from community ecology that is insensitive to the presence of rare types and provides a more intuitive explanation than does the Shannon index. SIDI can be calculated as follows:

$$\text{SIDI} = 1 - \sum_{i=1}^{m} {P_i}^2 \tag{9}$$

where P_i is the proportion of the landscape occupied by patch type i.

AI denotes the number of similar adjacencies of a certain type divided by the maximum possible number of similar adjacencies of this type, which is then multiplied by the landscape proportion of this type. Subsequently, the total value for all types is multiplied by 100 (conversion into a percentage), and AI has a value range of [0, 100] expressed as a percentage. For AI = 0, the patch types are decomposed to the greatest extent. With increasing landscape aggregation, AI increases. When the landscape comprises a single patch type, AI equals 100. AI represents the frequency of different patch types occurring side by side on the map, which can reflect the patch aggregation degree. The AI equation is

$$\text{AI} = \left[\sum_{i=1}^{m}\left(\frac{g_{ii}}{\max(g_{ii})}\right)P_i\right] \cdot (100) \tag{10}$$

where g_{ii} is the number of similar adjacencies between pixels of type i based on the single-counting method; $\max(g_{ii})$ is the maximum value of g_{ii}; and P_i is the landscape proportion of type i.

2.2.2. Class Level

At the class level, five landscape pattern indices, including the percentage of the landscape (PLAND), PD, LPI, largest shape index (LSI), and SPLIT, were selected, and their calculation methods and specific meanings are explained below.

PLAND is the percentage of the area of the corresponding patch type in the total landscape area, with a range of (0, 100) expressed as a percentage. When the number of corresponding patch types in the landscape decreases, PLAND approaches 0. When the whole landscape comprises a single patch type, PLAND = 100. PLAND quantifies the proportional abundance of each patch type in the landscape, which is an important measure of the landscape composition in many ecological applications. The PLAND calculation equation is

$$\text{PLAND} = P_i = \frac{\sum_{j=1}^{n} a_{ij}}{A} \cdot (100) \tag{11}$$

where P_i is the landscape proportion occupied by patch type i; a_{ij} is the area of patch ij (m^2); and A is the total landscape area (m^2).

PD is equal to the number of patches of the corresponding patch type divided by the total landscape area (m^2) and multiplied by 1,000,000 (converted to 100 hectares), and its value range is (0, +∞) in units of the patch number per 100 hectares. When each cell is an individual patch, PD is the largest, and the final cell size determines the maximum number of patches per unit area. PD represents the number of patches per unit area, which is helpful for comparing landscapes of different sizes. The formula of PD is

$$\mathrm{PD} = \frac{n_i}{A} \cdot (1{,}000{,}000) \tag{12}$$

where n_i is the number of patches i of the corresponding type, and A is the total landscape area (m^2).

LPI denotes the percentage coverage of the largest patch of the corresponding type in the total area, with a value range of (0, 100) expressed as a percentage. When the largest patch area of the corresponding type decreases, the LPI gradually approaches 0. When the whole landscape comprises individual patches of corresponding patch types, LPI = 100. This is a simple method to measure the dominant position. LPI can be calculated as

$$\mathrm{LPI} = \frac{\max\limits_{j=1}(a_{ij})}{A} \cdot (100) \tag{13}$$

where a_{ij} is the area of patch ij (m^2), and A is the total landscape area (m^2).

LSI is equal to 0.25 multiplied by the sum of the whole landscape boundary and the lengths (m) of all edge segments of the corresponding patch type in the landscape boundary, which is then divided by the square root (m^2) of the total landscape area. The LSI value range is [1, +∞] and is unitless. When the landscape contains a single square patch of the corresponding type, LSI = 1. With increasing edge lengths of the corresponding patch type in the landscape, LSI increases infinitely, which provides a standardized measure of the total edges or ED. The LSI calculation equation is

$$\mathrm{LSI} = \frac{0.25 \cdot \sum_{k=1}^{m} e_{ik}^{*}}{\sqrt{A}} \tag{14}$$

where e_{ik}^{*} is the total length (m) of lateral edges between patch types i and k, and A is the total landscape area (m^2).

SPLIT is determined as the square of the total landscape area (m^2) divided by the sum of the squares of all patch areas (m^2), and its value range is [1, n^2], where n is the total number of pixels in the area (dimensionless). When the landscape contains a single patch type, SPLIT = 1. As the area of the focal patch type gradually decreases or the focal patch type is subdivided into smaller patches, SPLIT increases. SPLIT is based on the regional distribution of accumulated patches and can characterize the degrees of separation between various patch types. The SPLIT equation is

$$\mathrm{SPLIT} = \frac{A^2}{\sum_{j=1}^{n} a_{ij}^2} \tag{15}$$

where a_{ij} is the area of patch ij (m^2), and A is the total landscape area (m^2).

3. Results and Discussion

3.1. Overview of the Study Area and Data

The BTH region is located between 113°27′~119°51′ E and 36°05′~42°40′ N (Figure 2). The terrain is inclined, exhibiting high elevations in the northwest and low elevations in the southeast, with complex and heterogeneous landforms, including plateaus, mountains, hills, basins, and plains. From northwest to southeast, the region is roughly divided into the Bashang Plateau, Yanshan-Taihang Mountains, and Haihe Plain. The BTH region experiences a temperate of a semihumid continental monsoon climate, with distinct winters and summers, and more than 67% of precipitation is concentrated in the summer. The BTH region covers an area of approximately 217,200 square kilometers, which accounts for 2.3% of the total land area in China. This area hosts a dynamic economy, the highest degree of economic openness, the strongest innovation ability, and the highest foreign population absorption level in China. However, many particulate matter indicators seriously exceed relevant standards due to the dense population and notable industrial agglomeration, and smog and other air pollution phenomena frequently occur in the BTH region due to its rapid economic development. Air pollution seriously threatens the ecological environment and human health and affects the residential quality of life. The pollution conditions in the BTH region are highly representative of those in China and worldwide.

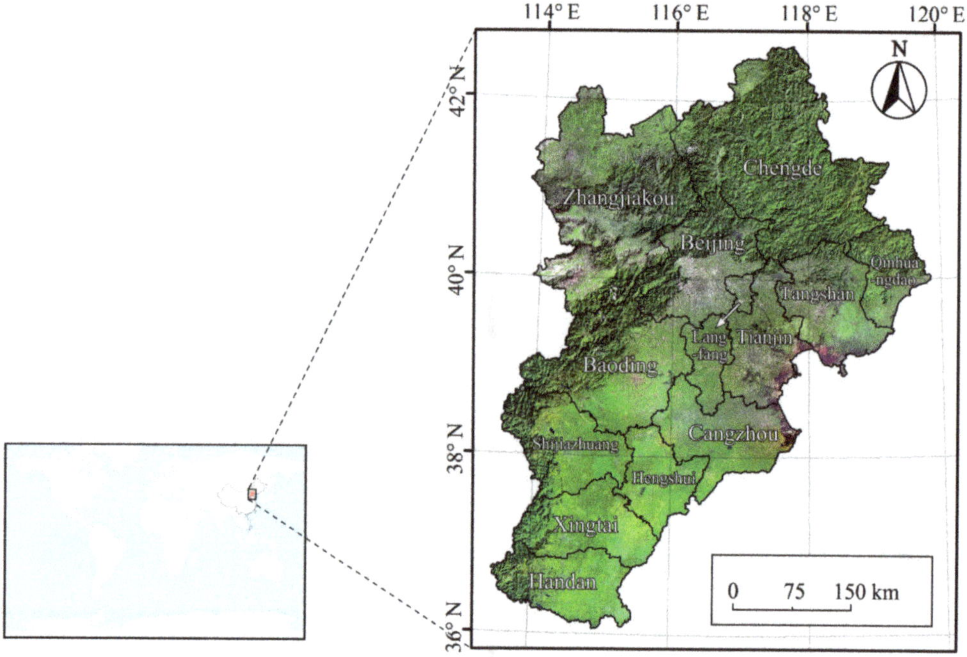

Figure 2. Geographical location of the study area.

3.1.1. Land Use Data

The land use/land cover data analyzed in this experiment included the C6 version of MCD12Q1, which is a land cover product of the Moderate Resolution Imaging Spectroradiometer (MODIS, https://modis.gsfc.nasa.gov/, accessed on 1 December 2021). The temporal resolution of the data was 1 year; the spatial resolution was 500 m; and a data mosaic of four scenes could cover the entire BTH region. Since 2001, relevant data were updated at annual intervals and stored in the HDF4 file format. The projection method is sinusoidal projection, which includes 8-day comprehensive MODIS observation results obtained from Terra and Aqua satellite MODIS sensors throughout the year. Combined with the characteristics of the surface reflectivity and surface temperature, the data were analyzed with a supervised decision tree classification algorithm based on a high-quality land cover training sample database. The land cover classification schemes for the dataset include the International Geosphere-Biosphere Programme (IGBP) scheme, University of Maryland (College Park, MD, USA), leaf area index (LAI)/fraction of the absorbed photosynthetically active radiation (FPAR), net primary production (NPP), plant functional type (PFT), and Food and Agriculture Organization-Land Cover Classification System (FAO-LCCS, https://www.fao.org/home/en/, accessed on 1 December 2021) (Rome, Italy). The PFT classification scheme, including water bodies, evergreen coniferous forestland, evergreen broad-leaved forestland, deciduous broad-leaved forestland, shrub/forestland, grassland, cereal crops, broad-leaved crops, built-up areas, snow and ice, bare land, and other types, was adopted in this experiment [36,37]. Many experiments show that the MCD12Q1 dataset has a high classification accuracy. For more information on the accuracy of the dataset products, please refer to Sulla-Menashe et al. [36,37]. Before conducting the experiment, we first reclassified the data and organized them into six types (Table 2).

Table 2. Land use types and descriptions after reclassification.

Land Use Type	Description
Water bodies	Permanent water coverage rates not lower than 60%.
Forestland	Tree cover not lower than 10%.
Grassland	Dominated by herbaceous annuals (less than 2 m) with no cultivation.
Farmland	Dominated by herbaceous annuals (less than 2 m) and no less than 60% cereal crops or broadleaf crops.
Construction land	Impervious surface coverage, including building materials, asphalt, and vehicles, not lower than 30%.
Unused land	Unvegetated barren (sand, rock, soil) with less than 10% vegetation.

3.1.2. Air Quality Monitoring Data

The air quality data obtained in this experiment included real-time monitoring data from monitoring stations. The China National Environmental Monitoring Center (Beijing, China) releases automatic hourly air quality monitoring results from all monitoring stations and cities in China to the public through the National Urban Air Quality Real-time Publishing Platform (http://106.37.208.233:20035/) accessed on 1 December 2021. In this experiment, 215 stations distributed in the BTH region and surrounding cities were selected (as shown in Figure 3). From 0:00 on 1 December 2014, to 23:00 on 31 December 2018, 35,808 h of hourly $PM_{2.5}$ and PM_{10} concentration data were acquired. First, the data were calculated and integrated. For the small amount of missing data, hourly average values on other days in the corresponding months were used to obtain $PM_{2.5}$ and PM_{10} values at each point. Then, the ordinary kriging interpolation method was applied to obtain the spatial distribution of particulate matter concentrations in each time interval within the study area, with a spatial resolution of 500 m. Moreover, the $PM_{2.5}/PM_{10}$ ratio at each spatial position and each region was calculated. Finally, the spatial mean values of particulate matter concentrations in the whole study area in each region and for each land use type at different time scales were obtained by the statistical partitioning method. The ordinary kriging method dynamically determines the values of variables according to an optimization criterion function in the interpolation process to ensure that the interpolation function is in the best state; this process takes into account both the positional relationship between the observed and estimated points and the relative positional relationship among the observed points and is more effective when the number of points is relatively sparse.

3.1.3. Accuracy Verification of the Optimization Model

To verify the accuracy of the improved model, both the traditional and improved models were adopted to fit the monthly mean values of $PM_{2.5}$ and PM_{10} concentrations in the whole study area and in cities. In addition, the goodness-of-fit R^2 value was compared between these two models, and the results are listed in Table 3. Among the $PM_{2.5}$ and PM_{10} fitting results, the average R^2 values of the traditional model are 0.51 and 0.49, respectively, and the average R^2 values of the improved model are 0.65 and 0.57, respectively, which indicate increases of approximately 0.14 and 0.08, respectively. The goodness of fit of the improved model is considerably higher than that of the traditional model, which verifies that the improved model achieves a higher fitting accuracy, can extract more information from the data, and is more suitable for quantitative analyses of the variation characteristics of particulate matter concentrations. In addition, compared to the traditional sine function, the improved function better quantifies the general trends, amplitudes, deformation degrees, deviation degrees, and other parameters. In addition to the particulate matter concentration, this method can be extended to other elements conforming to the periodic variation pattern involving a single peak and a single valley, such as the vegetation coverage, monthly mean temperature, and monthly mean precipitation. The corresponding patterns of variation can be examined based on the obtained fitting results.

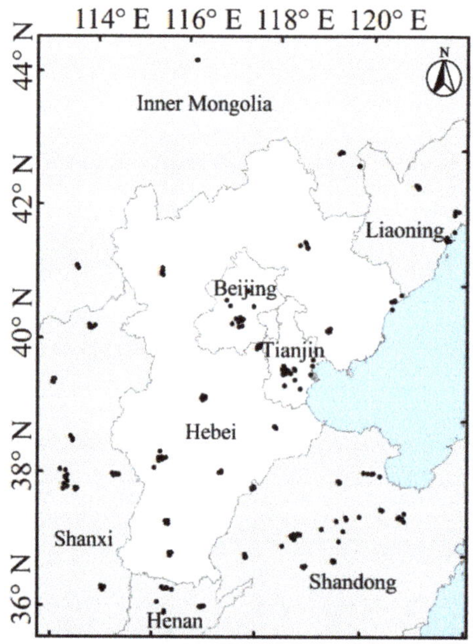

Figure 3. Spatial distribution of the air quality monitoring sites in the study area and surrounding areas.

Table 3. Comparison of R^2 between the traditional and improved models.

	$PM_{2.5}$		PM_{10}	
	The Traditional Model	**The Improved Model**	**The Improved Model**	**The Improved Model**
The BTH region	0.60	0.74	0.53	0.58
Beijing	0.36	0.50	0.29	0.32
Tianjin	0.50	0.67	0.51	0.56
Shijiazhuang	0.52	0.67	0.48	0.56
Tangshan	0.53	0.64	0.48	0.57
Qinhuangdao	0.55	0.56	0.60	0.64
Handan	0.50	0.77	0.53	0.68
Xingtai	0.64	0.81	0.63	0.68
Baoding	0.66	0.81	0.65	0.72
Zhangjiakou	0.08	0.18	0.19	0.34
Chengde	0.54	0.59	0.36	0.42
Cangzhou	0.58	0.76	0.60	0.71
Langfang	0.50	0.67	0.43	0.50
Hengshui	0.63	0.84	0.64	0.74

3.2. *Influence of Land Use on the Atmospheric Particulate Matter Concentration*

3.2.1. Spatiotemporal Characteristics of Land Use

The spatial distribution of land use in the study area for each year is shown in Figure 4a–d. The eastern and southern parts of the study area are on the Haihe Plain, and the construction land is concentrated in the built-up areas of the cities and the surrounding villages, while the agricultural land is mostly between the cities and villages. The northeastern and southwestern parts of the study area are in the Yanshan Mountain Range and Taihang

Mountain Range, and forestland is mostly distributed in areas with higher elevations and higher slopes, while grassland is distributed in areas with lower elevations and gentler slopes. The northwestern part of the study area is in the Bashang Plateau, and grassland is the most dominant land use type. The locations where land use changes occurred are shown in Figure 4e. Forestland, as the largest land use type with the largest new area, occurred mainly in the mountainous areas. The increase in grassland occurred mostly on the gentle slopes of the mountains. New farmland is mainly present in the northwestern Bashang Plateau, the eastern side of the Taihang Mountains, and the eastern coastal areas. The new construction land mainly occurs on the periphery of the built-up areas of the cities and is the spatial manifestation of urban expansion.

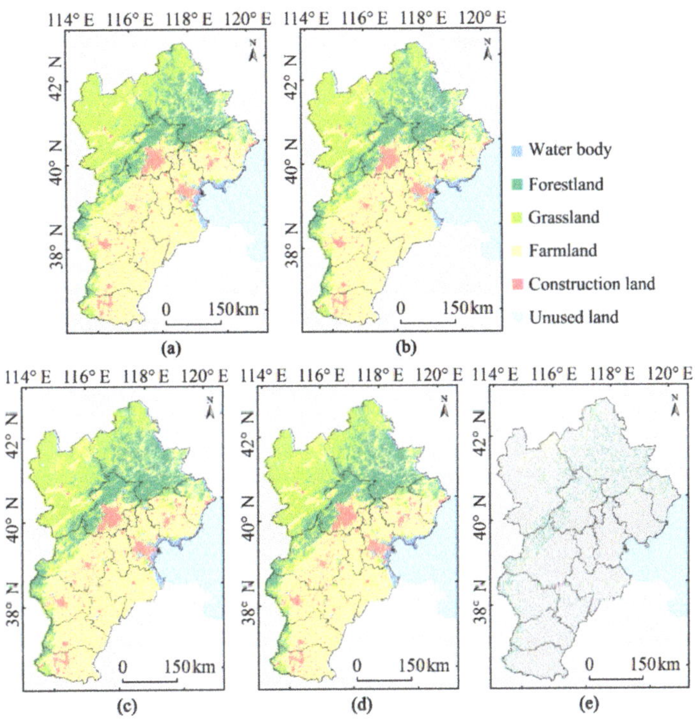

Figure 4. Spatial distributions of land use in 2015 (**a**), 2016 (**b**), 2017 (**c**), and 2018 (**d**) and a schematic representation of the land use changes that occurred (**e**).

The area of each land use type in descending order is farmland, grassland, forestland, construction land, water bodies, and unused land, among which the areas of farmland and grassland are decreasing; the areas of forestland and construction land are increasing at relatively fast and slow rates, respectively; and the area of water bodies and unused land is small and has remained relatively unchanged. The land use conversion matrix is shown in Table 4. Approximately 4.5% of the study area underwent land use change in mid-2015–2018, with grassland and forestland being the two relatively active types. A total of 8.8% of grassland was transformed, mainly to forestland and farmland. A total of 4.8% of forestland was transformed, mainly to grassland. A total of 2.6% of farmland was transformed into grassland, forestland, and construction land. Construction land is the most stable of all types and is generally not converted to other types. Smaller areas of water bodies and unused land underwent negligible conversion.

Table 4. Land use transfer matrix from 2015 to 2018.

		2018						
		Water Bodies	**Forestland**	**Grassland**	**Farmland**	**Construction Land**	**Unused Land**	**Total**
2015	**Water bodies**	2650	0	4	0	0	5	2658
	Forestland	1	30,066	1404	107	4	0	31,582
	Grassland	53	3404	57,858	2008	115	23	63,461
	Farmland	7	304	1861	98,473	447	0	101,091
	Construction land	0	0	0	0	16613	0	16,613
	Unused land	4	0	36	0	2	348	390
	Total	2714	33,773	61,163	100,588	17181	376	215,795

Unit: km^2

3.2.2. Comparison of Atmospheric Particulate Matter Concentrations between Various Land Use Types

The spatial distributions of each land use type in the study area from 2015 to 2018 were extracted, and the extracted distributions were superimposed onto the interpolation data of the monthly mean values of the particulate matter concentration in the corresponding year and month to determine the monthly mean particulate matter concentrations corresponding to each land use type. Moreover, the average values for the same month in each year were calculated. The results are shown in Figure 5. The results indicate obvious differences in particulate matter concentrations among the various land use types. Among the land use types, the particulate matter concentrations corresponding to construction land and farmland were higher than those corresponding to the other types. Among the various land use types, construction land encompasses areas with the highest human activity intensity and motor vehicle exhaust, road dust, industrial emissions, domestic emissions, fuel combustion-related emissions, and construction dust, all of which are closely related to particulate matter emissions. In addition, the building density and height in urban built-up areas are high. The resultant blocking and friction effects weaken regional winds, greatly affecting particulate matter diffusion [11,12]. The influence of farmland on particulate matter exhibited seasonal differences. During the crop growing season, farmland is mostly covered with vegetation, which inhibits the emission of surface particulate matter to a certain extent, while leaves generate a certain dust-retention effect. However, due to the relatively low heights and comparatively neat arrangements of crops, this dust-retention effect may be lower than those of forestland and grassland [19,20]. After harvest, the dust emitted during straw burning represents an important source of particulate matter [21,22]. Especially during the cold winter season, farmland lacking vegetation coverage is mostly bare, and a large amount of dust becomes exposed and is easily emitted into the atmosphere under windy weather conditions, resulting in a notable increase in the wintertime particulate matter concentration [19,20,38]. The particulate matter concentrations corresponding to forestland and grassland were significantly lower than those corresponding to the other land use types, and the intensity of particulate matter removal was positively related to the vegetation growth degree, which was considerably higher in summer than in winter. Vegetation stems, leaves, and other organs remove particles suspended in air via retention, attachment, and adhesion mechanisms [39]. First, the plant canopy can block airflow and locally reduce the wind speed, which can result in the deposition of atmospheric particles onto the surfaces of leaves. Moreover, the turbulence between branches and leaves is very high, which makes it easier for particles to collide with and contact branches and leaves, thus increasing the settling rate [40]. Second, the surface structures of plant leaves and bark are very rough, and certain groove-like tissues, cilia, waxy layers, etc. can intercept particles that become embedded into their rough surfaces, and the deposition effect remains relatively consistent [41]. Third, plant leaves often secrete sticky substances, which can cause particle adhesion to their surfaces. This dust retention effect remains the most stable and can withstand rain erosion. Vegetation can also create an environment conducive to particle sedimentation. Tree canopies generate shade effects, and leaves generate transpiration effects, which can achieve cooling effects. Moreover, vegetation can increase the relative air humidity and adjust the local microclimate, thus altering the viscosity and quality of particles, shortening the particle suspension time and accelerating the sedimentation process. Furthermore, vegetation effectively inhibits chemical reactions of particles and reduces the generation of secondary particles [42]. In addition, the PM_{10} concentrations in forestland and grassland areas increased the most in spring, which is mainly attributed to frequent dusty weather conditions. The effect of water on the particulate matter concentration is complicated. On the one hand, the particulate matter emission levels of water bodies were lower than those of the other types, and water simultaneously exerted a certain adsorption effect on suspended particulate matter. On the other hand, water evaporation can increase humidity, which can affect the secondary formation, accumulation, and diffusion of particulate matter [19,43–45]. The vegetation coverage in unused land areas is low, and bare soil or sand on the surface can easily be entrained by wind, resulting in particulate pollution. However, because the unused land area in the study region is small and mostly distributed in the coastal areas and northwestern Bashang Plateau, unused land areas are mostly surrounded by water bodies and grasslands. Therefore, the particulate matter concentration mostly varied between the levels corresponding to water bodies and grasslands.

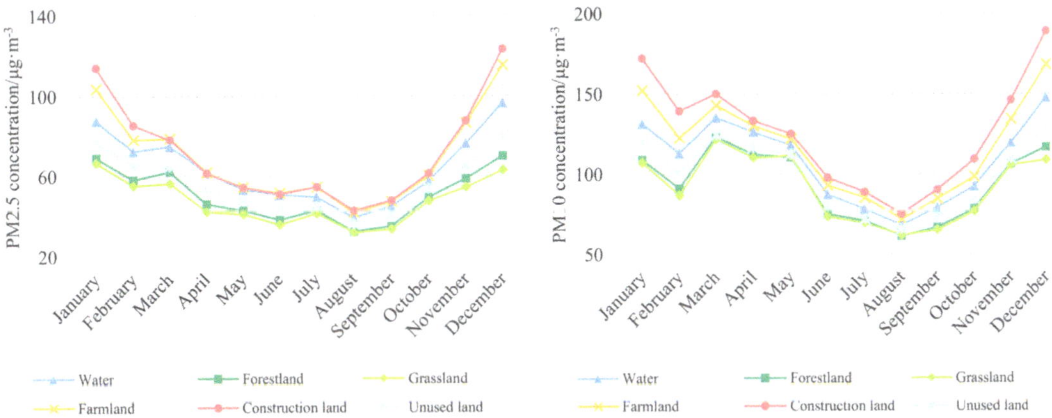

Figure 5. Monthly mean values of the $PM_{2.5}$ and PM_{10} concentrations corresponding to each land use type.

3.2.3. Variation Patterns of the Atmospheric Particulate Matter Concentration for Different Land Use Types

The analysis of the temporal variation in the particulate matter concentration corresponding to each land use type in the study area revealed that the particulate matter concentration corresponding to each land use type was similar to that in the whole study area, and an obvious periodic variation pattern occurred. Therefore, the improved sine function model was employed to fit the particulate matter concentration corresponding to each land use type, and the inherent relationship between the land use type and particulate matter concentration was explored by comparing the fitting parameters between various land use types. The results are shown in Figures 6 and 7.

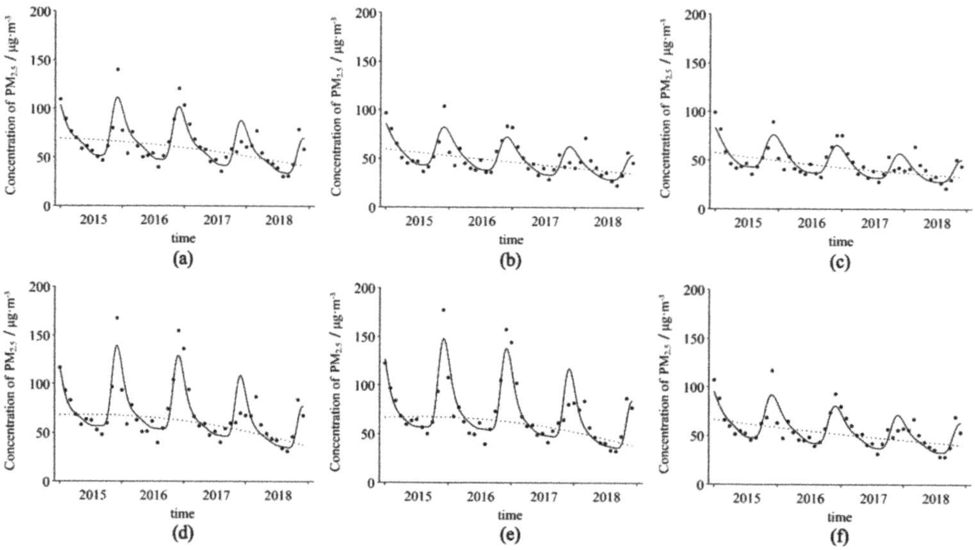

Figure 6. Fitting results of the monthly $PM_{2.5}$ concentration for each land use type: (**a**) water; (**b**) forestland; (**c**) grassland; (**d**) farmland; (**e**) construction land; and (**f**) unused land.

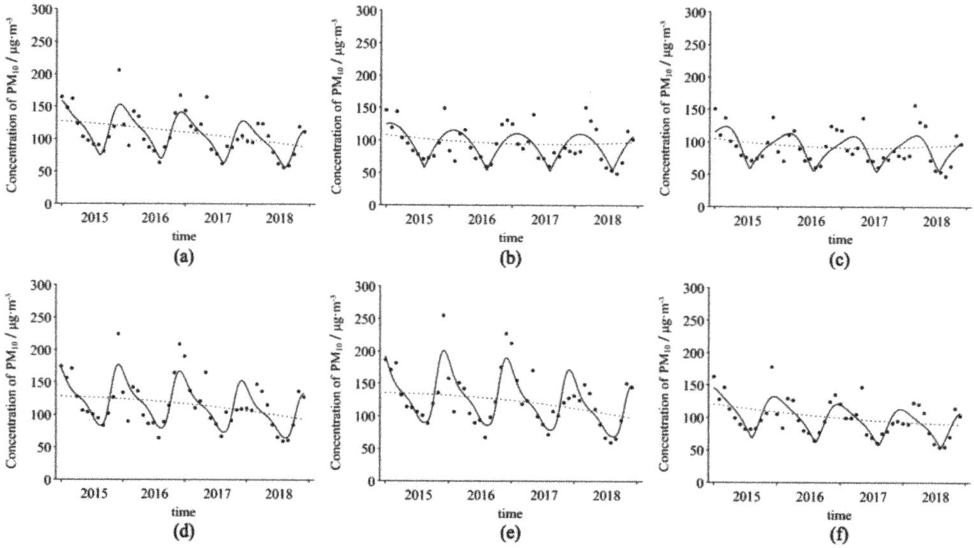

Figure 7. Fitting results of the monthly PM_{10} concentration for each land use type: (**a**) water; (**b**) forestland; (**c**) grassland; (**d**) farmland; (**e**) construction land; and (**f**) unused land.

Fitting parameters were separately obtained for the whole study area and each city, and the results are summarized in Table 5.

Table 5. Fitting parameters of the monthly PM concentration.

		R^2	Average Concentration	A_t	B_t	Turning Point	Overall Trend	Average Monthly Variation	C_t	D_t	E_t	F_t	G_t	H_t
$PM_{2.5}$	Water bodies	0.71	64	−0.0066	−0.13	Before 2015	Descends rapidly	−0.46	53.20	1.20	12	1.95	1.86	1.43
	Forestland	0.63	50	0.0008	−0.45	After 2018	Descends slowly	−0.41	46.92	0.97	12	1.77	2.09	1.00
	Grassland	0.63	47	0.0040	−0.62	After 2018	Descends slowly	−0.42	47.69	0.87	12	1.93	2.22	0.72
	Farmland	0.71	69	−0.0148	0.19	2015.06	Rises and then descends	−0.54	56.19	1.48	12	2.85	1.86	1.41
	Construction land	0.79	72	−0.0154	0.24	2015.07	Rises and then descends	−0.51	56.23	1.60	12	3.12	1.84	1.25
	Unused land	0.71	56	0.0026	−0.56	After 2018	Descends slowly	−0.43	52.46	0.99	12	1.81	2.00	1.19
PM_{10}	Water bodies	0.58	108	−0.0027	−0.37	Before 2015	Descends rapidly	−0.50	78.17	1.09	12	0.76	1.73	1.18
	Forestland	0.41	93	0.0082	−0.53	2017.08	Descends and then rises	−0.13	61.02	1.09	12	0.46	0.97	0.14
	Grassland	0.38	91	0.0101	−0.60	2017.06	Descends and then rises	−0.11	62.81	1.02	12	0.58	0.54	−0.86
	Farmland	0.57	117	−0.0080	−0.10	Before 2015	Descends rapidly	−0.49	84.77	1.15	12	1.17	1.74	1.46
	Construction land	0.66	126	−0.0102	−0.05	Before 2015	Descends rapidly	−0.55	91.18	1.26	12	1.37	1.76	1.42
	Unused land	0.56	99	0.0058	−0.69	After 2018	Descends slowly	−0.40	73.82	1.00	12	0.63	1.60	0.94

First, the goodness-of-fit R^2 value was evaluated. The average R^2 values of the $PM_{2.5}$ and PM_{10} concentrations corresponding to the six land use types are 0.70 and 0.53, respectively, verifying that the improved model can better reflect the change pattern of the monthly mean particulate matter concentration for each land use type. The fitting R^2 value of the $PM_{2.5}$ concentration for all land use types is above 0.6, and the fitting effect is good. The fitting R^2 value of the PM_{10} concentration exceeds 0.6 for construction land; is above 0.5 for water bodies, farmland, and unused land; and is below 0.5 for forestland and grasslands. Generally, the fitting R^2 value of the $PM_{2.5}$ concentration for all land use types is higher than that of the PM_{10} concentration, which once again demonstrates that the periodic fluctuation patterns of the $PM_{2.5}$ concentration were stronger than those of the PM_{10} concentration. Among all types, construction land exhibits the highest R values, which is attributable to the close relationship between the seasonal patterns of human activities. Especially after the heating period in winter, the particulate matter concentration corresponding to construction land sharply increased, resulting in the largest differences between high and low values for all land use types and the most notable periodicity. The fitting R^2 values for forestland and grassland are relatively low. On the one hand, periodic fluctuations were slightly limited in forestland and grassland areas due to the presence of fewer particulate matter emission sources and fewer anthropogenic emissions. On the other hand, particulate matter concentrations corresponding to forestland and grassland were significantly lower than those corresponding to the other land use types, and random factors imposed a great influence, producing certain randomness and consequently reducing the periodic fluctuation patterns, thus yielding slightly poor fitting effects.

The quadratic coefficient A_t and linear coefficient B_t reflect the overall change trends of particulate matter concentrations in the whole study area and in cities. The $PM_{2.5}$ concentrations corresponding to all land use types exhibited downward trends, with those of water bodies exhibiting an accelerated downward trend; those in forestland, grassland, and unused land all exhibiting gradual downward trends; and those in both construction land and farmland first increasing and then decreasing. Additionally, the axis of symmetry occurred in 2015, i.e., an accelerated downward trend was exhibited throughout most of the research period. In terms of the overall change trends of PM_{10} concentrations for various land use types, the PM_{10} concentrations corresponding to water bodies, farmland, and construction land decreased at an accelerated rate; the concentration corresponding to unused land decreased at a low rate; the concentrations in forestland and grassland areas first decreased and then increased; and the axis of symmetry occurred in 2017. However, the average PM_{10} concentration was the lowest among all types, with a limited decline and increase; thus, the PM_{10} concentrations basically remained at low levels. Overall, although the overall average concentrations in the study area exhibited downward trends, the change rates varied between different land use types. The particulate matter concentrations corresponding to construction land and farmland were relatively high but simultaneously decreased at higher rates, and the decrease rates were accelerated. The particulate matter concentrations in forestland and grassland areas were low; the decrease rates of the $PM_{2.5}$ concentration were decelerated; and the PM_{10} concentrations slightly increased after decreasing to a constant level. The particulate matter concentrations in areas containing water bodies and unused land were lower than those in farmland and construction land areas but higher than those in forestland and grassland areas, with the particulate matter concentration corresponding to water bodies rapidly decreasing, while the particulate matter concentration corresponding to grassland decreased gradually.

The amplitude D_t is the ratio of the focal concentration fluctuation amplitude to that of the overall concentration. Construction land and farmland exhibited significantly higher values than did the other land use types because these two types are the most affected by human activities. Therefore, these land use types exhibited large differences between winter and summer. Moreover, the amplitude of the $PM_{2.5}$ concentration for all land use types was obviously larger than that of the PM_{10} concentration, which also indicates that the $PM_{2.5}$ concentration is more closely related to human sources.

The deformation factor F_t reflects the durations of periods with high and low concentrations. The larger the value is, the shorter the durations of high and low concentrations are. The deformation factor F_t of the $PM_{2.5}$ concentration for each type was larger than 1, which suggests that the duration of low-value periods was longer than that of high-value periods, with the F_t values of construction land and farmland being significantly higher than those of the other land use types. This finding was attributable to the maximal intensity of human activities in construction land and farmland areas producing sharp increases in particulate matter concentrations in winter. This pattern was quite different from the low-value periods in summer. The relatively high values further shortened the duration of high-value periods. However, the F_t values of the PM_{10} concentration for water bodies, forestland, grassland, and unused land were smaller than 1, which indicates that the duration of high-value periods was longer than that of low-value periods for these land use types. This pattern occurred because the frequent dusty weather conditions in spring prolonged the duration of high-PM_{10}-concentration periods, and high values continued to occur in winter and spring, which imposed a particularly notable impact on water bodies, forestland, grassland, and unused land. Farmland and construction land areas are more affected by human activities, and the F_t values were larger than 1, i.e., the duration of low-value periods was increased.

The phase G_t reflects the time when a peak value occurs during a period. The larger the value, the earlier the peak and valley values are observed. There was little difference in the phases of $PM_{2.5}$ concentrations among different land use types, which indicates that the peak and valley values for each land use type occurred at similar times. Regarding the PM_{10} concentration model, the phases G_t of forestland and grassland were obviously lower than those of the other land use types, which was attributed to the low vegetation coverage in spring. In addition, frequent dust events slightly increased the PM_{10} concentration for all land use types, while the spring season exhibited relatively low vegetation coverage, which exerted a notable impact on vegetation types such as forestland and grassland. Therefore, the PM_{10} concentrations in forestland and grassland areas were higher than the notable peak values observed in March and April, while the peak values for the other land use types often occurred in December or January, resulting in higher values of the phase G_t in forestland and grassland areas.

The offset factor H_t controls the slight shift in the peaks and valleys from left to right. A positive value indicates a rightward shift in the valleys; a negative value indicates a leftward shift in the valleys; and the larger the absolute value is, the greater the shift. The shift factor H_t of the $PM_{2.5}$ concentration for each land use type was larger than 0, which indicates that for all land use types, each valley was far from the previous peak and that the next peak was closer. Notably, the concentration slowly decreased and subsequently increased rapidly. The PM_{10} concentration indicates that the H_t values in forestland and grassland areas were smaller than or close to 0, i.e., the peak values occurred later. In addition, the concentrations decreased faster and increased slower, which is similar to the reason why the G_t values of the PM_{10} concentrations of forestland and grassland were smaller than those of the other land use types. Among all types, the H_t value of farmland was the largest because farmland reduced particulate matter concentrations during the crop growth period. Hence, the particulate matter concentration exhibited a steady and gradual downward trend. However, after crops were harvested, surface dust became exposed due to the absence of vegetation coverage and was emitted into the air, and the particulate matter concentration rapidly increased.

3.3. *Influence of Landscape Pattern on the Atmospheric Particulate Matter Concentration*

According to the division of county-level administrative regions, the study area was divided into 200 county-level units, and the average $PM_{2.5}$ and PM_{10} concentrations throughout the whole year and during the four seasons of 2018 in each district and county were separately determined. The landscape pattern indices of each district and county at the landscape and class levels were obtained in Fragstats 4.2 software (University of Massachusetts Amherst, Amherst, MA, USA), and the influence of the landscape pattern on the atmospheric particulate matter concentration was evaluated by correlation analysis. Considering that water bodies and unused land accounted for only 1.3% and 0.2%, respectively, of the total area of the study region and imposed limited regulation effects on the large-scale particulate matter concentration, only the four most important land use types, namely, forestland, grassland, farmland, and construction land, were examined at the class level.

3.3.1. Atmospheric Particulate Matter Concentration Distribution

The average $PM_{2.5}$ and PM_{10} concentrations in all districts and counties throughout the whole year and in the four seasons were statistically analyzed. The results are shown in Figure 8. The results reveal that the overall particulate matter concentration exhibited a spatial distribution trend similar to that in the whole region, in which the annual average $PM_{2.5}$ concentration exhibited a trend of high values in the southeast and low values in the northwest. High values were mainly concentrated in the Shijiazhuang–Baoding and Handan–Xingtai areas, while the concentrations in Zhangjiakou, Chengde, and other cities in the north were low. During each season, the spatial distribution patterns of the seasonal average concentrations in winter and spring in all districts and counties were similar to those of the annual average concentrations, while in spring and summer, excluding the Shijiazhuang–Baoding and Handan–Xingtai areas, relatively high concentrations were observed in Beijing, Tianjin, Langfang, and other cities. The annual average trend of the PM_{10} concentration was similar to that of the $PM_{2.5}$ concentration, exhibiting the same distribution trend of high values in the southeast and low values in the northwest. The difference is that Tangshan city on the eastern coast also contained relatively high-concentration areas, with the lowest concentrations mainly concentrated in Zhangjiakou city and Chengde city in the north. The distribution pattern during each season was similar to that of the annual average concentration. Moreover, the frequent dusty weather conditions in spring caused a slight increase in the concentration in Northwest China, and simultaneously, the concentration in Tangshan increased, while relatively high-concentration areas moved slightly northwards.

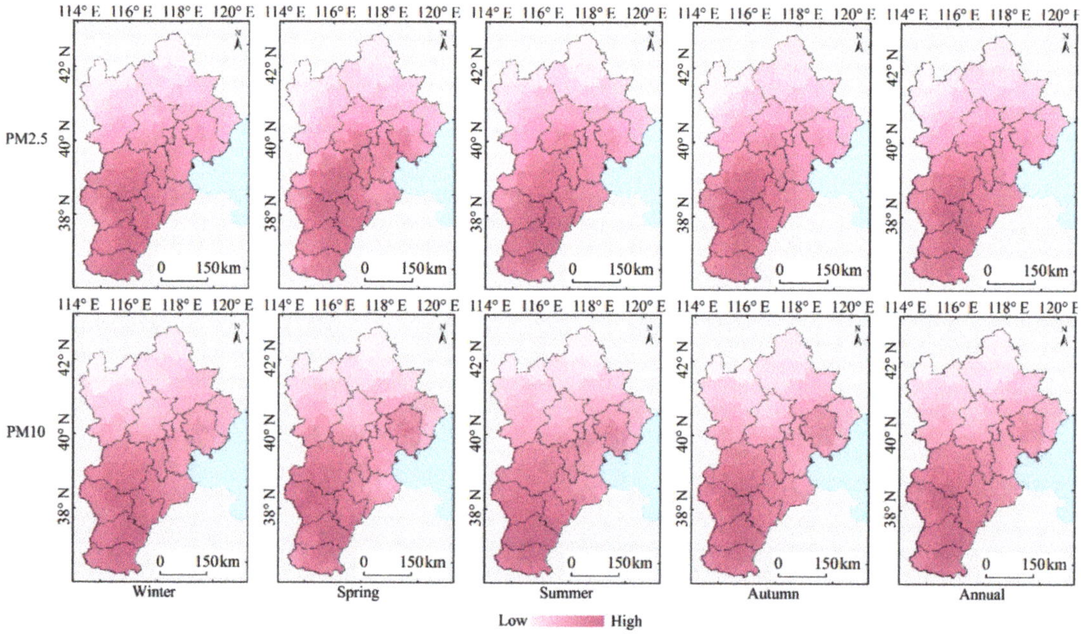

Figure 8. Spatial distributions of the $PM_{2.5}$ and PM_{10} concentrations at the county level.

3.3.2. Effects of Land Use Types on Atmospheric Particulate Matter Concentrations

The landscape pattern indices at the landscape level are shown in Figure 9, and the correlation between the particle concentration and indices at the landscape level is shown in Table 6. At the landscape level, each selected index exhibited a certain correlation with the particulate matter concentration, but the correlation degrees were lower than those at the class level, indicating that the distribution pattern of the whole landscape influenced the particulate matter concentration, but the effect degree was weaker than that observed in specific land cover types. Among the indices, LPI, AI, and CONTAG mostly exhibited positive correlations with the particulate matter concentration, while PD, ED, SPLIT, SHDI, and SIDI mostly showed negative correlations with the particulate matter concentration. The remaining indices exhibited weak correlations. The lower the aggregation degree is, the higher the fragmentation degree, the more complex the shape, and the richer the land use types are, the more closely the landscape resembles a natural landscape. Conversely, the opposite characteristics indicate an artificial landscape. Compared to artificial landscapes, natural landscapes contain fewer emission sources, and natural landscapes can simultaneously reduce and limit particulate matter concentrations. In a comparison of the seasons, the correlation of each index in winter and summer was generally higher than that in spring and autumn, which may be attributed to the high particulate matter concentration and notable spatial heterogeneity in winter. In contrast, in summer, the vegetation coverage was the highest; the reduction effect of vegetation on the particulate matter concentration was the strongest; and the landscape pattern distribution exerted a notable influence.

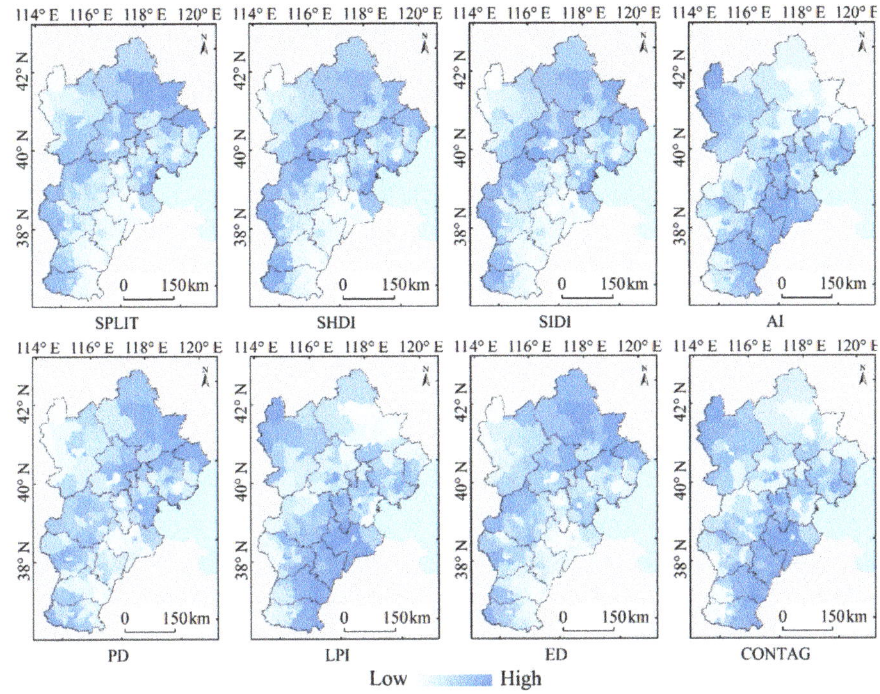

Figure 9. Spatial distributions of the landscape pattern indices at the landscape level.

Table 6. Correlations between the particle concentrations and indices at the landscape level.

	$PM_{2.5}$					PM_{10}				
	Winter	**Spring**	**Summer**	**Autumn**	**All**	**Winter**	**Spring**	**Summer**	**Autumn**	**All**
PD	−0.185 **	−0.108	−0.181 **	−0.158 *	−0.170 *	−0.155 *	−0.114	−0.144 *	−0.128	−0.143 *
LPI	0.401 **	0.368 **	0.407 **	0.372**	0.399 **	0.366 **	0.291 **	0.346 **	0.309 **	0.343 **
ED	−0.362 **	−0.307 **	−0.395 **	−0.322 **	−0.356 **	−0.331 **	−0.280 **	−0.358 **	−0.284 **	−0.320 **
CONTAG	0.249 **	0.200 **	0.251 **	0.195 **	0.234 **	0.227 **	0.154 *	0.214 **	0.168 *	0.203 **
SPLIT	−0.385 **	−0.387 **	−0.419 **	−0.386 **	−0.398 **	−0.343 **	−0.330 **	−0.357 **	−0.317 **	−0.342 **
SHDI	−0.446 **	−0.332 **	−0.403 **	−0.373 **	−0.415 **	−0.425 **	−0.324 **	−0.382 **	−0.362 **	−0.395 **
SIDI	−0.418 **	−0.320 **	−0.386 **	−0.349 **	−0.391 **	−0.395 **	−0.293 **	−0.355 **	−0.329 **	−0.364 **
AI	0.291 **	0.263**	0.333 **	0.265 **	0.293 **	0.254 **	0.198 **	0.281 **	0.205 **	0.241 **

* indicates a significant correlation at the 0.05 level, and ** indicates a significant correlation at the 0.01 level.

The landscape pattern indices of forestland, grassland, farmland, and construction land at the class level are shown in Figure 10, and the correlation between the particle concentration and indices at the class level is shown in Table 7. PLAND, PD, LPI, and LSI all exhibited negative correlations with the particulate matter concentration, with LSI showing the highest correlation. In contrast, SPLIT exhibited no significant correlation, indicating that the larger the area, density, dominance, and shape complexity of forestland patches were, the lower the particulate matter concentration was, and the complexity of the patch shape imposed the greatest influence on the particulate matter concentration. If the forestland area within a given region is large, this land use type can become the dominant type in the region. However, if the forestland area is concentrated and connected as one unit, this area can reduce and inhibit particles. Among all seasons, the correlation in summer was slightly higher than that in the other seasons, which was related to the high vegetation coverage in forestland areas in summer and the strongest particulate matter inhibition effect of lush trees. Moreover, the influence of forestland areas on the PM_{10} concentration exceeded that on the $PM_{2.5}$ concentration.

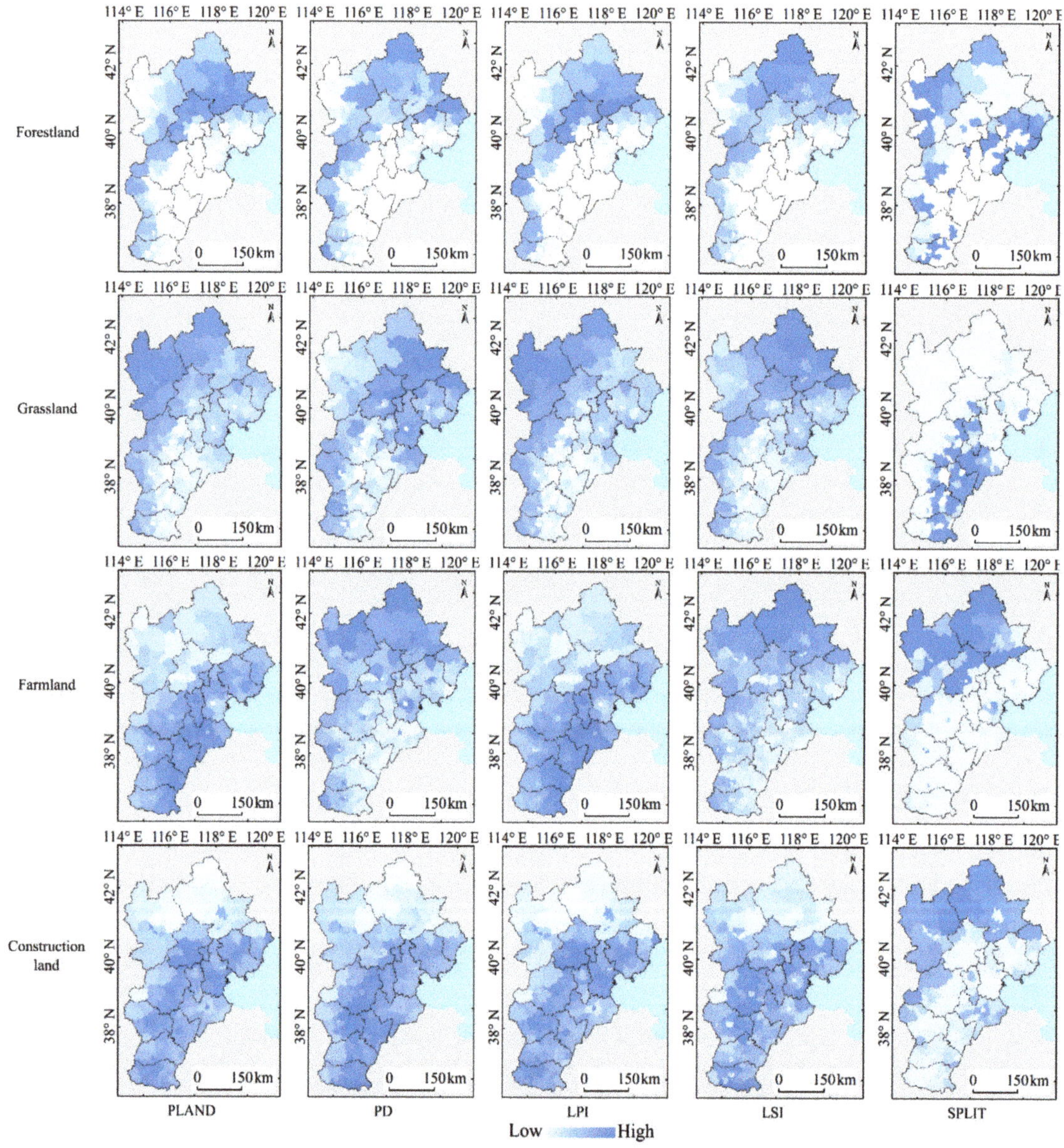

Figure 10. Spatial distributions of the landscape pattern indices at the class level.

Table 7. Correlations between the particle concentration and indices at the class level.

		$PM_{2.5}$					PM_{10}				
		Winter	Spring	Summer	Autumn	All	Winter	Spring	Summer	Autumn	All
Forestland	PLAND	−0.382 **	−0.346 **	−0.372 **	−0.359 **	−0.376 **	−0.389 **	−0.367 **	−0.424 **	−0.378 **	−0.393 **
	PD	−0.280 **	−0.289 **	−0.286 **	−0.269 *	−0.285 **	−0.255 *	−0.176	−0.255 *	−0.212	−0.235 *
	LPI	−0.270*	−0.195	−0.222 *	−0.229 *	−0.246 *	−0.293 **	−0.252 *	−0.306 **	−0.278 *	−0.288 **
	LSI	−0.464 **	−0.551 **	−0.530 **	−0.487 **	−0.502 **	−0.444 **	−0.468 **	−0.477 **	−0.439 **	−0.457 **
	SPLIT	0.099	0.113	0.088	0.070	0.095	0.106	0.053	0.094	0.070	0.088
Grassland	PLAND	−0.656 **	−0.791 **	−0.713 **	−0.713 **	−0.712 **	−0.623 **	−0.515 **	−0.592 **	−0.564 **	−0.598 **
	PD	−0.394 **	−0.259 **	−0.351 **	−0.333 **	−0.360 **	−0.369 **	−0.325 **	−0.348 **	−0.336 **	−0.357 **
	LPI	−0.588 **	−0.737 **	−0.648 **	−0.653 **	−0.647 **	−0.560 **	−0.462 **	−0.530 **	−0.508 **	−0.537 **
	LSI	−0.528 **	−0.505 **	−0.553 **	−0.515 **	−0.535 **	−0.509 **	−0.507 **	−0.527 **	−0.487 **	−0.514 **
	SPLIT	0.237 **	0.174 *	0.212 **	0.199 **	0.220 **	0.237 **	0.162 *	0.188 *	0.183 *	0.210 **
Farmland	PLAND	0.631 **	0.619 **	0.618 **	0.602 **	0.633 **	0.630 **	0.512 **	0.586 **	0.554 **	0.597 **
	PD	−0.267 **	−0.291 **	−0.278 **	−0.261 **	−0.276 **	−0.236 **	−0.138	−0.183 *	−0.170 *	−0.202 **
	LPI	0.626 **	0.613 **	0.614 **	0.597 **	0.627 **	0.621 **	0.508 **	0.576 **	0.543 **	0.588 **
	LSI	−0.507 **	−0.587 **	−0.594 **	−0.541 **	−0.549 **	−0.483 **	−0.510 **	−0.531 **	−0.471 **	−0.498 **
	SPLIT	−0.058	−0.014	−0.018	−0.031	−0.040	−0.081	−0.048	−0.068	−0.079	−0.075
Construction land	PLAND	0.084	0.242 **	0.195 **	0.188 **	0.150 *	0.026	0.058	0.083	0.046	0.044
	PD	0.555 **	0.503 **	0.505 **	0.538 **	0.548 **	0.551 **	0.536 **	0.510 **	0.524 **	0.545 **
	LPI	0.034	0.189 **	0.147 *	0.135	0.098	−0.021	0.015	0.043	0.002	−0.002
	LSI	0.388 **	0.365 **	0.344 **	0.371 **	0.383 **	0.370 **	0.335 **	0.294 **	0.324 **	0.349 **
	SPLIT	−0.378 **	−0.492 **	−0.458 **	−0.435 **	−0.428 **	−0.357 **	−0.390 **	−0.384 **	−0.356 **	−0.371 **

* indicates a significant correlation at the 0.05 level, and ** indicates a significant correlation at the 0.01 level.

PLAND and LPI reflect the proportion, dominance degree, and patch connectivity of grassland patches, respectively, and these indices decreased spatially from northwest to southeast. High values were mainly concentrated on the Bashang Plateau and Taihang Mountains, indicating that these areas contained high grassland proportions and high dominance degrees and good connectivity of grassland patches. PD reflects the density of grassland patches, and high PD values were mostly concentrated in the Yanshan Mountains and eastern coastal areas, indicating that although the grassland patches in these areas are small, many patches are present, and their distribution is fragmented. LSI reflects the shape of grassland patches, and high LSI values largely occurred in the Yanshan Mountains and Taihang Mountains, where grassland and forestland are staggered, with complex shapes, dense patches, and small patch sizes. High SPLIT values primarily occurred in the Haihe Plain in the southeast, where the grassland coverage is low, and the distribution is scattered.

Among the various indices for farmland patches, PLAND and LPI exhibited strong positive correlations with the particulate matter concentration, while PD and LSI showed negative correlations. SPLIT exhibited no significant correlation, which indicates that the higher the farmland proportion in the region, the higher the connectivity and the more complete the patches, the higher the particulate matter concentration, and corresponding regions mainly occurred in the Haihe Plain. However, the particulate matter concentration was low in areas with high densities, scattered distributions, and irregular shapes of farmland patches. Among all seasons, PLAND, PD, and LPI exhibited the highest correlations in winter or spring, and there was less vegetation coverage in farmland patches in winter and spring. In areas where farmland dominated, bare surfaces became the main emission particulate matter source, and the resultant correlation was high. LSI showed the highest negative correlation in summer, which may have occurred because the more complex the shapes of farmland patches are, the closer the landscape is to a natural landscape. These areas were dominated by mountains, and farmlands were mostly distributed among many forestlands and grasslands. In summer, the farmland coverage intensity increased. Moreover, the forestland and grassland in the surrounding areas played a notable role in blocking and settling particles, thus reducing the particulate matter concentration.

PLAND and LPI were mostly positively correlated with the $PM_{2.5}$ concentration but not significantly correlated with the PM_{10} concentration. Notably, the $PM_{2.5}$ concentration was higher in areas with high proportions of construction land, once again verifying that human activities comprise the main $PM_{2.5}$ source, while PM_{10} is more related to natural factors. PD and LSI were positively correlated with the particulate matter concentration, indicating that the denser the construction land patches and the more complex the shape are, the higher the particulate matter concentration is. A dense distribution of construction land patches can lead to the concentration of emission sources. In contrast, a complex shape indicates that the city is highly developed, and the human activity intensity is high, which produces higher particulate emissions and does not facilitate particulate matter diffusion. SPLIT was negatively correlated with the particulate matter concentration, which indicates that in areas where the distribution of construction land patches is scattered, construction land is mostly separated by other land use types, and it is difficult to establish centralized emission sources. Moreover, the other land use types limited particulate matter concentrations, and the particulate matter concentrations were therefore low. Among the seasons, PLAND, LPI, and SPLIT exhibited the highest correlations with the $PM_{2.5}$ concentration in spring, which demonstrates that the influence of construction land on the $PM_{2.5}$ concentration was the strongest in spring. In the other seasons, the $PM_{2.5}$ concentration was more notably influenced by other factors. PD and LSI showed the highest correlations with the particulate matter concentration in winter, which may be related to the overall high particulate matter concentration and highest spatial heterogeneity in winter.

4. Implications and Limitations

As urbanization continues to accelerate, the extent to which human activities contribute to particulate matter emissions is increasing. A large amount of harmful substances are emitted into the atmosphere, which poses a great danger to production and living activities and human health. The problem of atmospheric pollution is receiving more and more attention from relevant departments and individuals. China, especially the eastern part of the country, is one of the more serious regions in the world. Many cities have concentrations of atmospheric pollutants that far exceed the guideline values provided by the World Health Organization (Geneva, Switzerland). At the same time, with the rapid expansion of built-up areas, many cities are gradually approaching saturation in terms of space. Therefore, to control air pollution, a rational planning and layout of the limited space available is necessary. In this experiment, an attempt was made to establish the relationship between land use and particulate matter concentrations using geography, landscape ecology, and statistics. Firstly, the differences in the patterns of change in particulate matter concentrations on each land use type were analyzed, and then the relationship between landscape patterns and particulate matter concentrations was compared for each type. These analyses can provide some theoretical support for the optimization of land use structures. This can be useful for urban planning in the study area and can also provide a reference for other regions and cities. The results obtained from the experiments may be deficient due to limitations in the data sources or experimental methods. The spatial distribution of particulate matter concentrations is obtained by spatial interpolation of the Ordinary Kriging method based on the measured data at the stations. Due to the limited number of stations and the uneven spatial distribution, the interpolation results may have some errors. The experiment covers the period from 2015 to 2018, which is a limited time span. It should also be noted that the spatial and temporal characteristics of particulate matter concentrations are a complex system driven by a combination of influencing factors. Land use affects air quality both through anthropogenic emissions and agricultural activities that alter pollutant emission potential, and through interactions with the atmosphere, such as turbulence stimulation and wind speed reduction. In this manuscript, we have not explored the influence of meteorological factors in depth. All of the above will be focused on in subsequent experiments.

5. Conclusions

In this manuscript, we collated land use and atmospheric particulate matter concentration data pertaining to Beijing, Tianjin, and Hebei in China and analyzed the corresponding temporal and spatial changes and correlations. The conclusions are as follows:

(1) The optimized sine function model can better represent periodic changes in atmospheric particulate matter concentrations. The experimental results reveal that the concentrations corresponding to land use types with higher concentrations, such as construction land and farmland, generally exhibited rapid downward trends, while the concentrations corresponding to land use types with lower concentrations, such as forestland and grassland, generally exhibited gradual downward or consistent trends.
(2) The concentration gradually decreased in spring and summer and rapidly increased in autumn and winter. Human activities such as heating in winter in construction land areas and planting and harvesting of crops in farmland areas were the main factors affecting the concentration fluctuations. The concentrations in forestland and grassland areas were the lowest; the differences between high and low values were small; the fluctuation patterns of the particulate matter concentrations remained relatively uniform; and the peak times occurred slightly later than those of the other land use types. Natural sources greatly influenced the concentration fluctuations, among which the frequent dusty weather conditions in spring imposed a greater influence on forestland and grassland than on the other land use types.
(3) The landscape pattern associated with the land use type significantly affects the particulate matter concentration. Overall, the lower the aggregation degree, the higher the fragmentation degree, the more complex the shape, and the higher the landscape abundance, the lower the particle concentration is. Among the land use types, the density and diversity of forestland, grassland, farmland, and construction land patches contribute to a reduction in the particulate matter concentration. The higher the dominance of forestland and grassland patches in the landscape, the more favorable the setting is to a reduction in the particulate matter concentration.
(4) The protection of the atmosphere is essential in the context of high social and economic development. According to the experimentally observed results, the limiting effect of vegetation such as trees and grasses on the concentration of particulate matter is significant. Theoretically, increasing the percentage of vegetation can help control air pollution. However, in a highly developed urban city with high population density such as Beijing, the area of land that can be changed is very limited. Therefore, it is more important to adjust the rationality of land use in densely populated and motorized built-up areas. The area of urban green space can be appropriately increased to create better connectivity of vegetation in order to give full play to the limiting effect of green vegetation on particulate matter concentration. In addition, targeted selection of tree species for urban greening and increasing the rationality of the vertical level of vegetation are also key to improving the efficiency of urban land use and environmental protection functions.

Author Contributions: J.Y. is the corresponding author. H.Z. contributed to improving the methodology and wrote the manuscript. X.T. helped edit and improve the manuscript. G.W. contributed to testing the methodology. All authors have read and agreed to the published version of the manuscript.

Funding: This work was jointly supported by grants from the Application Demonstration System of GaoFen Remote Sensing Mapping of China (No. 42-Y30B04-9001-19/21); the Chinese Postdoctoral Science Foundation (No. 2021M693782).

Conflicts of Interest: The authors declare no conflict of interest.

References

1. Zhai, H.; Tang, X.; Wang, G.; Li, J.; Liu, K. Characteristic analyses, simulations and predictions of land use in poor mountainous cities: A case study in the central area of Chengde County, China. *Environ. Earth Sci.* **2018**, *77*, 585. [CrossRef]
2. Gong, J.; Xie, Y.; Jia, Z.; Qian, D. Recent progress in land use and cover change in Heihe River Basin. *J. Lanzhou Univ. (Nat. Sci.)* **2014**, *50*, 390–397, 407. [CrossRef]
3. Zhang, X. Simulation of land use change scenarios of Tangshan Coastal Zone Based on CLUE-S Model. Master's Thesis, Hebei Normal University, Shijiazhuang, China, 20 March 2008. [CrossRef]

4. Yan, J.; Zhai, H. Dynamic simulation of land use with CLUE-S model and change rate of land-sea gradient. *Int. J. Earth Sci. Eng.* **2016**, *9*, 1447–1453.
5. Yan, J.; Zhai, H.; Cheng, F. Spatio-temporal evolution of land use and its socio-economic effects in the Shandong Peninsula. *Int. J. Earth Sci. Eng.* **2015**, *8*, 1994–2003.
6. Wang, G.; Yang, K.; Yang, Y. The spatio-temporal variation of $PM_{2.5}$ pollution and the its correlation with the impervious surface expansion. *China Environ. Sci.* **2017**, *37*, 2470–2481.
7. Yang, K.; Wang, G.; Yang, Y.; Luo, Y. Association between impervious surface and $PM_{2.5}$ concentrations in Kunming, China. *Res. Environ. Sci.* **2017**, *30*, 499–509. [CrossRef]
8. Xu, S.; Zou, B.; Pu, Q.; Guo, Y. Impact analysis of land use/cover on air pollution. *J. Geo-Inf. Sci.* **2015**, *17*, 290–299. [CrossRef]
9. Fu, B. The spatial pattern analysis of agricultural landscape in the loess area. *Acta Ecol. Sin.* **1995**, *15*, 113–120. [CrossRef]
10. Tang, X.; Liu, H.; Li, J.; Xie, Z.; Zhao, W. Response analysis of haze/particulate matter pollution to Land Use/Cover in Beijing. *China Environ. Sci.* **2015**, *35*, 2561–2569. [CrossRef]
11. Shi, J.; Cui, L.; He, Q.; Sun, L. The changes and causes of fog and haze days in Eastern China. *Acta Geogr. Sin.* **2010**, *65*, 533–542. [CrossRef]
12. Shi, J.; Cui, L.; Zhou, W. Change trend of climatic factors in the Yangtze River Delta from 1959 to 2005. *Resour. Sci.* **2008**, *30*, 1803–1810.
13. Peng, W.; Zhou, J.; Luo, H.; Yang, C.; Zhao, J. Relationship between land use change and its urban air environmental impacts. *Res. Soil Water Conserv.* **2010**, *17*, 87–91.
14. Font, A.; Baker, T.; Mudway, I.S.; Purdie, E.; Dunster, C.; Fuller, G.W. Degradation in urban air quality from construction activity and increased traffic arising from a road widening scheme. *Sci. Total Environ.* **2014**, *497-498*, 123–132. [CrossRef] [PubMed]
15. Zhai, H.; Tang, X.; Wang, G.; Li, J.; Xing, L. Impact of impervious surface expansion on tropospheric NO_2 column density in Chinese megacities. *Acta Sci. Circumstantiae* **2019**, *39*, 797–808. [CrossRef]
16. Wei, J.; Sun, L.; Liu, S.; Duan, D.; Guo, Y.; Mi, X.; Tian, X.; Yu, H. Response analysis of particulate air pollution to Land-use and land-cover change. *Acta Ecol. Sin.* **2015**, *35*, 5495–5506. [CrossRef]
17. Mo, L.; Yu, X.; Zhao, Y.; Sun, F.; Mo, N.; Xia, H. Correlation analysis between urbanization and particle pollution in Beijing. *Ecol. Environ. Sci.* **2014**, *23*, 806–811. [CrossRef]
18. Sun, N. Using Satellite Remote Sensing Data for Monitoring PM_{10} Concentrating in the Pearl River Delta and Study on the Correlation between PM_{10} and Underlying Surface. Master's Thesis, China University of Geosciences, Beijing, China, 1 May 2013.
19. Lou, C.; Liu, H.; Li, Y.; Li, Y. Research on the response of air particles ($PM_{2.5}$, PM_{10}) to landscape structure: A review. *Acta Ecol. Sin.* **2016**, *36*, 6719–6729. [CrossRef]
20. Lou, C. Tempo-spatial characteristics of $PM_{2.5}$ at multi-scales and its influencing in the core of the Yangtze River Delta. Ph.D. Thesis, Nanjing Normal University, Nanjing, China, 19 May 2018.
21. Zhuang, Y.; Chen, D.; Li, R.; Chen, Z.; Cai, J.; He, B.; Gao, B.; Cheng, N.; Huang, Y. Understanding the Influence of Crop Residue Burning on $PM_{2.5}$ and PM_{10} Concentrations in China from 2013 to 2017 Using MODIS Data. *Int. J. Environ. Res. Public Health* **2018**, *15*, 1504. [CrossRef]
22. Zhuang, Y.; Li, R.; Yang, H.; Chen, D.; Chen, Z.; Gao, B.; He, B. Understanding Temporal and Spatial Distribution of Crop Residue Burning in China from 2003 to 2017 Using MODIS Data. *Remote Sens.* **2018**, *10*, 390. [CrossRef]
23. Li, J.; Wang, H.; Wang, G.; Guo, J.; Zhai, H. A new method of high resolution urban water extraction based on index. *Remote. Sens. Inf.* **2018**, *33*, 99–105. [CrossRef]
24. Jiang, Y.; Zhu, R.; Zhu, K.; Li, Z. Numerical simulation on the air pollution potential in the severe air pollution episodes in Beijing-Tianjin-Hebei Region. *Acta Sci. Circumstantiae* **2015**, *35*, 2681–2692. [CrossRef]
25. Wang, X.; Lang, J.; Cheng, S.; Chen, G.; Liu, X. Study on transportation of $PM_{2.5}$ in Beijing-Tianjin-Hebei (BTH) and its surrounding area. *China Environ. Sci.* **2016**, *36*, 3211–3217. [CrossRef]
26. Wang, Y.; Li, Y.; Qiao, Z.; Lu, Y. Atmospheric transmission rule on air pollution in Beijing-Tianjin-Hebei urban agglomeration: A comparative analysis of two emission inventories. *China Environ. Sci.* **2019**, *39*, 4561–4569. [CrossRef]
27. Yang, X.; Zhao, W.; Xiong, Q.; Wang, L.; Zhao, W. Spatio-temporal distribution of $PM_{2.5}$ in Beijing-Tianjin-Hebei (BTH) area in 2016 and its relationship with meteorological factors. *Ecol. Environ. Sci.* **2017**, *26*, 1747–1754. [CrossRef]
28. Liu, X.; Zheng, T.; Wan, Q.; Tan, H.; Deng, X.; Li, F.; Deng, T. Spatio-temporal characteristics of NO_2 in concentrated PRD Urban Districts and analysis of anthropogenic influences based on OMI remote sensing data. *J. Trop. Meteorol.* **2015**, *31*, 193–201. [CrossRef]
29. Wang, Y.; Jiang, H.; Zhang, X.; Zhou, G.; Yu, S.; Xiao, Z. Temporal-spatial distribution of tropospheric NO_2 in China using OMI satellite remote sensing data. *Res. Environ. Sci.* **2009**, *22*, 932–937. [CrossRef]
30. Zhang, Y.; Niu, Z.; Wang, L.; Chang, C. Study on tropospheric NO_2 change trend in cities using OMI satellite data. *Geogr. Geo-Inf. Sci.* **2008**, *24*, 96–99.
31. Xu, J.; Jiang, H.; Xiao, Z. Estimate PM_{10} concentration over the Yangtze Delta using remote sensing: Analysis of spatial and temporal variations. *Chin. J. Environ. Eng.* **2016**, *10*, 1349–1357. [CrossRef]
32. Xu, J.; Jiang, H. Estimation of $PM_{2.5}$ concentration over the Yangtze Delta using remote sensing: Analysis of spatial and temporal variations. *Environ. Sci.* **2015**, *36*, 3119–3127. [CrossRef]

33. Zhai, H.; Yao, J.; Wang, G.; Tang, X. Spatio-Temporal Characteristics and Variation Pattern of the Atmospheric Particulate Matter Concentration: A Case Study of the Beijing–Tianjin–Hebei Region, China. *Atmosphere* **2022**, *13*, 120. [CrossRef]
34. McGarigal, K.; Cushman, S.A.; Ene, E. Fragstats v4: Spatial Pattern Analysis Program for Categorical and Continuous Maps. Computer Software Program Produced by the Authors at the University of Massachusetts, Amherst. 2012. Available online: http://www.umass.edu/landeco/research/fragstats/fragstats.html (accessed on 8 May 2013).
35. Fu, B.; Chen, L. Landscape diversity types and their ecological significance. *Acta Geogr. Sin.* **1996**, *51*, 454–462. [CrossRef]
36. Sulla-Menashe, D.; Friedl, M.A. User Guide to Collection 6 MODIS Land Cover (MCD12Q1 and MCD12C1) Product. 2018. Available online: https://lpdaac.usgs.gov/documents/101/MCD12_User_Guide_V6.pdf (accessed on 8 December 2021).
37. Sulla-Menashe, D.; Gray, J.; Abercrombie, S.P.; Friedl, M.A. Hierarchical mapping of annual global land cover 2001 to present: The MODIS Collection 6 Land Cover product. *Remote Sens. Environ.* **2019**, *222*, 183–194. [CrossRef]
38. Li, J.; Wang, H.; Wang, G.; Zhai, H.; Han, M.; Cheng, Q. Plastic-mulched farmland extraction with multi-source satellite data. *Bull. Surv. Mapp.* **2018**, 78–82. [CrossRef]
39. Chen, X.; Jiao, Y.; Pei, T.; Zhou, Z. The effect of adsorbing fine particulate matter ($PM_{2.5}$) by garden plants: A review. *Chin. J. Ecol.* **2014**, *33*, 2558–2566. [CrossRef]
40. Matsuda, K.; Fujimura, Y.; Hayashi, K.; Takahashi, A.; Nakaya, K. Deposition velocity of $PM_{2.5}$ sulfate in the summer above a deciduous forest in central Japan. *Atmos. Environ.* **2010**, *44*, 4582–4587. [CrossRef]
41. Freer-Smith, P.H.; Holloway, S.; Goodman, A. The uptake of particulates by an urban woodland: Site description and particulate composition. *Environ. Pollut.* **1997**, *95*, 27–35. [CrossRef]
42. Zhao, C.; Wang, Y.; Wang, Y.; Zhang, H. Interactions between fine particulate matter ($PM_{2.5}$) and vegetation: A review. *Chin. J. Ecol.* **2013**, *32*, 2203–2210. [CrossRef]
43. Chen, A.; Lei, Y.; Sun, R.; Chen, L. How many metrics are required to identify the effects of the landscape pattern on land surface temperature? *Ecol. Indic.* **2014**, *45*, 424–433. [CrossRef]
44. Kang, X.; Cui, L.; Zhao, X.; Li, W.; Zhang, M.; Wei, Y.; Lei, Y.; Ma, M. Effect of wetlands on reducing atmospheric fine particles $PM_{2.5}$ in Beijing. *Chin. J. Ecol.* **2015**, *34*, 2807–2813. [CrossRef]
45. Zhu, C.; Zeng, Y.; Chen, Y.; Guo, H. Effects of urban lake wetland on air PM_{10} and $PM_{2.5}$ concentration: A case study of Wuhan. *Chin. Landsc. Archit.* **2016**, 88–93.

atmosphere

Article

Impact of the Restaurant Chimney Emissions on the Outdoor Air Quality

Mahmoud Fathy ElSharkawy * and Osama Ahmed Ibrahim

Department of Environmental Health, College of Public Health, Imam Abdul Rahman Bin Faisal University, Dammam 34212, Saudi Arabia; olabib@iau.edu.sa
* Correspondence: msharkawy@iau.edu.sa

Abstract: The emission of cooking fumes becomes a serious concern due to the fast development of the restaurant business because it harms the health of restaurant workers and customers and damages the outdoor air quality. This study was conducted to evaluate the impact of restaurant emissions on ambient air quality. Twenty restaurants with four different types of food cooking were selected in Dammam City, which represents a densely populated urban city in Saudi Arabia. Levels of five air pollutants were simultaneously measured in the restaurants' chimneys and in the surrounding ambient air. The highest mean levels of CO (64.8 $\pm$ 44.3 ppm), CO_2 (916.7 $\pm$ 463.4 ppm), VOCs (105.1 $\pm$ 61.3 ppm), NO_2 (4.2 $\pm$ 2.4 ppm), and SO_2 (8.0 $\pm$ 7.4 ppm) were recorded in chimneys of the grilling restaurants. Similarly, the highest levels of all pollutants were recorded in the areas adjacent to the grilling restaurants rather than other types.

Keywords: restaurants; chimney emission; combustion efficiency; ambient air pollution; emission standards

Citation: ElSharkawy, M.F.; Ibrahim, O.A. Impact of the Restaurant Chimney Emissions on the Outdoor Air Quality. *Atmosphere* **2022**, *13*, 261. https://doi.org/10.3390/atmos13020261

Academic Editors: Begoña Artíñano, Manousos Ioannis Manousakas, Anikó Angyal, Maria Gini and Elena Hristova

Received: 17 December 2021
Accepted: 31 January 2022
Published: 3 February 2022

1. Introduction

Worldwide, the cooking process consumes huge amounts of energy, especially in developing countries [1,2]. Several types of fuels are usually used for cooking including natural gas, charcoal, wood, kerosene, liquefied petroleum gas, electricity, biogas, and biomass [3,4]. Consequently, large amounts of harmful air pollutants and greenhouse gases are emitted daily during the cooking processes [5,6]. Restaurants represent the most important site for cooking where many local and foreign people tend to spend a lot of their time. The emission of cooking fumes has become more serious due to the fast development of the restaurant business [7]. The emitted pollutants from restaurants not only harm the health of restaurant workers and customers but also represent a great contributor to the outdoor air pollution levels [8–12]. Complaints against cooking fume/odor emissions from restaurants have been increasing and recorded in some areas of the world [13].

For several years, great concern has been given to the cooking fumes, particularly in the highly crowded cities where restaurants are usually located in densely populated areas that are very close to residential and other sensitive buildings [13]. Emission of pollutants from restaurants results from heating and cooking operations where several types of food are cooked and different types of fuels are used [14]. The amounts and composition of pollutants emitted from those sources depend greatly on the cooking materials, cooking styles, and even cooking fuel [15]. For example, charcoal is used extensively for barbecuing in most restaurants in the world because it has high heating value, is cheap compared to other types of fuels, can be easily stored, and gives a unique flavor and texture to the food [16–18]. Charcoal contains various types of organic and inorganic compounds such as hydrocarbons, sulfur, water, and oxygen along and numerous trace elements [19–21]. Therefore, the combustion of charcoal creates a considerable amount of airborne toxic elements both in the solid and gaseous states. The coal-tars and soot (fine black particulate matter) have been documented as human carcinogens since the late 1700s [22,23]. Several previous studies revealed that the combustion of charcoal is considered a potential source of volatile organic compounds (VOCs) and polycyclic aromatic hydrocarbon (PAHs) [24,25] that have several adverse health effects including carcinogenicity in addition to its contribution to the formation of photochemical ground-level ozone [26,27]. Sulfur—as a component of fuels that occurs primarily in coal, petrol, kerosene, and diesel—can produce sulfur dioxide gas (SO_2) when combusted during the cooking process or any high-temperature combustion. The presence of SO_2 in the air leads to irritations of the mucous membranes and the eyes, as well as chronic bronchitis [28]. Combustion of charcoal is also considered a source of carbon monoxide (CO) in the air [29] that is considered a chemical asphyxiant for humans [30]. The other cooking activities in restaurants, such as charbroiling, frying, and baking, are also considered sources of the same or different air pollutants that make significant contributions to both indoor and outdoor air pollution [31,32].

Unfortunately, a lot of the exhaust outlets of ventilation ducting systems in restaurants are always not located at favorable locations, in particular in densely populated urban regions near the sensitive receptors, such as residential premises, schools, or clinics [13]. Effective control measures should be taken from the formal governmental agencies with the cooperation of owners and operators of the restaurants to ensure that no visible cooking fumes nor objectionable odor would be emitted causing any harm or forms of pollution. In this regard, appropriate high-performance air pollution control equipment must be installed at the kitchen ventilation system of the food premises for treating cooking fume emissions before being discharged to the outdoor environment as well as it is considered a cost-effective way to reduce indoor air pollution and the related health problems [33,34]. As a general guideline, the control equipment of the restaurant must be installed directly above the stoves and cooking appliances and properly connected with the exhaust ducts to prevent cooking fume from leaking through possible cracks. Moreover, the ducts must be connected with exhaust fans of adequate capacity [35]. The range hood is one of the most common types of ventilation [36]. It has the advantage of providing constant ventilation for the smoke to escape [37]. The more effective type of cooker-hood is the one that extracts the contaminated air from the cooking zone and ejects it to the ambient environment [38]. Chimneys, that must be extended to above the roof of the restaurant, are more effective because they largely prevent the smoke from entering the kitchen or any other internal site of the restaurant [39,40]. Additionally, the chimney plays an active role in the performance of the stove and in reducing emissions by influencing the overall air-to-fuel ratio and subsequently the production of CO and/or particulate matter (PM) [41]. Additionally, chimneys keep flue gas separated from ambient conditions, providing a longer residence time of the gas within a heated environment [42].

Numerous previous studies have been conducted concerning the impact of cooking emissions on the indoor environment of restaurants in developed and developing countries [43–45]. Despite their importance, data on the impact of cooking emissions on the direct surrounding environment and air pollution levels are still very scarce. This study was conducted to fill this gap by studying the impact of the restaurant emissions on the outdoor ambient air pollution in a densely populated urban city representing a developing country. It was conducted to quantify emissions of different pollutants from the chimneys of various restaurants and simultaneously levels of the same pollutants in the ambient air. It was aiming also to guide the owners and operators of restaurants, food businesses, and corresponding governmental agencies in helping them understand and apply the best practical control measures to minimize these emissions, thereby preventing air pollution problems.

2. Materials and Methods

2.1. Study Area and Period

This study was conducted in Dammam City in the Eastern province of the Kingdom of Saudi Arabia (KSA). Generally, KSA is characterized by the presence of a wide variety and large number of restaurants in all cities, particularly those with high dense populations or visitors such as Mecca, Riyadh, and Dammam because of the large numbers of immigrants, foreign workers, and people visiting the country to perform Hajj and Omrah. Dammam is considered one of the most important cities in the Kingdom. It is the capital, the major seaport, and one of the most populated cities in the Eastern Province of KSA. It is also a major administrative center for the Saudi oil industry (Aramco) and about 40% of the industrial activity of KSA is located in it. There is an increase in the migration of people to Dammam city for obtaining jobs and studying because of the presence of a large number of industries, universities, and different governmental centers. Due to the rapid population growth in the city, there is an increasing demand for food and, consequently, the number of restaurants is also increasing. All restaurants in Dammam are located close to the residential premises, schools, hospitals, and other sensitive receptors.

A variety of cooking methods are used in restaurants, but it differs from one to another according to the type of food that characterizes each restaurant. For example, the main cooking methods of some restaurants include stir frying, simmering, steaming, roasting, smoking, and stewing, while in other restaurants the main cooking methods are grilling, broiling, and deep frying. Twenty restaurants were selected in Dammam for this study representing four different types of food cooking; grilling (such as chicken or meat grilled on charcoal), frying (such as fried chicken), cooking (such as cooked rice and vegetables), and baking (such as pizza and pastry). Five restaurants from each type were selected and the twenty restaurants were contributed to the same criteria and specifications, except the type of food cooking. All selected restaurants were located on the ground floor of a residential building of 3–4 floors with a chimney extended to above the building roof. They are installed in densely populated areas adjacent to moderate traffic activity streets (about 500 cars/h) and far from any other air pollution sources such as industrial activity or any other restaurants. All restaurants have nearly the same size and number of cooking appliances. The objective of this selection was to remove any factor that could affect the results of a comparison between the emission of pollutants from the restaurants' chimneys and the ambient levels of the same pollutant. For confidence, confirmation, and comparison, a building with the same characteristics was selected in an area far from any restaurants but with the same traffic activity. This area was considered a "control area".

Climatically, Dammam has a hot desert climate. The winter temperatures range from mild to warm, while the summer temperatures are extremely hot, usually exceeding 40 °C (104 °F) for about six months. Rainfall in Dammam is generally sparse and usually occurs in small amounts in December. Heavy thunderstorms are not uncommon in winter. For this reason, this study was conducted during the six warm and hot months (April–August) of the year 2019.

2.2. Measurement of Chimney Emissions

The cooking fumes of all selected restaurants were extracted through an exhaust hood and then discharged into the atmosphere near the surrounding neighbors. Owners of restaurants did not want to make a hole in the chimney. Therefore, the sampling probe was placed directly near the outlet center of the outdoor chimney and paralleled with the direction of the chimney. Generally, for measuring the quality of the combustion of the cooking tools, the probe of a flue gas analyzer (electronic sensor) was applied to the cooking tool chimney where levels of gaseous pollutants can be measured directly on site. The advantage of the electronic sensors is that they can be used for a long time and their usage lies in real-time measurement. The air pollution content is determined using sensors where there are whole ranges of measuring principles that can be employed such as flame or photo-ionization detection for organic species, chemiluminescence for oxides of nitrogen, non-dispersive infrared for carbon monoxide, Fourier transform infrared for sulfur dioxide, etc. A digital readout indicates the measured value at the spot. These devices need to be calibrated before each monitoring session. The calibration occurs through a test gas of known pollutant concentration [42].

During our study, the Lancom 4 Portable Flue Gas Analyzer was used for measuring levels of combustion gases and cooking vapors in the exhaust chimneys of all restaurants. This analyzer meets the US EPA CTM 034 reference method, and it has data acquisition and analysis software. The analyzer is composed of a monitor, a probe hose of 3 m length, and a probe pipe of 0.3 m length. Its monitor can read up to 17 measurement parameters and it is a useful tool to observe trends. Moreover, it is free from any bias that can be caused by substances in the waste gas. The measurement specifications of this analyzer are illustrated in Table 1. The Quality Management System of Land Instruments International is approved to BS EN ISO 9001 for the design, manufacture, and on-site servicing. For quality assurance and quality control (QA/QC), the analyzer was recently calibrated by the manufacturer themselves with a certificate of conformity and calibration No. 21572853, and calibration before and after measurement using standards that are traceable to certified reference materials was conducted. Carbon dioxide (CO_2), carbon monoxide (CO), nitrogen dioxide (NO_2), and sulfur dioxide (SO_2) were representing the combustion gases, while the volatile organic compounds (VOCs) were representing cooking vapors. Concentrations of these five pollutants were measured in parts per million (ppm). The exhaust measurements were carried out at least twice during the peak cooking period. A real-life photograph of the sampling setup while sampling was being undertaken is shown in Figure 1.

Table 1. Measurement specifications of the Lancom 4 Portable Flue Gas Analyzer.

Sensor	Detection Limit	Full Scale Range	Upscale Repeatability	Resolution
O_2	0.2%	0 to 30% v/v	±1%	0.1% v/v
CO (low)	2 ppm	0 to 6000 ppm	CO (low)	2 ppm
CO (high)	20 ppm	0 to 10%	±2% *	0.1 ppm
SO_2	2 ppm	0 to 4000 ppm	±2% *	0.1 ppm
NO	2 ppm	0 to 5000 ppm	±2% *	0.1 ppm
NO_2	2 ppm	0 to 1000 ppm	±2% *	0.1 ppm
Hydrocarbons (C_xH_y)	(Application dependent)	0 to 5% v/v	±4% *	0.1% v/v
Flue Gas/Ambient Temperature		Measured		
Draft		±50 hPa/20 "Water Gauge"		
Flow (velocity)		1 to 50 m/s		

* Calibration per ASTM D-6522 or LAND factory procedure.

Figure 1. A real-life photograph of the sampling setup (authors personal contribution).

For each one of the four types of restaurants (grilling, frying, cooking, and baking), the measurements were conducted for inside all chimneys of each restaurant that were installed on the restaurant's building roof and simultaneously from the ambient air outside the stack. Some restaurants have only one chimney while others have two chimneys. For each chimney, at least three measurements for each pollutant were performed for 2 h. Each one of the 20 selected restaurants was visited twice, and the monitoring process was performed during the evening period because this period represents the rush hours and maximum activity of cooking and food preparation inside each type of restaurant.

2.3. *Measurement of Outdoor Ambient Air Pollution*

Simultaneously with the chimney measurements, levels of the same pollutants in the outdoor air were measured at 10–20 m from the chimney in the downwind direction to study the effect of emitted pollutants in the ambient air levels. The selection of this distance was based on the actual presence of inhabitants' rooms on the residential building roof at 10 m from the chimney in some of the selected restaurants' buildings. The above five air pollutants were directly measured by the Gray Wolf's DirectSense® (Shelton, CT, USA) mobile PC-based products, AdvancedSense™ (Shelton, CT, USA) meters, and Wolf Pack™ (Shelton, CT, USA) area monitor. This monitor is composed of multi-gas detectors, and it is equipped with a wireless radio frequency modem that allows the unit to communicate and transmit readings and other information on a real-time basis with a remotely located base controller. Reliably measure key specific pollutants (VOCs, CO, O_3, NO_2, NH_3, HCHO, etc.; choose from 25+ gas sensors), as well as particulate, ventilation rates (CO_2 and airflow), differential pressure (DP), and more. High-performance, fast-response instrumentation for consistent use over portable, long-term, and continuous testing applications. In stand-alone operation, it is a rugged, weather-resistant, portable monitor that can run over 24 h on either rechargeable lithium-ion or alkaline batteries. The probe dimensions are 2 in. (5 cm) diameter × 12.5 in. (30 cm) length. Concentrations of the five pollutants were also measured in ppm. The measurement specifications of this gas detector are illustrated in Table 2. For quality assurance and quality control (QA/QC), the detector was recently calibrated by the manufacturer with a certificate of conformity and calibration No. 03-1291. Similarly, for each measuring point in the ambient around the chimney, at least three measurements for each pollutant were undertaken for 2 h.

Table 2. Measurement specifications of the Gray Wolf's DirectSense Gas Detector.

Parameter	Range	Limit of Detection	T90 Response	Sensor Drift
SO_2	0–20.0 ppm	0.2 ppm	<25 s	<2% per mo
NO_2	0–20.0 ppm	0.1 ppm	<20 s	<2% per mo
NO	0–200 ppm	1 ppm	<20 s	<2% per mo
CO	0–500 ppm	1 ppm	<35 s	<2% per mo
CO_2	0 to 10,000 ppm	±3% rdg ±50 ppm	<25 s	<2% per mo
TVOCs	0 to 10,000 ppm	0.1 ppm	<25 s	<2% per mo

2.4. *Measuring of Meteorological Factors*

So far as the dispersion of pollutants from a chimney is concerned, temporal wind distribution is the most important factor for the concentration buildup of air pollutants in the surrounding air basin. The most important meteorological condition in this study was the prevailing wind at the time of measurement in the study area. Before conducting any measurements, the prevailing wind direction was recorded by the Kestrel 4500 electronic weather station (Kestrelmeters, Boothwyn, PA, USA). This tool calculates crosswind and headwind/tailwind regarding a user-set target heading and stores the information along with all the other environmental readings in its 1400 data point memory. The smoke exhaust is mixed fast with ambient air, in which high temperature and relative humidity (RH) is not a major issue in the measurements. However, wind speed, temperature, and RH were also measured by the same tool.

3. Results and Discussion

3.1. Chimney Emissions

Figure 2 represents the mean levels of the measured air pollutants in the chimney exhaust of the four methods of food cooking (grilling, frying, cooking, and baking). The highest levels of all pollutants were emitted from the grilling chimneys followed by frying and baking while the lowest levels were emitted from the cooking ones. Inside the grilling chimneys, the highest mean levels ± standard deviation (SD) of CO, CO_2, VOCs, NO_2, and SO_2 were (64.8 ± 44.3 ppm), (916.7 ± 463.4 ppm), (105.1 ± 61.3 ppm), (4.2 ± 2.4 ppm), and (8.0 ± 7.4 ppm), respectively, while in the cooking chimneys the lowest mean levels were (8.3 ± 4.4 ppm), (555.2 ± 108.7 ppm), (17.7 ± 7.1 ppm), (1.4 ± 0.6 ppm), and (1.2 ± 1.0 ppm), respectively. In most restaurants, food is prepared under high temperatures when grilled or fried whereas most fire-based cooking is based on the combustion of various fuel types (e.g., coal, natural gas, liquefied petroleum gas (LPG), and electrical energy) [3,46]. The most important chemical processes during the high-temperature treatment of food are the degradation of sugars, pyrolysis of proteins and amino acids, and the degradation of fats [44]. Several previous studies reported that the burning of charcoal is the major source of emission of air pollutants and offensive odorants in the atmosphere [16,20,47]. For example, a recent study was conducted to quantify and characterize the gaseous emissions from charcoal combustion in a brick barbecue grill revealed that emissions of CO, CO_2, NOx, acid gases, NH_3, and VOCs from the combustion of charcoal were higher than those of the other fuels and appliances [16]. Another study which was conducted in Portugal to assess levels of VOCs in the exhaust stacks on the roofs of a university canteen and a charcoal-grilled chicken restaurant concluded that the cooking fumes of the barbecued chicken contribute to emissions of VOCs higher than those of the university canteen [16]. Although the frying pan is different from the charcoal-burner, it can be used to heat food for high temperatures, and consequently, excess air pollutants, such as PM, CO, and VOCs, are released in the atmosphere. Numerous previous studies revealed that emission of pollutants from frying food on a hot steel pan and broiling food on steel bars above a charcoal burner was always higher than those of any other methods of cooking [12,48,49]. The results of my study are quite like most of these studies.

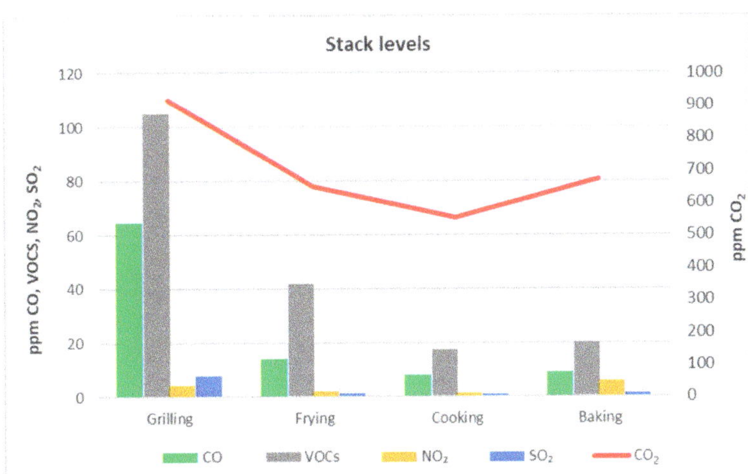

Figure 2. Mean levels of air pollutants in different restaurant chimneys.

Combustion efficiency is influenced by the fuel quality and the combustion chamber characteristics of a stove. A simple determination of the combustion efficiency can be conducted with the calculation of the CO/CO_2 ratio. It is known that the mass of pollutants can be related to the mass of burnt fuel or the ratio between CO and CO_2. The value of 0.1 or lower for this ratio is a good indication of the combustion efficiency [28]. In the present study, the CO/CO_2 for grilling, frying, cooking, and baking was 0.071, 0.022, 0.015, and 0.013, respectively, which reflects the good quality of combustion chambers in all selected restaurants.

Applying the independent t-test for comparing statistically between means of pollutants, indicated that there is a significant difference ($p < 0.05$) between mean levels of all pollutants, except CO_2, emitted from grilling and the other three methods of cooking. As for CO_2, there is a significant difference ($p < 0.05$) between grilling and both frying and cooking, while there is no significant difference ($p > 0.05$) between grilling and baking. The presence of the absence of statistical differences for the other three methods differs from one pollutant to another as shown in Table 3. This means that the emission of air pollutants from the grilling process is much higher than those of the other cooking methods, and it reflects the great contribution of the grilling process in emitting air pollutants from restaurants with comparing to other types of food cooking.

Table 3. Independent *t*-test for mean levels of pollutants in chimney exhaust.

Pollutant	Food Preparation Type	Frying	Cooking	Baking
CO	Grilling	0.000 *	0.000 *	0.005 *
	Frying		0.023 *	0.146
	Cooking			0.943
CO_2	Grilling	0.027 *	0.008 *	0.252
	Frying		0.193	0.760
	Cooking			0.236
VOCs	Grilling	0.002 *	0.000 *	0.003 *
	Frying		0.000 *	0.004 *
	Cooking			0.899
NO_2	Grilling	0.008 *	0.002 *	0.035 *
	Frying		0.444	0.000 *
	Cooking			0.017 *
SO_2	Grilling	0.001 *	0.001 *	0.049 *
	Frying		0.970	0.947
	Cooking			924

* The mean difference is significant at the 0.05 level.

Unfortunately, there are no emission standards for restaurants' chimneys. The emission standards promulgated in the U.S. by the EPA, Europe, and some countries of Asia are standards intended to control air pollution from several industries [50–53]. From these standards, we selected the nearest industries to restaurants such as coal-fired power plants and municipal waste combustors (MWCs) to compare the results of my study. The purpose of this comparison is a trial to set a range of safe limits for protecting people's health against restaurant fumes. Table 4 indicates the results of our study compared with the selected emission standards for only three pollutants: CO, NO_2, and SO_2. I did not find any standards for VOCs and CO_2 for the same industries. All mean levels of my study were much lower than the selected standards.

Table 4. Mean levels of pollutants in chimneys compared to emission standards.

Pollutant	Mean Concentration of the Study (ppm)				Standard (ppm)
	Grilling	Frying	Cooking	Baking	
CO	64.8	14.3	8.3	9.3	100
NO_2	4.2	2.0	1.4	6.0	50–150
SO_2	8.0	1.4	1.2	1.3	30–80

3.2. Outdoor Air Quality

Figure 3 represents mean levels of air pollutants in the outdoor air at the surrounding areas of the selected chimneys, in addition to the control area. Similarly, with the chimneys results, the highest levels of CO (5.4 ± 1.4 ppm), CO_2 (427.1 ± 86.8 ppm), VOCs (0.31 ± 0.23 ppm), NO_2 (0.044 ± 0.029), and SO_2 (0.18 ± 0.07 ppm) were emitted from the grilling chimneys followed by frying and baking while the lowest levels were emitted from the cooking chimneys. It is shown that Figure 2 has completely the same trend as Figure 1, which indicates the direct effect of restaurant chimney exhaust in the adjacent outdoor air quality levels. It can be confirmed by the lowest levels of all pollutants that were recorded in the control area as shown in Figure 2. This means that any negative or positive change in the combustion efficiency or the internal cooking process of any restaurant will be accompanied by the same change in the outer atmosphere. No doubt, this conclusion will help the decision-makers and regulators to effectively inspect the cooking emissions from restaurants.

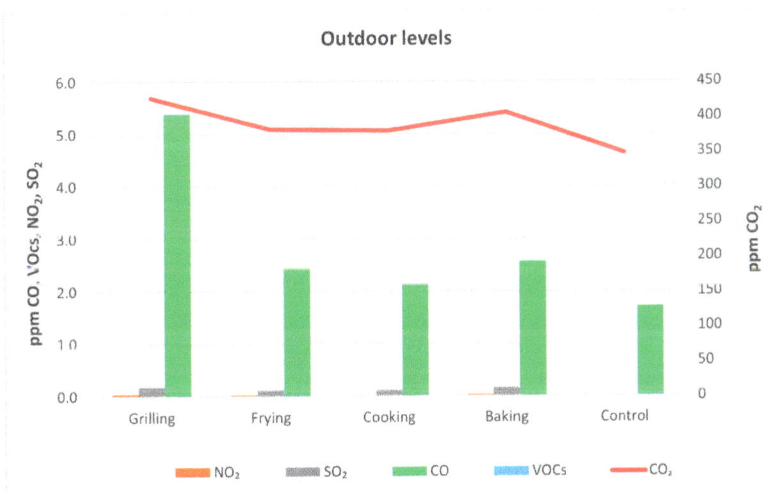

Figure 3. Mean levels of air pollutants in the surrounding areas outdoor of restaurants.

Statistically, the one-way ANOVA test was used to compare means of pollutants at the outdoor air for the four types of chimneys and the control area as shown in Table 5. Similarly, and surprisingly there is nearly the same statistical significance of chimneys results. Significant differences ($p < 0.05$) between mean levels of all pollutants, except CO_2 and SO_2, emitted from grilling were found with the other three methods of cooking. It confirms again the role of the grilling process in polluting the indoor and outdoor air. On the other hand, except for grilling restaurants, there is no significant difference ($p > 0.05$) between the control area and areas of frying, cooking, and baking restaurants. This can be explained by the considerable emission of the studied pollutants from the traffic activity.

Mean concentrations of CO, NO_2, SO_2, and VOCs of the outdoor air during this study were compared with their Air Quality Guidelines (AQG) as adopted by the Saudi

Environmental Law [54] and the WHO guidelines [55] and presented in Table 6. Levels of CO and NO_2 were lower in their AQGs. Levels of VOCs were higher than their AQGs in areas of grilling and frying restaurants, while levels of SO_2 exceeded their AQGs in areas of grilling and baking restaurants. This can be easily clarified by the presence of sulfur in both charcoal and diesel fuel which is usually used in baking stoves.

Table 5. One-way ANOVA test for mean levels of pollutants in the outdoor ambient air.

Pollutant	Food Preparation Type	Frying	Cooking	Baking	Control
CO	Grilling	0.000 *	0.000 *	0.000 *	0.000 *
	Frying		0.965	0.976	0.934
	Cooking			0.999	0.961
	Baking				0.968
CO_2	Grilling	0.009	0.002 *	0.132	0.000 *
	Frying		0.569	0.870	0.044 *
	Cooking			0.573	0.118
	Baking				0.082
VOC	Grilling	0.000 *	0.000 *	0.000 *	0.000 *
	Frying		0.033 *	0.124	0.085
	Cooking			0.998	0.934
	Baking				0.948
NO_2	Grilling	0.004 *	0.001 *	0.007 *	0.000 *
	Frying		0.677	0.000 *	0.073
	Cooking			0.000 *	0.145
	Baking				0.000 *
SO_2	Grilling	0.000 *	0.000 *	0.008	0.001 *
	Frying		0.993	0.987	0.555
	Cooking			982	556
	Baking				0.649

* The mean difference is significant at the 0.05 level.

Table 6. Mean levels of pollutants in ambient air compared to their AQGs.

Pollutant	Mean Concentration of the Study (ppm)					AQG (ppm)
	Grilling	Frying	Cooking	Baking	Control	
CO	5.0	2.0	2.0	3.0	2.0	9
NO_2	0.04	0.03	0.01	0.02	0.01	0.08
SO_2	0.18	0.12	0.11	0.15	0.01	0.14
VOCs	0.31	0.26	0.12	0.11	0.08	0.24

However, exact quantification of the contribution of restaurants' emissions to outdoor air is very scarce. Few previous studies were conducted to study the impact of cookstove smoke on ambient air quality. For example, a field study was conducted in four randomly selected households in two rural locations of southern Nepal during April 2017. This study revealed that 66% of particulate matter is less than 2.5 microns ($PM_{2.5}$) and 80% of the black carbon emissions from biomass cookstoves directly escape into ambient air [56]. Another study was also conducted in rural Nepal revealed that a range of 6–58% of the particulate matter emitted from the open design cookstoves is liberated to the outdoor atmosphere [57].

Cooking emissions are produced from the stove used for cooking and the emissions produced by cooking the food itself. Characteristics of both stove and the food being cooked influence cooking emissions type and concentration levels. Emissions from the stove can vary significantly depending on the fuel source [58]. For example, gas burners produce higher particle concentrations, formaldehyde (HCHO), CO, and NO_2 when compared to electric stoves [59,60]. Solid fuel combustion in cookstoves emits a complex mixture of particulate and gaseous species, some of these pollutants contribute to levels of commonly regulated pollutants in the ambient environment [61]. Many biomass fuels and coal also contain low concentrations of chlorine that lead to low levels of emissions of dioxins and furans [62]. Besides stove and fuel source characteristics, the type of food, method of cooking, and cooking temperature can also impact the type and intensity of the cooking emission. For example, high-heat cooking activities such as broiling and frying can produce acrolein, polycyclic aromatic hydrocarbons (PAHs), and particulates, while it has also been demonstrated that the process of charbroiling and the practice of cooking fatty foods (such as high-fat hamburgers) yield higher particle emission concentrations compared to lower-heat cooking and low-fat foods [63,64].

Access to clean cooking fuels and technologies is essential for maintaining human health and achieving environmental sustainability, particularly in developing counties. A recent study has been conducted to for the first time the environmental sustainability of household cooking, focusing on remote communities in developing countries in the South-east Asia-Pacific (SEAP) region and considering both life cycle and local impacts. To guide rural development policies, the impacts of the following cooking fuels were considered: liquefied petroleum gas, kerosene, wood, charcoal, crop residues, biogas, and electricity. Results of the study revealed that biogas from manure is environmentally the most sustainable cooking fuel, while fuelwood is the best option for climate change, with relatively low other impacts, apart from freshwater eutrophication. Cooking using electricity is the worst option since it is typically generated from diesel in off-grid communities. LPG and kerosene have higher resource depletion and land use impacts compared to biomass fuels derived from waste. Solid biomass fuels (fuelwood, charcoal, and crop residues) have high freshwater eutrophication, terrestrial ecotoxicity, and human toxicity. In addition, direct emissions from their combustion cause significant local health and environmental impacts [65].

Many intervention strategies can be used to effectively mitigate the emissions of pollutants from restaurants and protect people's health against the restaurant fumes both inside the restaurant and in its chimneys before discharging their contaminants to the ambient air. For example, separate exhaust systems must be provided to those cooking operations giving rise to oily fume and strong odor emissions and treat the emissions with separate control equipment such as venturi and activated carbon. Control equipment must be installed directly above the stoves and properly connected with the exhaust ducts to prevent cooking fume from leaking through possible cracks. For exhaust outlets near the sensitive receptors, the air pollution problem would still exist even after the application of advanced control technologies. To avoid air nuisance likely caused to the air-sensitive receivers, the owners and operators of the restaurants and food business should refrain from choosing these sites for their business. Suitable siting or positioning of the outlet of the exhaust system is of paramount importance to avoid causing or contributing to air pollution. The exhaust outlet of the restaurant chimney must be installed as high as possible for upward discharge.

4. Conclusions

The wide and fast spread of restaurants in all urban areas of the world cannot be dispensed or neglected, particularly in densely populated areas. The emission of pollutants from the restaurant chimneys has a considerable and direct effect on the outdoor ambient air, particularly the grilling process that emits pollutants at a much higher rate than those of the other food cooking methods used in restaurants. Any negative or positive change in the combustion efficiency or the internal cooking process of restaurants will be accompanied by the same change in the outer atmosphere. Fortunately, the combustion chambers and processes in all selected restaurants for this study were working efficiently, and most of the emitted pollutants were lower than their standards. The result of this study is expected to help the decision-makers and regulators to effectively inspect the emissions of pollutants from restaurants for protecting people's health against restaurant fumes and helping the restaurants' owners to take the correct actions for reducing levels of air pollution both inside the restaurant and in its chimneys before discharging their contaminants to the outer atmosphere. Furthermore, more studies must be conducted to separately study the effect of each type of fuel that is used in restaurants on the outdoor air quality.

Author Contributions: Methodology, M.F.E. and O.A.I.; formal analysis, M.F.E. and O.A.I.; data curation, M.F.E.; writing—original draft preparation, M.F.E.; writing—review and editing, M.F.E.; supervision, M.F.E.; project administration, M.F.E. All authors have read and agreed to the published version of the manuscript.

Funding: This research received no external funding.

Data Availability Statement: Not applicable.

Conflicts of Interest: The authors declare no conflict of interest.

Ethical Approval: The approval to conduct this study was obtained from the Municipality of the Eastern Province, Saudi Arabia. Each restaurant owner was provided with information about the study and the purpose of the study before conducting any sampling step. All the aspects of the subjects were kept confidential and used only for the study purpose.

References

1. Pol torak, A.; Wyrwisz, J.; Moczkowska, M.; Marcinkowska-Lesiak, M.; Stelmasiak, A.; Rafalska, U. Microwave vs. convection heating of bovine gluteus Medius muscle: Impact on selected physical properties of final product and cooking yield. *Int. J. Food Sci. Technol.* **2015**, *50*, 958–965. [CrossRef]
2. Bansal, M.; Saini, R.P.; Khatod, D.K. Development of cooking sector in rural areas in India—A review. *Renew. Sustain. Energy Rev.* **2013**, *17*, 44–53. [CrossRef]
3. İşler, A.; Karaosmanoğlu, F. Traditional cooking fuels, ovens and stoves in Turkey. In Proceedings of the Thirty-First IAEE International Conference, Istanbul, Turkey, 16–17 June 2008.
4. Ramanathan, R.; Ganesh, L.S. A multi-objective analysis of cooking energy alternatives. *Energy* **1994**, *19*, 469–478. [CrossRef]
5. Afrane, G.; Ntiamoah, A. Analysis of the life-cycle costs and environmental impacts of cooking fuels used in Ghana. *Appl. Energy* **2012**, *98*, 301–306. [CrossRef]
6. World Bank. *Household Cookstoves, Environment, Health, and Climate Change: A New Look at an Old Problem*; World Bank: Washington, DC, USA, 2011.
7. Jian-li, W.; Jun-bo, Z.; Mao-chu, G.; Zhi-min, L.; Ming, Z.; Yao-qiang, C. Remove cooking fume using catalytic combustion over Pt/La-Al_2O_3. *J. Environ. Sci.* **2007**, *19*, 644–646.
8. Zhang, Z.; Zhao, Y.; Zhou, M.; Tao, P.; Li, R. Measurement of indoor air quality in Chinese charcoal barbecue restaurants. *Proc. Eng.* **2017**, *205*, 887–894. [CrossRef]
9. Guo, H. Source apportionment of ambient volatile organic compounds from petroleum and non-petroleum emissions. *J. Pet. Environ. Biotechnol.* **2012**, *3*, e112. [CrossRef]
10. Kim, K.H.; Jahan, S.A.; Kabir, E. A review of diseases associated with household air pollution due to the use of biomass fuels. *J. Hazard. Mater.* **2011**, *192*, 425–431. [CrossRef]
11. Bhanarkar, A.D.; Majumdar, D.; Nema, P.; George, K.V. Emissions of SO_2, NOx and particulates from a pipe manufacturing plant and prediction of impact on air quality. *Environ. Monit. Assess.* **2010**, *169*, 677–685. [CrossRef] [PubMed]
12. Lee, S.C.; Li, W.; Chan, L.Y. Indoor air quality at restaurants with different styles of cooking in metropolitan Hong Kong. *Sci. Total Environ.* **2001**, *279*, 181–193. [CrossRef]
13. Pang, S.; Wong, A. Challenges on the control of cooking fume emissions from restaurants. In Proceedings of the Better Air Quality in Asian and Pacific Rim Cities (BAQ 2002), Hong Kong, 16–18 December 2002.
14. Alves, C.A.; Evtyugina, M.; Cerqueira, M.; Nunes, T.; Duarte, M.; Vicente, E. Volatile organic compounds emitted by the stacks of restaurants. *Air Qual. Atmos. Health* **2015**, *8*, 401–412. [CrossRef]
15. Kabir, E.; Kim, K.H. An investigation on hazardous and odorous pollutant emission during cooking activities. *J. Hazard. Mater.* **2011**, *188*, 443–454. [CrossRef] [PubMed]
16. Vicente, E.D.; Vicente, A.; Evtyugina, M.; Carvalho, R.; Tarelho, L.A.C.; Oduber, F.I.; Alves, C. Particulate and gaseous emissions from charcoal combustion in barbecue Grills. *Fuel Proc. Technol.* **2018**, *176*, 296–306. [CrossRef]
17. Lahiff, M.; Rehfuess, E.A.; Mishra, V.; Smith, K.R. Solid fuel use for household cooking: Country and regional estimates for 1980–2010. *Environ. Health Perspect.* **2013**, *121*, 784–790.
18. Johnson, E. Charcoal versus LPG grilling: A carbon-footprint comparison. *Environ. Impact Assess. Rev.* **2009**, *29*, 370–378. [CrossRef]
19. Taner, S.; Pekey, B.; Pekey, H. Fine particulate matter in the indoor air of barbeque restaurants: Elemental compositions, sources and health risks. *Sci. Total Environ.* **2013**, *454–455*, 79–87. [CrossRef]

20. Rahman, M.M.; Kim, K.H. Release of offensive odorants from the combustion of barbecue charcoals. *J. Hazard. Mater.* **2012**, *215-216*, 233–242. [CrossRef] [PubMed]
21. Susaya, J.; Kim, K.; Ahn, J.; Jung, M.; Kang, C. BBQ charcoal combustion as an important source of trace metal exposure to humans. *J. Hazard. Mater.* **2010**, *176*, 932–937. [CrossRef]
22. Lewtas, J. Air pollution combustion emissions: Characterization of causative agents and mechanisms associated with cancer, reproductive, and cardiovascular effects. *Mut. Res.* **2007**, *636*, 95–133. [CrossRef]
23. Fitzer, E.; Ochling, K.; Boehm, H.P.; Marsh, H. Recommended terminology for the description of carbon as a solid. *Pure Appl. Chem.* **1995**, *67*, 473–506. [CrossRef]
24. Kabir, E.; Kim, K.-H.; Ahn, J.-W.; Hong, O.-F.; Sohn, J.R. Barbecue charcoal combustion as a potential source of aromatic volatile organic compounds and carbonyls. *J. Hazard. Mater.* **2010**, *174*, 492–499. [CrossRef]
25. Pandey, S.K.; Kim, K.-H.; Kang, C.H.; Jung, M.C.; Yoon, H. BBQ charcoal as an important source of mercury emission. *J. Hazard. Mater.* **2009**, *162*, 536–538. [CrossRef]
26. Hsieh, L.T.; Yang, H.H.; Lin, Y.C.; Tsai, C.H. Levels and composition of volatile organic compounds from the electric oven during roasting pork activities. *Sustain. Environ. Res.* **2012**, *2*, 17–24.
27. Huang, Y.; Ho, S.S.H.; Ho, K.F.; Lee, S.C.; Yu, J.Z.; Louie, P.K.K. Characteristics and health impacts of VOCs and carbonyls associated with residential cooking activities in Hong Kong. *J. Hazard. Mater.* **2011**, *186*, 344–351. [CrossRef]
28. Usinger, J. *Manual on Simple Monitoring Techniques for the Control. of Indoor Air and Combustion Quality Standards in Developing Countries*, 2nd ed.; Deutsche Gesellschaft für Technische Zusammenarbei: Wiesbaden, Germany, 1999.
29. Chao, C.Y.; Wong, K.K. Residential indoor PM_{10} and $PM_{2.5}$ in Hong Kong and the elemental composition. *Atmos. Environ.* **2002**, *36*, 265–277. [CrossRef]
30. Sweeney, L.M.; Sommerville, D.R.; Goodwin, M.R.; James, R.A.; Channel, S.R. Acute toxicity when concentration varies with time: A case study with carbon monoxide inhalation by rats. *Regul. Toxicol. Pharm.* **2016**, *80*, 102–115. [CrossRef]
31. Kleeman, M.J.; Schauer, J.J.; Cass, G.R. Size and composition distribution of fine particulate matter emitted from wood burning, meat charbroiling, and cigarettes. *Environ. Sci. Technol.* **1999**, *33*, 3516–3523. [CrossRef]
32. Nolte, C.G.; Schauer, J.J.; Cass, G.R.; Simoneit, B.R.T. Highly polar organic compounds present in meat smoke. *Environ. Sci. Technol.* **1999**, *33*, 3313–3316. [CrossRef]
33. Maes, W.H.; Verbist, B. Increasing the sustainability of household cooking in developing countries: Policy implications. *Renew. Sustain. Energy Rev.* **2012**, *16*, 4204–4221. [CrossRef]
34. Rehfuess, E.; Tzala, L.; Best, N.; Briggs, D.J.; Joffe, M. Solid fuel use and cooking practices as a major risk factor for ALRI mortality among African children. *J. Epid. Community Health* **2009**, *63*, 887–892. [CrossRef]
35. MacCarty, N.; Still, D.; Ogle, D. Fuel use and emissions performance of fifty cooking stoves in the laboratory and related benchmarks of performance. *Energy Sustain. Develop.* **2010**, *14*, 161–171. [CrossRef]
36. Ballard-Tremeer, G.; Mathee, A. *Review of Interventions to Reduce the Exposure of Women and Young Children to Indoor Air Pollution in Developing Countries*; WHO/USAID Global Consultation: Washington, DC, USA, 2000.
37. Budds, J.; Biran, A.; Rouse, J. *What's Cooking? A Review of Health Impacts of Indoor air Pollution and Technical Interventions for Its Reduction*; Leicser; Water and Environmental Health at London and Loughborough (WELL): London, UK, 2001.
38. Probert, D.; Newborough, M. Designs, thermal performances and other factors concerning cooking equipment and associated facilities. *Appl. Energy* **1985**, *21*, 81–222. [CrossRef]
39. Kshirsagar, M.P.; Kalamkar, V.R. A comprehensive review on biomass cook stoves and a systematic approach for modern cook stove design. Renew. *Sustain. Energy Rev.* **2014**, *30*, 580–603. [CrossRef]
40. Parmigiani, S.P.; Vitali, F.; Lezzi, A.M.; Vaccari, M. Design and performance assessment of a rice husk fueled stove for household cooking in a typical sub-Saharan setting. *Energy Sustain. Develop.* **2014**, *23*, 15–24. [CrossRef]
41. Xu, Z.; Sun, D.-W.; Zhang, Z.; Zhu, Z. Research developments in methods to reduce carbon footprint of cooking operations: A review. *Trends Food Sci. Technol.* **2015**, *44*, 49–57. [CrossRef]
42. Prapas, J.; Baumgardner, M.E.; Marchese, A.J.; Willson, B.; DeFoort, M. Influence of chimneys on combustion characteristics of buoyantly driven biomass stoves. *Energy Sustain. Develop.* **2014**, *23*, 286–293. [CrossRef]
43. El Sharkawy, M.; Javed, W. Study of indoor air quality level in various restaurants in Saudi Arabia. *Environ. Prog. Sustain. Energy* **2018**, *37*, 1713–1721. [CrossRef]
44. Svendsen, K.; Jensen, H.N.; Sivertsen, I.; Sjaastad, A.K. Exposure to cooking fumes in restaurant kitchens in Norway. *Ann. Occup. Hyg.* **2002**, *46*, 395–400. [PubMed]
45. Chiang, C.M.; Lai, C.M.; Chou, P.C.; Li, Y.Y. The influence of an architectural design alternative (transoms) on indoor air environment in conventional kitchens in Taiwan. *Build. Environ.* **2000**, *35*, 579–585. [CrossRef]
46. Environment Protection Agency (EPA). *Air Emissions Monitoring Guidance Note #2 (AG2)*; EPA, Office of Environmental Enforcement (OEE): Wexford, Ireland, 2007.
47. Huang, H.L.; Lee, W.M.G.; Wu, F.S. Emissions of air pollutants from indoor charcoal barbecue. *J. Hazard. Mater.* **2016**, *302*, 198–207. [CrossRef]
48. Sung-OK, B.; Kim, Y.-S.; Perry, R. Indoor air quality in homes, offices and restaurants in Korean urban areas–indoor/outdoor relationships. *Atmos. Environ.* **1997**, *31*, 529–544.

49. Benfenati, E.; Pierucci, P.; Niego, D. A case study of indoor pollution by Chinese cooking. *Toxic Environ. Chem.* **1998**, *65*, 217–224. [CrossRef]
50. Central Pollution Control Board. *Guidelines for Continuous Emission Monitoring Systems*; East Arjun Nagar: Delhi, India, 2018.
51. The Ministry of Environment. Revised Standards for Coal-Based Thermal Power Plants, India. 2017. Available online: http://www.indiaenvironmentportal.org.in/content/423111/revised-standards-for-coal-based-thermal-power-plants (accessed on 10 May 2021).
52. Sloss, L.L. Emissions and effects of air toxics from coal combustion: An overview. In Proceedings of the Managing Hazardous Air Pollutants: State of the Art, Washington, DC, USA, 13–15 July 1993.
53. Licata, A.; Hartenstein, H.U.; Terracciano, L. Comparison of U.S. EPA and European Emission Standards for Combustion and Incineration Technologies. Columbia. 1997. Available online: http://www.seas.columbia.edu/earth/wtert/newwtert/Research/sofos/nawtec/nawtec05/nawtec05-48.pdf (accessed on 10 May 2021).
54. *Royal Commission for Jubail and Yanbu*; Royal Commission Environmental Regulations: Jubail, Kingdom of Saudi Arabia, 2015; Volume 1.
55. World Health Organization. *Guidelines for Air Quality*; WHO: Geneva, Switzerland, 2001.
56. Adhikari, S.; Mahapatra, P.S.; Pokheral, C.P.; Puppala, S.P. Cookstove smoke impact on ambient air quality and probable consequences for human health in rural locations of southern Nepal. *Int. J. Environ. Res. Public Health* **2020**, *17*, 550. [CrossRef] [PubMed]
57. Soneja, S.I.; Tielsch, J.M.; Curriero, F.C.; Zaitchik, B.; Khatry, S.K.; Yan, B.; Chillrud, S.N.; Breysse, P.N. Determining particulate matter 1 and black carbon exfiltration estimates for traditional cookstove use in rural Nepalese village households. *Environ. Sci. Technol.* **2015**, *49*, 5555–5562. [CrossRef]
58. Wong, G.W.; Brunekreef, B.; Ellwood, P.; Anderson, H.R.; Asher, M.I.; Crane, J.; Lai, C.K.W. Cooking fuels and prevalence of asthma: A global analysis of phase three of the international study of asthma and allergies in childhood (ISAAC). *Lancet Respir. Med.* **2013**, *1*, 386–394. [CrossRef]
59. Amouei Torkmahalleh, M.; Gorjinezhad, S.; Unluevcek, H.S.; Hopke, P.K. Review of factors impacting emission/ concentration of cooking generated particulate matter. *Sci. Total Environ.* **2017**, *586*, 1046–1056. [CrossRef]
60. Garrett, M.H.; Hooper, M.A.; Hooper, B.M.; Abramson, M.J. Respiratory symptoms in children and indoor exposure to nitrogen dioxide and gas stoves. *Am. J. Respir. Crit. Care Med.* **1998**, *158*, 891–895. [CrossRef] [PubMed]
61. Cardenas, B.; Umlauf, G.; Maiz, P.; Mariani, G.; Skejo, H.; Blanco, S. PCDD/F, PCB and HCB emissions from rural wood stoves: A preliminary evaluation in Mexico. *Organohalog. Compd.* **2010**, *72*, 1511–1513.
62. Northcross, A.L.; Katharine, H.S.; Canuz, E.; Smith, K.R. Dioxin inhalation doses from wood combustion in indoor cookfires. *Atmos. Environ.* **2012**, *49*, 415–418. [CrossRef]
63. Abdullahi, K.L.; Delgado-Saborit, J.M.; Harrison, R.M. Emissions and indoor concentrations of particulate matter and its specific chemical components from cooking: A review. *Atmos. Environ.* **2013**, *71*, 260–294. [CrossRef]
64. Seltenrich, N. Take Care in the Kitchen: Avoiding Cooking-Related Pollutants. Environ. *Health Perspect.* **2014**, *122*, A154–A159. [CrossRef] [PubMed]
65. Aberilla, J.M.; Gallego-Schmid, A.; Stamford, L.; Azapagic, A. Environmental sustainability of cooking fuels in remote communities: Life cycle and local impacts. *Sci. Total Environ.* **2020**, *713*, 136445. [CrossRef] [PubMed]

atmosphere

MDPI

Article

Black Carbon in Bulgaria—Observed and Modelled Concentrations in Two Cities for Two Months

Elena Hristova [1,*], Emilia Georgieva [1], Blagorodka Veleva [1], Nadya Neykova [1], Stela Naydenova [2], Lenia Gonsalvesh-Musakova [3], Rozeta Neykova [1] and Anton Petrov [1]

1 Department of Meteorology, National Institute of Meteorology and Hydrology, 1784 Sofia, Bulgaria; emilia.georgieva@meteo.bg (E.G.); blagorodka.veleva@meteo.bg (B.V.); nadya.neykova@meteo.bg (N.N.); rozeta.neikova@meteo.bg (R.N.); anton.petrov@meteo.bg (A.P.)
2 Department of Ecology and Environmental Protection, Assen Zlatarov University, 8000 Burgas, Bulgaria; steltion@gmail.com
3 Chemistry Department, Assen Zlatarov University, 8000 Burgas, Bulgaria; lenia_gonsalvesh@abv.bg
* Correspondence: elena.hristova@meteo.bg

Abstract: Black carbon (BC) is one of the particulate matter (PM) components that both affects human health and contributes to climate change. In this study, we present the preliminary results of the investigation of BC concentrations in $PM_{2.5}$ for two Bulgarian cites—Sofia and Burgas. The parallel $PM_{2.5}$ samplings were organized in October 2020 and January 2021. The Multi-Wavelength Absorption Black carbon Instrument (MABI) was used for the evaluation of light-absorbing carbon. In addition, we compared the observed BC and $PM_{2.5}$ values to modelled ones and analyzed the spatial distribution over the country, using data from advanced operational chemical transport models (CTM)—the European (regional) air quality system established at the Copernicus Atmosphere Monitoring Service (CAMS). Generally, the observed BC and $PM_{2.5}$ values were higher in January than in October for both cities. In October, the model underestimated the observed BC concentrations (Sofia—2.44 $\mu g.m^{-3}$, Burgas—1.63 $\mu g.m^{-3}$) by 17% and 51%. In January 2021, the observed monthly BC concentrations were higher (Sofia—3.62 $\mu g.m^{-3}$, Burgas—1.75 $\mu g.m^{-3}$), and the bias of the model was less than that in October, with an overestimation of 22% for Sofia. The relative bias for $PM_{2.5}$ in October (17% for Sofia and −6% for Burgas) was less than the relative bias in January when the model underestimated $PM_{2.5}$ monthly mean concentrations by 20% (Sofia) and 42% (Burgas). In addition, we also elaborate on two episodes with high observed BC concentrations in view of the meteorological conditions.

Keywords: black carbon; $PM_{2.5}$; urban air pollution; modeling of BC; CAMS

Citation: Hristova, E.; Georgieva, E.; Veleva, B.; Neykova, N.; Naydenova, S.; Gonsalvesh-Musakova, L.; Neykova, R.; Petrov, A. Black Carbon in Bulgaria—Observed and Modelled Concentrations in Two Cities for Two Months. *Atmosphere* **2022**, *13*, 213. https://doi.org/10.3390/atmos13020213

Academic Editors: Yongming Han and Takuma Miyakawa

Received: 19 December 2021
Accepted: 26 January 2022
Published: 28 January 2022

1. Introduction

Air pollution with particulate matter (PM) is still a big problem in the large urban agglomerations in the world as well as in most of the cities of Bulgaria [1]. The urban air quality is determined by many factors: air masses, characteristic local flow and meteorology, low dispersion ability in built-up environments, the concentration of emission sources of different types, and various chemical processes. The PM chemical composition, thus, reflects site-specific complex processes. In the last decade, the studies on the chemical composition of PM in urban areas were widely developed and were associated mainly with the investigation of its effects on human health and understanding of its sources [2–5].

Studies on the chemical composition of PM in Bulgaria are extremely important. On one side, regular air quality stations of the national monitoring network often report high PM concentrations, whereas, on the other side, there is a lack of highly specialized monitoring sites like the ones of the EMEP (European Monitoring and Evaluation Programme) network [6]. Studies on PM speciation were conducted during field campaigns in the past years for Sofia only [7–9], however, without analysis for a significant class of pollutants such as carbon and carbonaceous compounds [10].

The carbonaceous fraction (elemental carbon (EC) and organic carbon (OC)) is a prime constituent of atmospheric aerosol. It generally contributes 20–45% of $PM_{2.5}$ (fine particles with an aerodynamic diameter of $\leq$2.5 µm) and 20–30% of PM_{10} (particles with an aerodynamic diameter of $\leq$10 µm) across Europe [5,11]. The examination of carbon particles in urban areas is a matter of great importance for determining the contribution of transportation and domestic heating to air pollution, and thus providing a basis for mitigation measures towards air quality improvement.

There is no universally accepted definition of the term "black carbon" (BC) [12]. The difference between BC and EC is in the measuring method used—BC concentrations are obtained via the optical method, whereas EC—via the thermo-optical method. Not only the shape but also the chemical and physical properties of BC particles can affect the measurements of the optical instruments. Despite that, the average values of BC and EC in urban conditions obtained with different methods appear to be similar [13,14]. Therefore, as we go through the text, we will refer to them using the term BC.

Black carbon, a component of particulate matter, has significant effects on climate and human health [11–18]. BC is typically formed through the incomplete combustion of fossil fuels, biofuel, and biomass and is emitted by anthropogenic and natural sources. It consists of pure carbon in several forms, and the relevant particle size fraction can include known carcinogens and other toxic species [19]. BC stays in the atmosphere for periods varying from days to weeks [11]. Sources of BC can differ widely across regions. The main sources are on-road diesel vehicles, non-road transport, and residential combustion [14–27]. In general, higher concentrations of BC are reported during winter, likely due to higher energy consumption, unfavourable meteorological conditions, and the occasional but significant influence of fires at specific locations during the spring and the summer [21–27].

Countries around the world have not yet defined an air quality standard for black carbon concentrations and mandatory monitoring. This might explain the relatively low number of urban observational sites, despite some environmental agencies and institutions that monitor BC concentration levels [20]. In Europe, BC measurements were carried out during intensive measurement campaigns at rural sites of the EMEP network [6]. Such observational data represent snapshot information from a set number of locations and for limited periods, but they provide valuable information for evaluating modelling results.

Numerical models that represent the atmospheric dynamics and the chemical transformation of air pollutants are powerful tools to complement sparse data from air quality monitoring networks. The models allow one to analyze the spatial and temporal distribution of pollutants and estimate the effects of different mitigation scenarios for air quality improvement. Chemical transport models (CTM) simulating BC concentrations were applied both at a global scale [28,29] and at a regional scale (over Europe) [30,31]. The uncertainties in modelling results are mainly associated with those in BC emissions, estimated to be a factor of two to five on a regional scale [32]. Another important source of uncertainty is linked to parameterizations of chemical mechanisms describing the black carbon ageing process [33]. The performance of models for BC concentrations in Europe was studied in model intercomparison exercises, comparing results from several CTMs to observations at background sites of the EMEP network [34,35]. The underestimation of EC in $PM_{2.5}$ was approximately 20–60% for 2005 [34] and a maximum of 60% for the intensive measurement campaigns in different seasons in the years 2006–2009 [35]. Modelling studies for other parts of the world also showed underestimation of BC by CTMs, for example, in South Africa, mean monthly BC concentrations were underestimated by approximately 50% [36]; in Japan, seasonal BC concentrations observed at urban sites were underestimated by a factor from 0.3 to 0.6; and in South Asia, daily BC concentrations were underestimated by CTM on average by a factor of 4 at urban sites [37,38]

Studies on BC concentration in Bulgaria are limited in number and are not executed regularly. Only a few studies on monitoring BC concentration in $PM_{2.5}$ in Bulgaria are available, conducted within the framework of the project "CARBOAEROSOL" [39,40], with an application of methods for estimation of BC from different sources.

In this study, we present and discuss preliminary results on BC concentrations in fine particulate matter. They were observed for months in two main Bulgarian cities: Sofia, found in a valley, and Burgas, located on the Black Sea coast. Furthermore, we compare the observed BC and $PM_{2.5}$ values to modelled ones and analyze the spatial distribution over the country, using data from an advanced operational CTM, which is the European (regional) air quality system established at the Copernicus Atmosphere Monitoring Service (CAMS) [41]. Two episodes with high observed BC concentrations are discussed in view of the meteorological conditions.

2. Materials and Methods

2.1. Sampling Sites Description

The $PM_{2.5}$ sampling field campaigns were organized in two large cities—Sofia and Burgas (Figure 1)—in the framework of the project "CARBOAEROSOL". Sofia, located in the western part of the country, is the capital and the largest city in Bulgaria, with a population of 1.3 million inhabitants [42]. It is located in the semi-closed Sofia valley at approximately 550 m a.s.l., surrounded by different mountains, where the topography prevents the dispersion of pollutants and determines unfavourable air quality conditions [10]. The climate is continental with a mean temperature of −0.6 °C in January and a maximum of 21.1 °C in July.

Figure 1. Topographic map of Bulgaria with locations of the sampling sites (Sofia—red, Burgas—blue), adapted from https://www.worldometers.info/maps/bulgaria-map/ (accessed on 29 September 2021).

Burgas, located on the Black Sea coast in eastern Bulgaria, is a regional and municipal center. It is the fourth largest town in Bulgaria, housing 277,922 inhabitants [42]. The climate is moderately continental with a clearly expressed sea influence and frequent sea breeze, most notable during the warm period of the year. Mean monthly temperatures are positive all over the year; the mean wind speed is $\geq$4.0 ms^{-1} with a maximum of 4.8 ms^{-1} in March. The city is surrounded by four large water bodies with different salinity, which determines the high relative humidity throughout the year. The winter is mild, usually without any snow. During the winter season, the average air temperature is 4.6 °C, and that of the sea water is 7.4 °C. Due to the influence of the sea, autumn at this site is long and significantly warmer than in inland regions, whereas spring tends to be cold and arrives a month later. These more dynamic atmospheric conditions affect the processes of atmospheric chemistry. The fewer residential, transport, and industrial sources are expected to lead to lower $PM_{2.5}$, BC, and nitrogen dioxide (NO_2) concentrations [43,44].

2.2. $PM_{2.5}$ and BC Emissions

Annual emissions of $PM_{2.5}$ and BC in Bulgaria reported to the Convention on Long-Range Transboundary Air Pollution (CLRTAP) [45] allow the analysis of the trend both of national totals and the contribution by different sectors (Figure 2). The $PM_{2.5}$ emissions in 2019 were 30 kt, which is slightly lower than in the previous few years. The main contributors were the sectors "other stationary combustion" (residential heating), with a share of 77%, road transportation, with a share of 7%, and industry (8%). The BC emissions in 2019 were 3.63 kt, representing 12% of the annual $PM_{2.5}$ emissions. The annual BC emissions in the period 2008–2019 exhibited little variation, with a maximum of 4.13 kt in 2012. In 2019, 63% of the BC emissions were attributed to residential heating, 33% to road transport. The emissions are unevenly distributed throughout the year. Although the residential sources prevail during winter months (November to February), the transport emissions are almost constant.

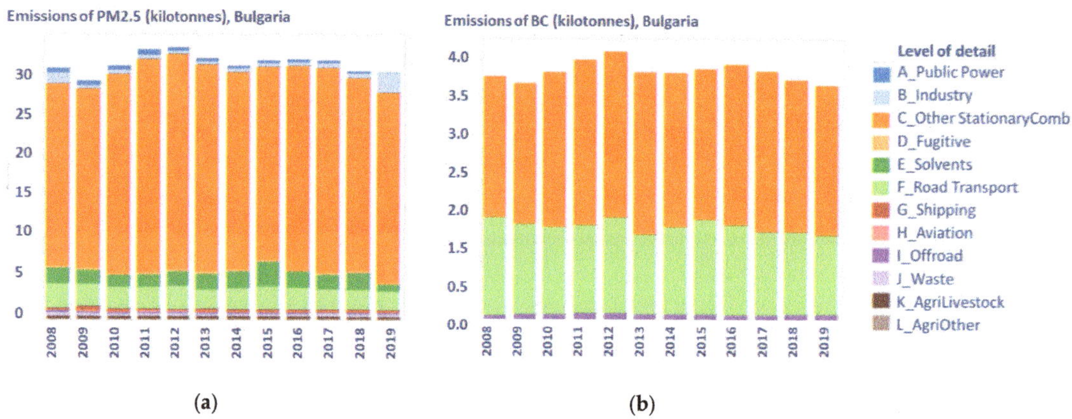

Figure 2. Annual emissions (kilotons) of $PM_{2.5}$ (**a**) and BC (**b**) in Bulgaria; national totals and the contribution by different sectors, adapted from https://www.ceip.at/data-viewer (accessed on 3 October 2021).

There have been no reliable estimations of emissions of particulate matter at city levels so far. Some estimates were made at the two municipalities concerning dispersion modelling for air quality planning. For example, the contribution from residential heating to $PM_{2.5}$ emissions in the Sofia region (about 20 km) in 2018 was estimated to be 73% [46]. For Burgas, the share of residential emissions to $PM_{2.5}$ in 2014 was estimated to be 68% [47].

A recent study for Sofia, based on the Positive Matrix Factorization (PMF) receptor model for source contribution of different groups of emission sources to PM_{10} concentrations [10], indicated 23% for biomass burning and 9% for transport on an annual basis for 2019. Another recent study on the determination of the local pollution of PM and PAH in the Sofia municipality by the Gaussian dispersion-based ADMS-Urban model showed that domestic heating contributed 59% of the $PM_{2.5}$ concentrations [48].

2.3. Sampling Procedure, Equipment, and Black Carbon Analysis

The parallel 24 h $PM_{2.5}$ sampling was performed at the Central Meteorological Observatory, Sofia (42.655 N, 23.384 E, at 586 m a.s.l.) and at "Prof. Dr Assen Zlatarov" University, Burgas (42.314 N, 27.264 E, at 30 m a.s.l.) during autumn (October 2020) and winter (January 2021). Both sampling sites can be classified as 'urban background'. A total of 69 daily $PM_{2.5}$ samples on polytetrafluoroethylene membrane filters (PTFE) Whatman® (47 mm) were collected with low volume air samplers following the EN-12341 standard (38.3 l min^{-1}). Although the PTFE filters are optically thinner and less homogeneous than the fiber media, they avoid interference from adsorbed organic gases that is associated with quartz and glass fiber media. This provides flexibility to perform subsequent analysis of the same filters using other techniques such as X-ray fluorescence, ICP, IC, etc. The samplers were installed on a grass field, and the sampling heads were at 2 m. a.g.l. Before and after sampling, the filters were weighed by analytical balances (in Sofia with Mettler Toledo, AG135 and in Burgas with KERN & Sohn GmbH, ABT 100-5M). $PM_{2.5}$ mass concentrations were determined from the weight differences and sampled air volume.

All collected $PM_{2.5}$ samples were analyzed for light-absorbing carbon (LAC) with the Multi-Wavelength Absorption Black carbon Instrument (MABI) at the National Institute of Meteorology and Hydrology (NIMH). This instrument consists of an optical module with a multi-wavelength light source (7 LEDs), a sampling holder, and a photodetector. Moreover, opaque glass is used in the MABI units to scatter the scattered light back through the filter to the detector. The MABI instrument measures light absorption at seven wavelengths, spanning from ultraviolet to infrared (405 nm, 465 nm, 525 nm, 639 nm, 870 nm, 940 nm, and 1050 nm). This allows the differentiation of the contributions from biomass burning (BC_{bb}) and fossil fuels combustion (BC_{ff}) based on different light absorption from different type particles at different wavelengths. The values obtained by subtracting the BC (1050 nm) data from BC (450 nm) data represent mainly BC from biomass burning [49,50]

The MABI is a research instrument developed at the Australian Nuclear Science and Technology Organisation that does not perform LAC or BC calculation [49,50]. This is done by the user, which ensures full control, understanding, and interpretation of the obtained BC values. MABI scans are initially performed on unexposed "blank filters" to determine the absorption at each wavelength from the filter substrate. These data for blank filters are referred to as *Io*. The MABI scans are then repeated on the same filters after exposure and the collection of particulate matter, i.e., "exposed filters", which represent the absorption at each wavelength from both the collected particles and the filter substrate. This transmission data are referred to as *I* [49,50]. Both of these values along with the filter area (10.75 cm^2) and sampled air volume (~55 m^3) are used to determine the black carbon light absorption coefficient (b_{abs}) in (Mm^{-1}]) at each wavelength (Equation (1)):

$$b_{abs} = 10^2 \frac{A}{V} ln\left(\frac{I_o}{I}\right) \quad (1)$$

where

A—filter collection area in cm^2;

V—volume of air sampled through the filter in m^3;

Io—measured light transmission through blank (unexposed) filter;

I—measured light transmission through an exposed filter.

The obtained values for b_{abs} (λ = 639 nm) in this study are in the range 6.74–45.10 Mm^{-1} for Sofia and 1.19–24.3 Mm^{-1} for Burgas.

The LAC mass concentration in [$ng.m^{-3}$] is determined using a mass absorption coefficient (ε) in ($m^2.g^{-1}$) at each wavelength (Equation (2)):

$$LAC\left(ng.m^{-3}\right) = \frac{10^5 A}{\varepsilon V} ln\left(\frac{I_o}{I}\right) \quad (2)$$

The mass absorption coefficient (ε) is a function of wavelength (λ): $\varepsilon = a^*\lambda^{-b}$ and was obtained following the procedure presented by [49,50]. A plot of the log of all wavelength data against the log of all λ = 639 nm data is made to determinate the values of ε for each wavelength (example in Figure S1, Supplementary). According to authors [49], the mass absorption coefficient ε at λ = 639 is standard baseline data with a known value—6.6 $m^2.g^{-1}$. The received values of ε as a function of λ for Sofia and Burgas are presented in Figure S2 in the Supplementary. These values are in the range of 4.9–8.4 $m^2.g^{-1}$, which is in correspondence with literature for fine particulate size fraction $PM_{2.5}$ [49,50].

The light source calibration was made automatically in two calibration cycles. During measurements, the calibration was checked at regular intervals.

Filter-based optical methods for estimation of BC concentration suffer from two major artifacts—the filter loading (or shadowing) effect and the multiple scattering effect [51–59]. The filter loading effect (FLE) is a bias that affects the apparent concentrations relative to the ambient ones. The multiple scattering from aerosols on the filter and from the filter material leads also to attenuation, which might be misinterpreted as due to absorption.

Different approaches have been developed over time for compensating for these effects, proposing various empirically estimated correction procedures [53–62]. The correction algorithms depend on the wavelength and make use of parameters that account for instrument details. As there is no worldwide consensus on the most representative correction procedure [60], inter-comparison workshops and studies applying different laboratory and field instruments propose algorithms for widely used instruments (e.g., for AE-31, PSAP).

Most of the correction algorithms were developed specifically for quartz or glass fiber filters. Correcting for loading effect is treated as quite straightforward [57], and it has already been implemented in an aethalometer using a dual-spot technology [58].

In our study we used a relatively new research type instrument (MABI) with a PTFE filter. There are still a limited number of published studies with this device [49,50], and, to our knowledge, they do not have details as to whether it is necessary to apply a filter loading correction.

In a recent work on MABI measurements [49], comparison of data from MABI with data from a MAGEE AE-33 aethalometer (instrument with automatic dual-spot 'loading compensation') for almost one year was addressed. The correlation coefficient of R^2 = 0.96 was reasonable for these two quite different instruments measuring the fine $PM_{2.5}$ black carbon concentrations at the same site. On average, the AE_33 aethalometer estimates were 12% higher than the MABI estimates. Furthermore, they quoted each MABI LAC measurement as having an error of ±14% at 639 nm.

The precision of the measurements in our study was verified by conducting repeated measurements on 20 samples. The results showed the difference between measurements on the same filter was <0.1% for all readings of *Io* and *I*. The minimum detection limit of LAC ranges from 60 to 70 $ng.m^{-3}$. Considering the presented information in [49] and the [60] approach, the raw data could be further modified if researchers agree upon a correction method that could be used in combination with the MABI instrument. Therefore, in the next sections we will present and discuss raw BC (uncorrected) data.

2.4. Modelling Systems

We analysed data provided by the European (regional) air quality system established at the Copernicus Atmosphere Monitoring Service [63]. The system, further denoted as CAMS-ENS, is based on an ensemble approach, including results from nine state-of-the-art chemistry transport models [63]. All models use common data regarding meteorology, emissions, and boundary conditions. The meteorological driver is the Integrated Forecast system of the European Centre for Medium-range Weather Forecasts (ECMWF-IFS), [64]. The individual models use common anthropogenic emission input, the CAMS-REG-AP emission inventory, developed under the Copernicus Global and Regional emissions service [65]. The reference emission input (CAMS-REG-v.4.2) is for the year 2017, with upgrades in version 4.2 for 2018 and 2019 [66]. There are ongoing activities on including emissions of EC in the $PM_{2.5}$ fraction, still facing difficulties due to differences in reporting wood-burning emissions by several countries [67]. All models use common wildfire emission input, provided by the Global Fire Assimilation System at CAMS. The individual models use various chemistry algorithms and parameterizations and assimilation of in situ observations; further details can be found in [67]. The CAMS-ENS results (analysis and 4-day forecast hourly values) are available on a grid of $0.1° \times 0.1°$ at eight vertical levels (from the surface to 5000 m) and refer to more than 20 species. For this study, we downloaded and assessed 1-h analysis data for PM_{10}, $PM_{2.5}$, and total elementary carbon from the CAMS atmosphere data store [68].

To investigate the effects of long-range transport, we considered back-trajectories from the Hybrid Single-Particle Lagrangian Integrated Trajectories (HYSPLIT) trajectory model [69,70]. For analysis of selected episodes, back-trajectories arriving at the sampling sites at 500 m, 1500 m, and 3000 m a.g.l. were calculated for three days, using GFS0p25 meteorological input with resolution $0.25° \times 0.25°$.

2.5. Data Analyses

We analyzed data for $PM_{2.5}$ and BC concentrations for October 2020 and January 2021 in Sofia and Burgas. We selected these months to investigate the variability of concentrations from the two cities under different thermal and dynamic conditions. Moreover, October is the month before the start of the heating season, whereas January is in the middle of it. The data for observed black carbon concentrations presented in our study are a sum of fossil fuels (BC_{ff}) and biomass burning (BC_{bb}) and are marked as BC.

Furthermore, the obtained data for $PM_{2.5}$ and BC concentrations, together with basic meteorological parameters provided by NIMH for both cities, were processed with statistical software R and StatSoft v.6.0.

The observed BC and $PM_{2.5}$ mean concentrations were compared with modelled ones for October 2020 and January 2021. Additionally, they were compared for two selected episodes with elevated BC concentrations. For briefly commenting on the synoptic situation during the two selected episodes, we analysed the spatial distribution of some meteorological variables from the global reanalysis dataset of the National Centers for Environmental Prediction/National Center for Atmospheric Research (NCEP/NCAR) [71]. NCEP/NCAR reanalysis products have been widely used to investigate large-scale meteorological conditions and transport paths linked to air pollution episodes. This reanalysis dataset was selected as it provides a possibility for fast visualisation of composite maps. The composite (mean) maps can be produced for selected dates and hours, facilitating evidence of large scale differences in the meteorological fields for low and high pollution days, or for periods in different seasons. We produced composite maps of the geopotential height and the wind at 700 hPa for each of the two selected episodes using the web tool at the National Oceanic and Atmospheric Administration, Earth System Research Laboratories (NOAA/ESRL) [72].

3. Results

3.1. Observed $PM_{2.5}$ and BC Concentrations in Sofia and Burgas

The comparison of daily $PM_{2.5}$ mass concentrations in Sofia and Burgas for October 2020 and January 2021 is presented in Figure 3. The number of the $PM_{2.5}$ samples in Burgas (10) is fewer than that in Sofia (19) for January due to technical problems during the experimental campaign. The $PM_{2.5}$ concentrations in Sofia were in the range 4.9–25.5 $\mu g.m^{-3}$ in October and from 5.1 to 46.9 $\mu g.m^{-3}$ in January. For Burgas, the $PM_{2.5}$ concentrations varied from 3.3 to 27.1 $\mu g.m^{-3}$ in October and from 2.2 to 30.4 $\mu g.m^{-3}$ in January.

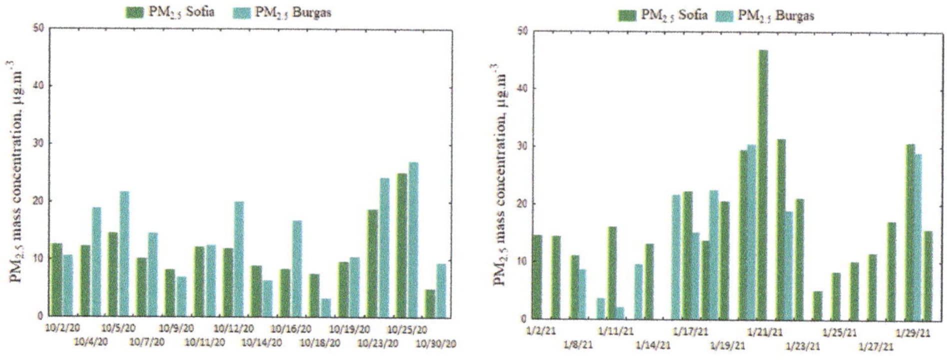

(**a**) (**b**)

Figure 3. Daily $PM_{2.5}$ mass concentration ($\mu g.m^{-3}$) in Sofia and Burgas for October 2020 (**a**) and January 2021 (**b**).

The analysis showed that 30% of the measured daily $PM_{2.5}$ concentrations in Sofia and 43% in Burgas exceeded the most recent WHO air quality guideline level (15 $\mu g.m^{-3}$) [19]. In October, a higher mean $PM_{2.5}$ concentration (14.5 $\mu g.m^{-3}$) was observed in Burgas than in Sofia (11.8 $\mu g.m^{-3}$). This is probably due to the fewer rainy days and monthly accumulated precipitation in Burgas (12 mm) in comparison to Sofia (46.7 mm). In January, a higher mean $PM_{2.5}$ concentration (18.6 $\mu g.m^{-3}$) was observed in Sofia than in Burgas (16.2 $\mu g.m^{-3}$). Although some daily $PM_{2.5}$ concentrations in Sofia were higher than the WHO guideline value, the monthly mean $PM_{2.5}$ concentration (18.6 $\mu g.m^{-3}$) was lower than the mean value for January 2019 (22.8 $\mu g.m^{-3}$), the period before the COVID-19 pandemic [27]. The highest daily $PM_{2.5}$ concentrations in Sofia were observed on 21 and 22 January, when the air temperature was around zero and the wind speed was very low (0.2–0.5 ms^{-1}). The highest $PM_{2.5}$ concentration in Burgas was measured on 20 January, when the air temperature was 1.9 °C, and the wind speed was 2.9 ms^{-1}.

Figure 4 presents the variation in BC concentrations obtained in October and in January, as well as the $BC/PM_{2.5}$ ratio.

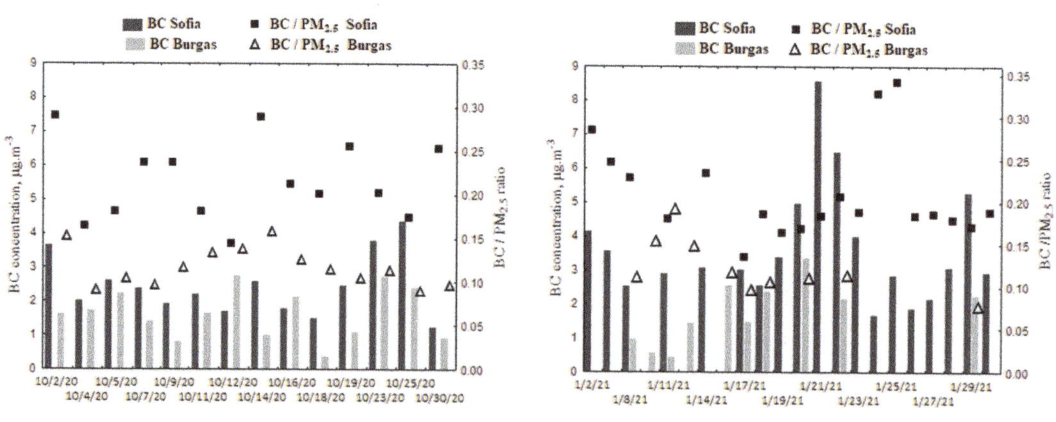

(**a**) (**b**)

Figure 4. Daily BC concentrations ($\mu g.m^{-3}$) (bars) and $BC/PM_{2.5}$ ratio (symbols) for October 2020 (**a**) and January 2021 (**b**).

Generally, the derived BC values are higher for January than those for October. The BC concentrations in Sofia samples varied from 1.25 to 4.36 $\mu g.m^{-3}$ for October and from

1.67 to 8.58 $\mu g.m^{-3}$ for January. The BC concentrations in Burgas samples ranged from 0.37 to 2.76 $\mu g.m^{-3}$ in October and from 0.42 to 3.36 $\mu g.m^{-3}$ in January. Based on the measured $PM_{2.5}$ and BC concentrations, the derived $BC/PM_{2.5}$ ratio for Sofia varies from 17% to 29% and for Burgas, from 1% to 17% in October. This ratio in January covers a wider range in Sofia (14–34%), reflecting the increased impact of combustion sources on the mass concentration of PM and more variable meteorological/synoptic situations. For Burgas the ratio is limited to a narrower range (9–11%) in January, probably because of the similar meteorological conditions in terms of dynamical mixing (high mean diurnal wind speed) for measurement days.

The monthly mean BC concentration for Sofia was lower in October (2.4 $\mu g.m^{-3}$) than in January (3.6 $\mu g.m^{-3}$). During the latter, the number of days with inversion was high, and there was an influence from biomass burning for heating. There was no significant difference between the mean BC concentration in Burgas in October (1.63 $\mu g.m^{-3}$) and January (1.75 $\mu g.m^{-3}$). This is most likely due to the more intense dispersion of air pollutants and the higher wind speed in Burgas during winter. Moreover, October 2020 was characterized by lower wind speed than the multi-annual (1981–2010) value (norm): in Burgas, it was 3.3 ms^{-1} (compared to the norm of 4.6 ms^{-1}); and in Sofia, it was 1.0 ms^{-1} (compared to the norm of 1.2 ms^{-1}). The monthly sum of precipitation was 30.2 mm in Burgas (less than the norm of 57 mm), whereas in Sofia monthly amount of rainfall was slightly higher—60.4 mm with a standard amount of 50 mm. The air temperatures in October 2020 in Burgas and Sofia were above the norm with 3.3 °C and 1.4 °C, respectively. In addition, January 2021 was warmer than usual: 2.8 °C above the norm in Burgas and 2 °C above the norm in Sofia. It was more humid, which is depicted by the precipitation sum—above the norm by 432% in Burgas and 335% in Sofia. The mean wind speed was near the norm for both cities. These specific meteorological conditions led to smaller than expected variations in daily and mean concentrations.

Because the BC and $PM_{2.5}$ concentrations depend on the meteorological conditions, a statistical analysis involving main meteorological parameters (temperature, wind speed, relative humidity) was carried out. The correlations between black carbon concentration from fossil fuel (BC_{ff}) and biomass burning (BC_{bb}), $PM_{2.5}$ mass concentration, and the noted meteorological parameters for both sites are illustrated in Figure 5. A high correlation between BC_{ff} and $PM_{2.5}$ in Sofia was observed, with a statistically significant correlation coefficient equal to 0.9. A moderately high correlation (0.7) between BC_{ff} and $PM_{2.5}$ was observed in Burgas. There were significant negative correlations (−0.5) between BC and the temperature (T) and between BC_{ff} and wind speed (WS) in Sofia, as expected as high concentrations of pollutants are observed in cold and calm conditions. There is a significant negative correlation between BC_{bb} concentration and temperature in Sofia (−0.5) and Burgas (−0.6) due to enhanced biomass burning during colder days.

The comparison between the obtained values for BC in Sofia and Burgas and those reported for other European countries is presented in Table 1. It was found that the obtained values for BC in Sofia were close to those measured in Madrid and Athens (January 2020) and lower than those obtained in Zabrze, Poland. The mean BC concentrations in Burgas were generally lower compared to the values recorded at most urban or urban background stations. In contrast, they were closer to those for Helsinki and Athens (October 2020). It should be noted that the applied method and instrument for estimation of BC concentrations in our study are relatively new. According to the authors who created this device, there is a possible underestimation of BC concentration by approximately 12% [49]. The possible underestimation of BC concentration might lead to a higher $BC/PM_{2.5}$ ratio in some particular days with high $PM_{2.5}$ concentrations in January in Sofia, not more than 5–6% of what was presented originally in Figure 4. New research and results from comparative measurements with different instruments are required for more accuracy in these estimates.

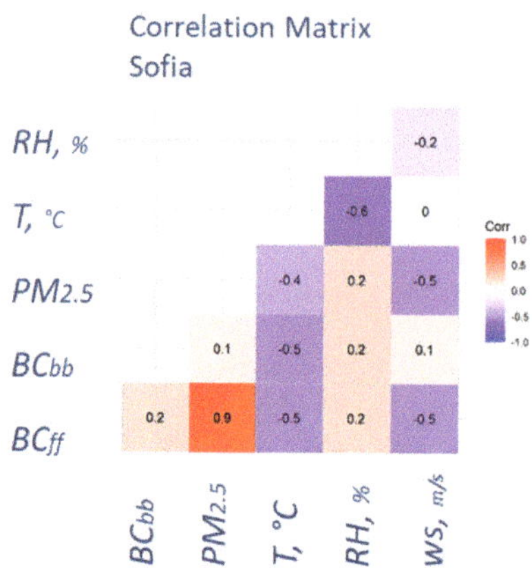

(**a**)

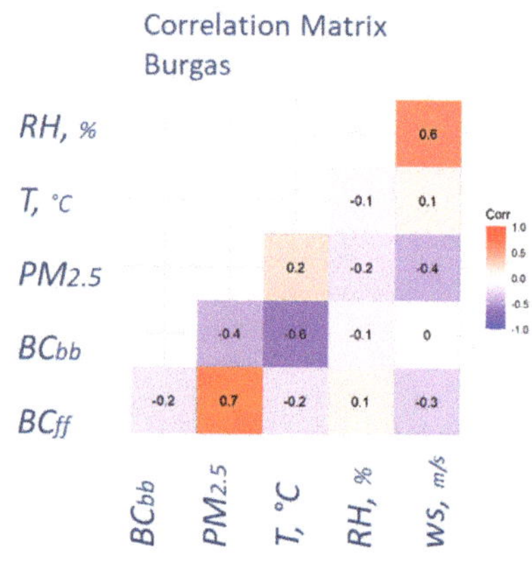

(**b**)

Figure 5. Correlation coefficients between $PM_{2.5}$, BC_{ff}, and BC_{bb} concentrations and some meteorological parameters (T—temperature, RH—relative humidity, ws—wind speed) for (**a**) Sofia and (**b**) Burgas; data for October 2020 and January 2021; *p*-value < 0.01.

Table 1. Mean BC values for different European sities.

Place	Period	Mean BC, $\mu g.m^{-3}$	Source
Zabrze, Poland (urban)	2019	3.4	[21]
Zabrze, Poland (urban)	2020	2.9	[21]
Zabrze, Poland (urban)	October 2020	4.4	[21]
Zabrze, Poland (urban)	January 2020	6.7	[21]
Athens, Greece (suburban)	October 2013 January 2014	1.8 3.7	[22]
Helsinki, Filand (urban)	October 2015–May 2017	1.7	[23]
Paris, France (urban)	September 2009– September 2010	3.0	[24]
London, UK (urban)	2009	2.0	[24]
Madrid, Spain, (urban background)	Autumn 2015 Winter 2014–2015	3.8 2.4	[25]
Madrid, Spain (urban traffic)	Autumn 2015 Winter 2014–2015	4.9 4.2	[25]
urban station in Germany	2005–2014	2.1	[26]
Ostrava, Czech Republic (urban)	2012–2014	3.5	[27]
Sofia, Bulgaria (urban background)	October 2018	3.8	[40]
Sofia, Bulgaria	January 2019	3.7	[40]
Sofia, Bulgaria (urban background)	October 2020 January 2021	2.4 3.6	This study
Burgas, Bulgaria (urban background)	October 2020 January 2021	1.6 1.8	This study

3.2. $PM_{2.5}$ and BC Concentrations: Observed vs. Model

In this section, we present the results from a comparison of observed and modelled BC and $PM_{2.5}$ concentrations for October 2020 and January 2021. Moreover, we elaborate on selected episodes with elevated values of observed BC concentration. As we noted in Section 2.5, the designation "BC" refers to observed BC and modelled total EC by CAMS-ENS. We must keep in mind that there is some uncertainty in the measurement of BC, estimated at ±14%, which we assume will have a lower effect in averaging the data, compared to single values.

3.2.1. Monthly Mean $PM_{2.5}$ and BC Concentrations: Observed vs. Model

The comparison for monthly mean BC and $PM_{2.5}$ concentrations at the two sampling sites for October 2020 and January 2021 is presented in Figure 6. The observed BC concentrations in October (Sofia—2.44 μg.m^{-3}, Burgas—1.63 μg.m^{-3}) are underestimated by the model, respectively, by 17% and 51%. In January 2021, the observed monthly BC concentrations are higher (Sofia—3.62 μg.m^{-3}, Burgas—1.75 μg.m^{-3}), and the bias of the model is less than in the autumn month. An overestimation of 22% for Sofia modelled BC in January compared to the observed might be partly a result of the higher $PM_{2.5}$ concentration and bias in measured BC concentrations because of the filter loading effect.

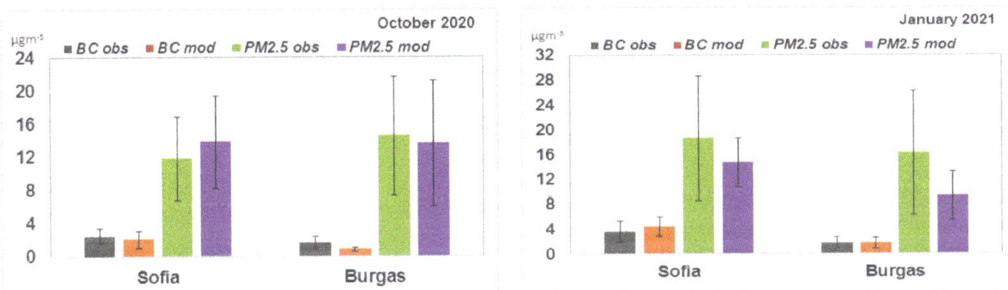

(a) (b)

Figure 6. Monthly mean BC and $PM_{2.5}$ concentrations (μg.m^{-3}) at Sofia and Burgas observed and modelled by CAMS-ENS; for October 2020 (**a**), for January 2021 (**b**); the whiskers represent the standard deviations.

It has to be noted that model values are representative of the computational grid (about 10 × 10 km) and background concentrations in urban environments located in regions with complex topography, whereas monitored concentrations reflect local emission sources and meteorological and dispersion conditions. The relative bias for $PM_{2.5}$ in October (17% for Sofia and −6% for Burgas) was less than the relative bias in January when the model underestimated $PM_{2.5}$ monthly mean concentrations by 20% (Sofia) and 42% (Burgas). The $PM_{2.5}$ underestimation and, at the same time, BC overestimation in the winter month at the Sofia sampling site suggest that model BC emissions from residential combustion around the capital are high. The underestimation of BC at both locations during autumn, with biases higher than for $PM_{2.5}$ values, suggest that BC emitted from all types of transportation might be underestimated. The observations and the model results show that BC and $PM_{2.5}$ concentrations were higher in Sofia than in Burgas.

The spatial distribution of monthly mean BC and $PM_{2.5}$ concentrations based on CAMS-ENS data is shown over part of the Balkans in Figure 7 for October 2020 and Figure 8 for January 2021.

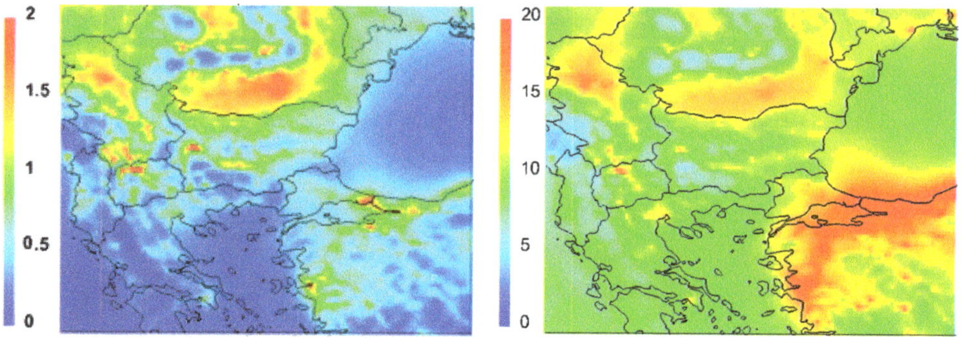

(a) (b)

Figure 7. CAMS-ENS monthly mean concentrations ($\mu g.m^{-3}$) in October 2020: BC (**a**), $PM_{2.5}$ (**b**).

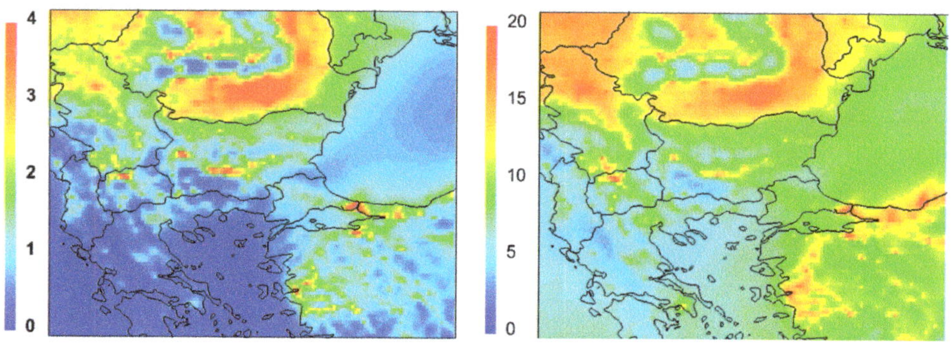

(a) (b)

Figure 8. CAMS-ENS monthly mean concentrations ($\mu g.m^{-3}$) in January 2021; BC (**a**), $PM_{2.5}$ (**b**).

The regions with higher BC concentrations in both months were mainly outside of the country. The hot spots in the country were around Sofia and in the lowlands of southern Bulgaria, especially in the winter month. Elevated BC concentrations were also noted in the northern part of the country, in the Lower Danube plain, covering wider areas in October. The $PM_{2.5}$ pattern was similar to the one for BC, although some new features were noted. The model simulated higher $PM_{2.5}$ concentrations in October than in January in the region south-east of Bulgaria. This area includes the megalopolis of Istanbul and the industrialized area of Turkey along the northern Aegean coast. The higher $PM_{2.5}$ concentrations were most likely linked to the influence of emission sources in the region and the atmospheric dynamics, more easily noticed in October than under winter conditions.

Model concentrations depend on many factors—emission sources, chemistry mechanisms, and meteorological processes. The black carbon in atmospheric particulates is inert and directly linked to emission sources. The spatial distribution of $PM_{2.5}$ emissions used by CASM-ENS is shown in Figure 9. The emission data are available in the Emissions of Atmospheric Compounds and Compilation of Ancillary Data (ECCAD) at [73]. The maps in Figure 9, produced by the ECCAD data tools, refer to the 2019 and CAMS-REG-AP-v4.2 emission inventory on a $0.05° \times 0.1°$ grid. The annual total emissions of $PM_{2.5}$ were mainly due to residential combustion, especially in the lowland regions in the northern and southern parts of the country.

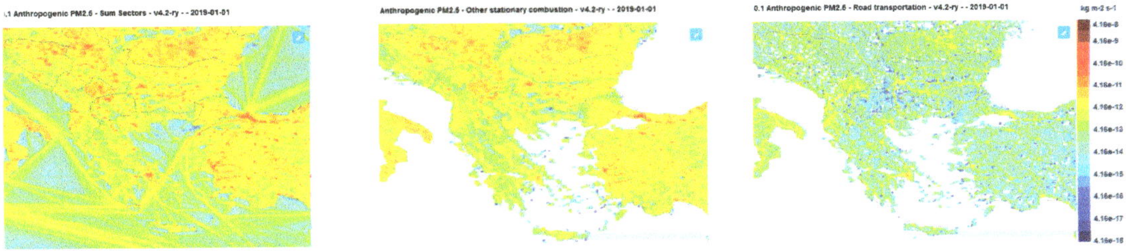

(**a**) (**b**) (c)

Figure 9. $PM_{2.5}$ anthropogenic emissions in 2019 by CAMS-REG-AP inventory: sum of all sectors (**a**), from other stationary combustion (residential) (**b**), from road transportation (**c**).

3.2.2. Daily Mean $PM_{2.5}$ and BC Concentrations: Observed vs. Model

Further, we discuss the comparison between daily BC and $PM_{2.5}$ concentrations at the two sites. Figure 10 shows the daily variability of BC and $PM_{2.5}$ in October, whereas Tables 2 and 3 present some main statistical parameters.

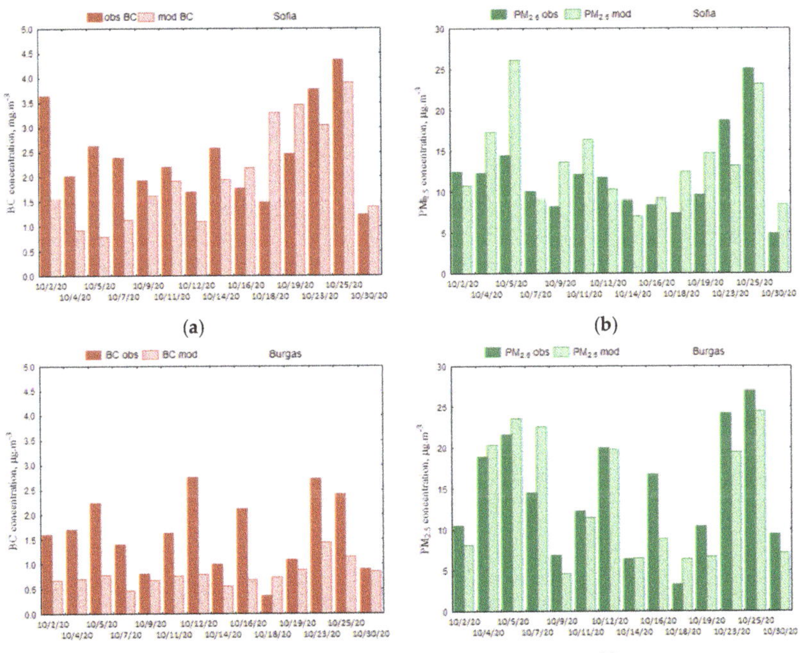

Figure 10. Five daily mean concentrations (μg.m^{-3}) observed (solid fill) and modelled (hatched) for October 2020 in Sofia (**a**,**b**) and in Burgas (**c**,**d**).

In Sofia, the observed BC daily concentrations ranged from 1.3 to 4.4 μg.m^{-3}, whereas the simulated BC ranged from 0.8 to 3.9 μg.m^{-3}, with a correlation coefficient equal to 0.42. For $PM_{2.5}$, the observed values were between 4.9 and 25.1 μg.m^{-3}; the modelled ones between 7.1 and 26.2 μg.m^{-3}, with a correlation coefficient equal to 0.65. In comparison, the correlation coefficients in Burgas were higher—0.51 for BC and 0.87 for $PM_{2.5}$.

Table 2. Statistical parameters for BC and $PM_{2.5}$ in October 2020.

October 2020	n	Mean_OBS	Mean_MOD	MBE	RMSE	Corr	FGE	NMB %
BC_Sofia	14	2.44	2.02	−0.42	1.09	0.42	0.43	−17.21
PM2.5_Sofia	14	11.76	13.77	2.00	4.77	0.65	0.31	17.09
BC_Burgas	14	1.63	0.80	−0.83	1.04	0.51	0.67	−50.92
PM2.5_Burgas	14	14.47	13.6	−0.87	3.81	0.87	0.26	−6.01

Mean observed and modelled values: MB—mean bias error, RMSE—root mean square error, Corr—correlation coefficient, FGE—fractional gross error, NMB—normalized mean bias. Units: $\mu g.m^{-3}$.

Table 3. Statistical parameters for BC and $PM_{2.5}$ in January 2021.

January 2021	n	Mean_OBS	Mean_MOD	MBE	RMSE	Corr	FGE	NMB %
BC_Sofia	19	3.62	4.67	1.05	1.85	0.46	0.39	29.01
$PM_{2.5}$_Sofia	19	18.58	14.71	−3.87	8.80	0.65	0.28	−20.83
BC_Burgas	10	1.75	1.70	−0.04	0.55	0.80	0.25	−2.86
$PM_{2.5}$_Burgas	10	16.16	9.32	−6.84	10.07	0.70	0.47	−42.33

Mean observed and modelled values: MBE—mean bias error, RMSE—root mean square error, Corr—correlation coefficient, FGE—fractional gross error, NMB—normalized mean bias. Units: $\mu g.m^{-3}$.

The daily values in January (Figure 11) were, in general, higher at both sites. In Sofia, the observed BC and $PM_{2.5}$ were in the interval 1.7–8.6 $\mu g.m^{-3}$ and 5.1–46.9 $\mu g.m^{-3}$, whereas the model ones were in the range 2.6–6.7 for BC and 7.5–21 $\mu g.m^{-3}$ for $PM_{2.5}$. BC was overestimated for almost all days in Sofia and half of the days in Burgas. It is interesting to note that on the days with the highest observed concentration of BC and $PM_{2.5}$, the modelled values are lower. The correlation coefficients in Sofia were almost the same as in October: −0.46 for BC and 0.65 for $PM_{2.5}$. The correlation coefficients in Burgas were higher (0.78 for BC and 0.7 for $PM_{2.5}$). However, one must interpret the numbers for Burgas with caution due to the limited number of daily values.

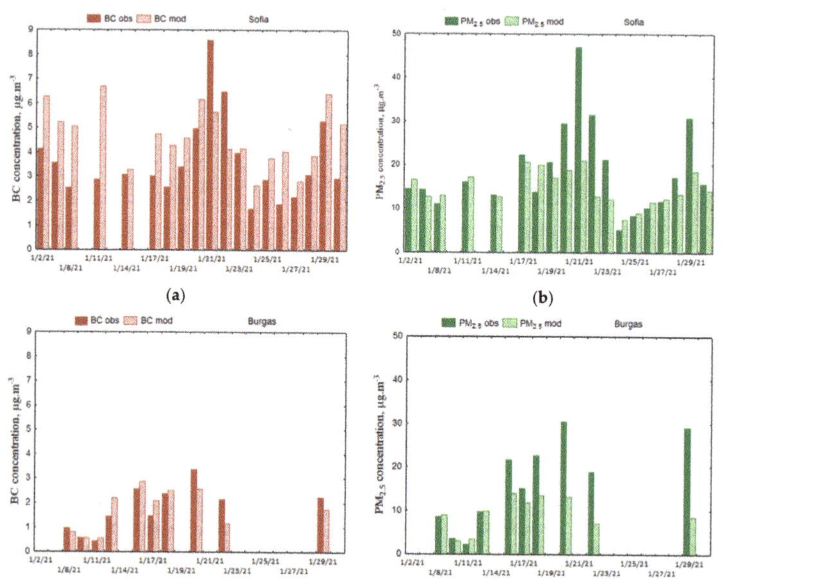

Figure 11. BC and $PM_{2.5}$ daily mean concentrations ($\mu g.m^{-3}$) observed (solid fill) and modelled (hatched) for January 2021 in Sofia (**a**,**b**) and in Burgas (**c**,**d**).

3.2.3. Daily Mean $PM_{2.5}$ and BC Concentrations: Observed vs. Model for Selected Episodes

Episode 1: 23–25 October 2020

Relatively high concentrations of $PM_{2.5}$ and BC were observed at both sites on 23 and 25 October 2020 and captured in the model in Figure 12. The CAMS-ENS model simulated an increase in the concentrations on 25 October, especially for $PM_{2.5}$ in Sofia.

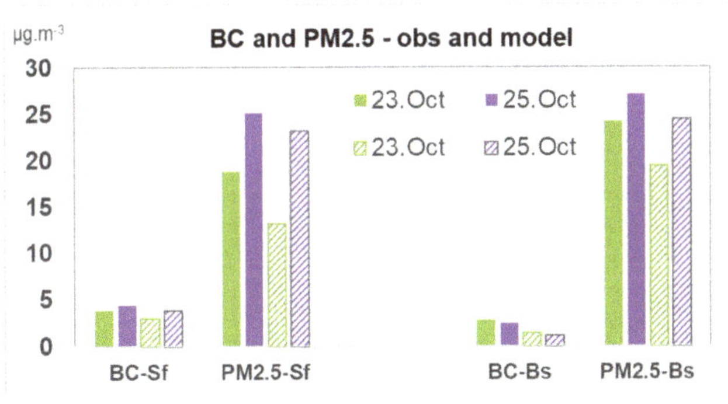

Figure 12. BC and $PM_{2.5}$ daily mean concentrations ($\mu g.m^{-3}$) observed (solid fill) and modelled (hatched) on 23 October 2020 (green) and 25 October 2020 (purple).

The synoptic situation in the period 19–24 October 2020 was characterized by high pressure over southern and eastern Europe, favouring the accumulation of pollutants [74].

On 25 October, a perturbation, associated with the passage of a low-pressure center from the region of central Europe towards the Balkans, led to an upper air flow from the south-west to Bulgaria, as shown on the reanalysis maps for the geopotential height and wind at 700 hPa (Figure 13).

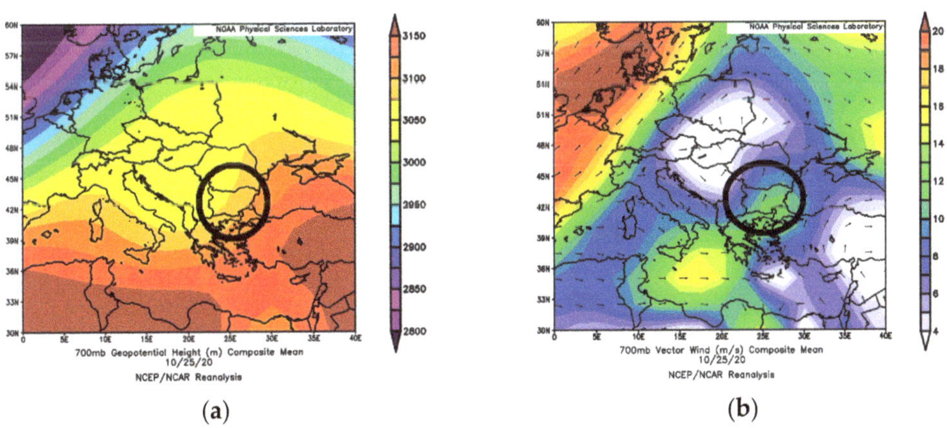

Figure 13. NCEP/NCAR Reanalysis for 25 October 2020: (**a**) geopotential height (m) at 700 hPa, (**b**) wind vectors and speed (ms^{-1}) at 700 hPa. Image provided by the NOAA/ESRL Physical Sciences Laboratory, Boulder Colorado from their Web site at http://psl.noaa.gov/ (accessed on 15 October 2021). Bulgaria is marked with a circle.

HYSPLIT back trajectories (Figure 14) indicated that the upper airflow (3000 m a.g.l.) was from the southern direction at both sites. However, at lower altitudes, due to the persistent high-pressure conditions, the winds were low, and temperature inversions on the surface and aloft were observed in Sofia in the morning and dissolved by midday.

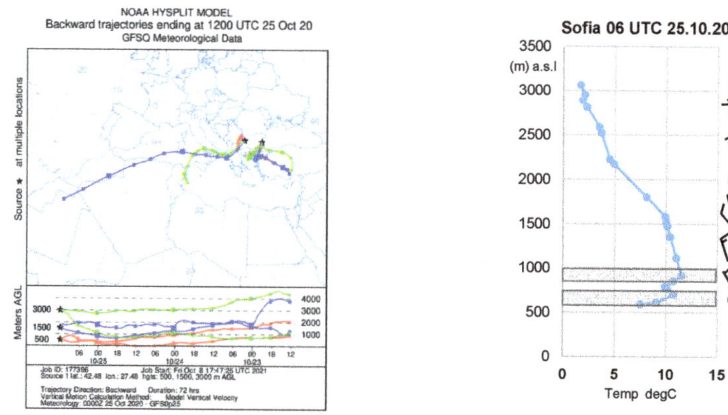

(**a**) (**b**)

Figure 14. HYSPLIT back-trajectories on 25 October 2010 (**a**); vertical profile of temperature at 06 UTC at NIMH-Sofia (**b**). Barbs indicate wind speed and direction; grey shaded rectangles are highlight inversion layers.

$PM_{2.5}$ concentrations simulated by CAMS-ENS in the morning of 25 October 2020 (at 06UTC) were evident at a height of about 1000 m a.g.l (Figure 15a). Along with the plain areas, high $PM_{2.5}$ values were noted south-west and south-east (Black Sea) of the country.

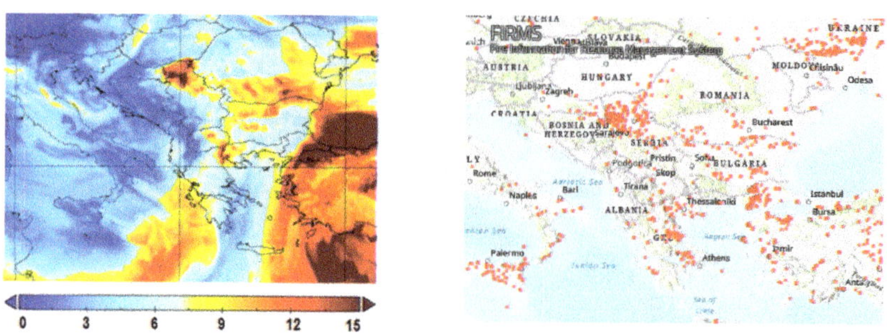

(**a**) (**b**)

Figure 15. $PM_{2.5}$ ($\mu g.m^{-3}$) from CAMS-ENS on 25 October 2020 06UTC, at height 1000 m a.g.l. (**a**); fires in the period 23–25 October 2020 as detected by satellites, NASA-FIRMS service- created from https://earthdata.nasa.gov/firms (accessed on 30 September 2021) (**b**).

Wildfires in the period 23–25 October 2020 were located south and west of Bulgaria, as detected by satellites and mapped by the NASA FIRMS (Fire Information for Resource Management System) service [73] (Figure 15b).

The analyses above suggest that the increased concentrations on 25 October 2010 were most likely due local emission sources in anti-cyclone atmospheric conditions over the country and an approaching perturbation from the west and the south-west, which led to a flow from regions affected by wildfires.

Episode 2: 20–21 January 2021

This episode was characterized by high observed BC and $PM_{2.5}$ concentrations in Sofia, with a maximum for the month recorded on 21 January 2021. The observations in Burgas showed the highest values on 20 January 2021 (Figure 16). Contrary to the previous case, the modelled $PM_{2.5}$ concentrations were much lower than the observed ones. The observed daily mean BC and $PM_{2.5}$ concentrations in Sofia on 21 January 2021 reached values by more than 50% higher than on 20 January 2021, whereas the model indicated less than 10% change for these two days.

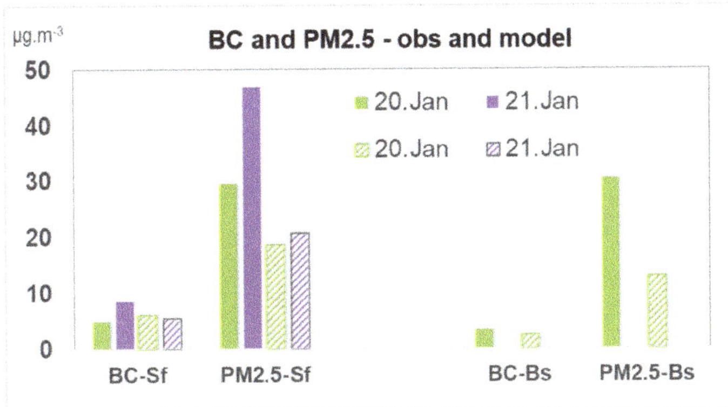

Figure 16. BC and $PM_{2.5}$ daily mean concentrations ($\mu g.m^{-3}$) observed (solid fill) and modelled (hatched) on 20 January 2021 (green) and 21 January 2021 (purple).

The synoptic situation in the period 15–19 January 2021 was characterised by a deep upper trough, intensifying over northern and eastern Europe and leading to a deep surface cyclone with a centre north-east of Bulgaria (southern Ukraine) [75]. The cold spell led to daily mean temperatures in Sofia during this period between −2 °C and −7 °C. On 20 January 2021 and 21 January 2021, a ridge of high pressure from northern Africa extended towards the country; the upper air winds (3000 m a.g.l) were from the west and the north-west (Figure 17). This led to milder conditions and an increase in the daily temperatures of about 1 °C in Sofia. It also resulted in temperature inversions, affecting the air quality in the Sofia valley.

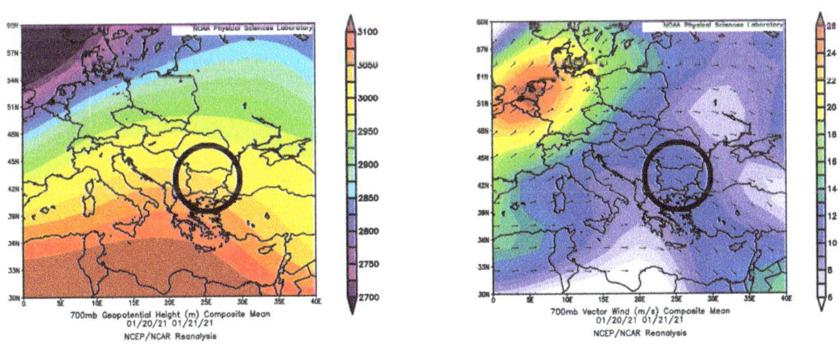

(a) (b)

Figure 17. NCEP/NCAR reanalysis composite map for 20–21 January 2021: (**a**) geopotential height (m) at 700 hPa, (**b**) wind vectors and speed (ms^{-1}) at 700 hPa. Image provided by the NOAA/ESRL Physical Sciences Laboratory, Boulder Colorado from their Web site at http://psl.noaa.gov/ (accessed on 15 October 2021). Bulgaria is marked with a circle.

The sounding data for 21 January 2021 in Sofia indicate an inversion on 06 UTC, not only on the surface but also aloft (Figure 18). This elevated inversion layer remained through noon-time.

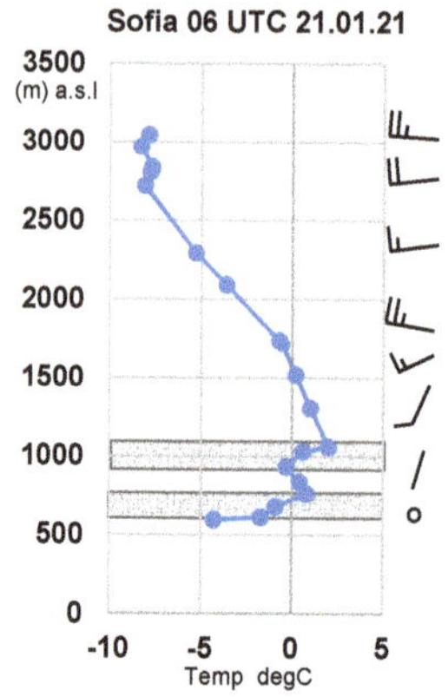

(a)

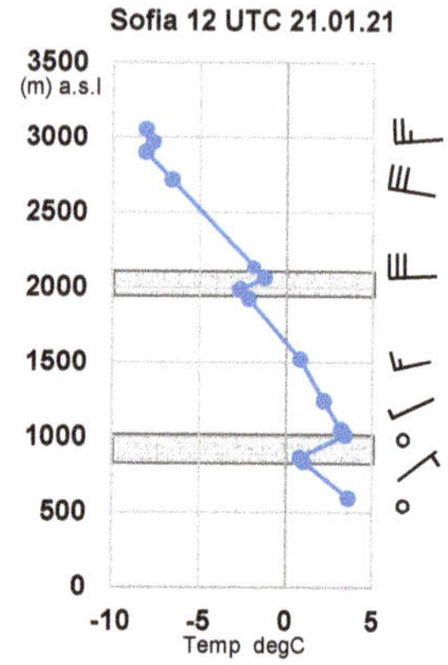

(b)

Figure 18. Vertical profile of temperature at NIMH Sofia on 21 January 2021: (**a**) 06 UTC, (**b**) 12UTC. Barbs indicate wind speed and direction; grey shaded rectangles highlight inversion layers.

The bias between model and observed $PM_{2.5}$ concentrations is approximately −55% at both sites, indicating that the model has difficulties in capturing the increase in concentrations under low wind conditions.

Modelled $PM_{2.5}$ concentrations at 06 UTC on 21 January 2021 (Figure 19) showed higher values in Sofia and over the lower plains of central Bulgaria at approximately 50 m a.g.l. At a height of about 100 m, the accumulation of pollutants around Sofia was not evident.

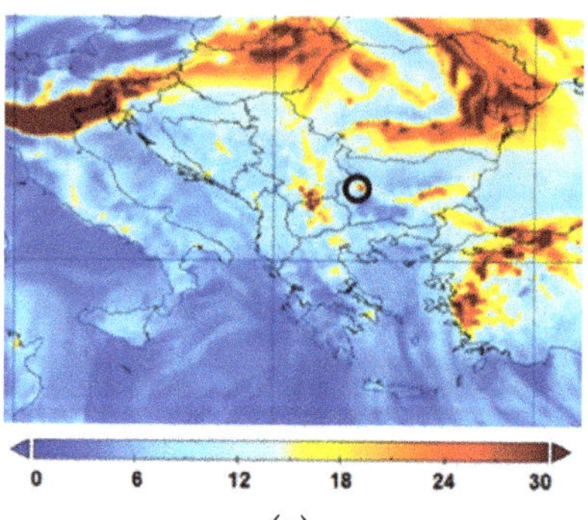

(a)

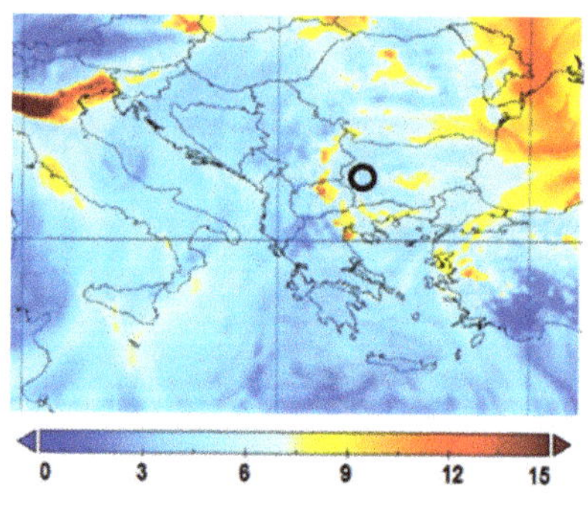

(b)

Figure 19. $PM_{2.5}$ (µg.m^{-3}) from CAMS-ENS on 21.02.2021 at 06 UTC: (**a**) at 50 m a.g.l, (**b**) at 500 m a.g.l. Sofia is marked with a circle.

4. Discussion and Conclusions

For the first time in Bulgaria, fine particles ($PM_{2.5}$) were analyzed for BC content. The reported results refer to data from field campaigns (October 2020 and January 2021) carried out in the cities of Sofia and Burgas. The mean contribution of BC to $PM_{2.5}$ was estimated to be about 20% in Sofia and 11% in Burgas. However, on some days, BC concentrations might have contributed up to 34% of $PM_{2.5}$ in Sofia and up to 19% of $PM_{2.5}$ in Burgas. The mean BC concentrations for the two months were 3.03 µg.m^{-3} in Sofia and 1.69 µg.m^{-3} in Burgas. The higher concentrations in Sofia, the largest city in the country, are linked both to more emissions (predominantly from road transport and biomass burning for heating) and to the particular topographic and meteorological conditions in the two cities. In general, wind speeds in Sofia are lower than at seaside Burgas. This affected the dispersion, especially in winter. At the same time, during the winter, the lower temperatures in Sofia, compared to Burgas, led to more emissions from household heating. The effects of the local and regional meteorological conditions on BC and $PM_{2.5}$ concentrations were investigated for two episodes with high daily mean BC concentrations (approximately 4.4–8.6 $\mu g.m^{-3}$ in Sofia). The analysis showed that whereas the episode in October might have been influenced by regional airflow from areas with wildfires, the one in January was associated with temperature inversions.

The comparison of the obtained BC results in Sofia and those reported for other European cites shows values close to those measured in urban background stations in Madrid [25], in a suburban station of Athens [22], and in Ostrava [27]. These values are lower than those obtained in urban background stations in Zabrze, Poland [21]. The mean BC concentrations in Burgas were generally lower compared to the values recorded at most urban or urban background stations and closer to those for Helsinki [23] and Athens [22].

The maps for the spatial distribution of BC and $PM_{2.5}$ over the country, constructed using model data by CAMS-ENS [41], revealed common features for the two months in the areas with elevated concentrations around Sofia, the Thracian lowlands of Southern Bulgaria, and the Danube plain in Northern Bulgaria. These areas correspond to the emissions zones where combustion from stationary sources is a prevailing contributor to anthropogenic $PM_{2.5}$ emissions. It must be noted that the vast areas with elevated concentrations are outside of the country.

The comparison between observed and modelled concentrations showed a mean bias of approximately 21% (in absolute value) for $PM_{2.5}$, with the highest underestimation by—42%—occurring in Burgas during January. In general, BC was underestimated by the model, especially in autumn, by approximately—34%. Other studies on evaluation of model performance for carbonaceous aerosol in Europe [35] have revealed an underestimation of seasonally averaged BC concentrations in the $PM_{2.5}$ mass fraction by a maximum of 60%. The observations in [35] had a mean value of 1.5 $\mu g.m^{-3}$ in the 2007 winter campaign and approximately 1.4 $\mu g.m^{-3}$ in the 2008 autumn campaign. These values are lower than in our study, but they are representative of background conditions as obtained at sites of the EMEP network.

The reported results and analysis can be considered as the first step Bulgaria is taking towards gaining a better understanding of BC concentration in $PM_{2.5}$ in the urban areas of the country. We have applied a variety of tools, analyzing observational data, modelling results, and meteorological parameters. They proved to be helpful in the investigation of elevated BC and PM concentrations during particular episodes.

However, more observational data for different seasons and further studies are needed to draw firm conclusions. A comparative measurement for BC concentration with MABI and AE33 on the territory of Sofia is planned.

We believe that the presented data contribute to filling in the gaps relevant to BC observations in Southeastern Europe.

Supplementary Materials: The following are available online at https://www.mdpi.com/article/10.3390/atmos13020213/s1, Figure S1. Example of log plot of wavelength data against log of all λ = 639 nm data; Figure S2. Mass absorption coefficients vs. wavelength (a) Sofia and (b) Burgas.

Author Contributions: Conceptualization, E.H. and E.G.; methodology, E.H., E.G. and B.V.; formal analysis, E.H., E.G., B.V., N.N., L.G.-M. and S.N.; writing—original draft preparation, E.H., E.G. and B.V.; visualization, E.H., E.G., N.N., R.N. and A.P.; project administration, E.H. All authors have read and agreed to the published version of the manuscript.

Funding: The research work was funded by Bulgarian National Science Fund through contract № КП-06-Н 34/9, 19 December 2019.

Institutional Review Board Statement: Not applicable.

Informed Consent Statement: Not applicable.

Data Availability Statement: The data are available on request from the corresponding author.

Acknowledgments: We acknowledge the use of data and/or imagery from NASA's Fire Information for Resource Management System (FIRMS), part of NASA's Earth Observing System Data and Information System (EOSDIS). We also acknowledge the Global Emission InitiAtive data portal (ECCAD) for the archiving and distribution of the emissions data. We are grateful to Maria Hristova, a student at the American College in Sofia, for her help with the English language editing.

Conflicts of Interest: The authors declare no conflict of interest.

References

1. European Environment Agency. *Air Quality in Europe—2020 Report*; Publications Office of the European Union: Luxembourg, 2020; ISBN 978-92-9480-292-7. [CrossRef]
2. Samek, L.; Stegowski, Z.; Furman, L.; Fiedor, J. Chemical content and estimated sources of fine fraction of particulate matter collected in Krakow. *Air Qual. Atmos. Health* **2017**, *10*, 47–52. [CrossRef] [PubMed]
3. Manousakas, M.; Diapouli, E.; Papaefthymiou, H.; Migliori, A.; Karydas, A.G.; Padilla-Alvarez, R.; Bogovac, M.; Kaiser, R.B.; Jaksic, M.; Bogdanovic-Radovic, I.; et al. Source apportionment by PMF on elemental concentrations obtained by PIXE analysis of PM10 samples collected at the vicinity of lignite power plants and mines in Megalopolis, Greece. *Nucl. Instrum. Methods Phys. Res.* **2015**, *349*, 114–124. [CrossRef]
4. Almeida, S.M.; Manousakas, M.; Diapouli, E.; Kertesz, Z.; Samek, L.; Hristova, E.; Šega, K.; Padilla Alvarez, R.; Belis, C.A.; Eleftheriadis, K. The Iaea European Region Study GROUP, Ambient particulate matter source apportionment using receptor modelling in European and Central Asia urban areas. *Environ. Pollut.* **2020**, *266*, 115–199. [CrossRef] [PubMed]
5. Putaud, J.-P.; Van Dingenen, R.; Alastuey, A.; Bauer, H.; Birmili, W.; Cyrys, J.; Flentje, H.; Fuzzi, S.; Gehrig, R.; Hansson, H.C.; et al. A European aerosol phenomenology 3: Physical and chemical characteristics of particulate matter from 60 rural, urban, and kerbside sites across. *Eur. Atmos. Environ.* **2010**, *44*, 1308–1320. [CrossRef]
6. Meteorological Synthesizing Centre-West of EMEP (MSC-W); Chemical Co-ordinating Centre (CCC); Centre on Emission Inventories and Projections (CEIP). *Transboundary Particulate Matter, Photo-Oxidants, Acidifying and Eutrophying Components*; EMEP Status Report 1/2021; Norwegian Meteorological Institute: Oslo, Norway, 2021; ISSN 1504-6109 (print).
7. Hristova, E.; Veleva, B. Variation of air particulate concentration in Sofia, 2005–2012. *Bulg. J. Meteorol. Hydrol.* **2013**, *18*, 47–56.
8. Veleva, B.; Hristova, E.; Nikolova, E.; Kolarova, M.; Valcheva, R. Statistical evaluation of elemental composition data of PM10 air particulate in Sofia. *Int. J. Environ. Pollut.* **2015**, *57*, 175–188. [CrossRef]
9. Hristova, E.; Veleva, B.; Georgieva, E.; Branzov, H. Application of Positive Matrix Factorization Receptor Model for Source Identification of PM10 in the City of Sofia, Bulgaria. *Atmosphere* **2020**, *11*, 890. [CrossRef]
10. Sandradewi, J.; Prévôt, A.S.H.; Szidat, S.; Perron, N.; Alfarra, M.R.; Lanz, V.A.; Weingartner, E.; Baltensperger, U. Using aerosol light absorption measurements for the quantitative determination of wood burning and traffic emission contributions to particulate matter. *Environ. Sci. Technol.* **2008**, *42*, 3316–3323. [CrossRef] [PubMed]
11. Gerlos-Nijland, M.; Lanki, T.; Salonen, R.; Cassee, F. *Health Effects of Black Carbon*; WHO Regional Office for Europe, The WHO European Centre for Environment and Health: Bonn, Germany, 2012.
12. Petzold, A.; Ogren, J.A.; Fiebig, M.; Laj, P.; Li, S.-M.; Baltensperger, U.; Holzer-Popp, T.; Kinne, S.; Pappalardo, G.; Sugimoto, N.; et al. Recommendations for reporting "black carbon" measurements. *Atmos. Chem. Phys.* **2013**, *13*, 8365–8379. [CrossRef]
13. Hitzenberger, R.; Petzold, A.; Bauer, H.; Ctyroky, P.; Pouresmaeil, P.; Laskus, L.; Puxbaum, H. Intercomparison of Thermal and Optical Measurement Methods for Elemental Carbon and Black Carbon at an Urban Location Environ. *Sci. Technol.* **2006**, *40*, 6377–6383. [CrossRef]
14. Peralta, O.; Ortínez-Alvarez, A.; Basaldud, R.; Santiago, N.; Alvarez-Ospina, H.; de la Cruz, K.; Barrera, V.; de la Luz Espinosa, M.; Saavedra, I.; Castro, T.; et al. Atmospheric black carbon concentrations in Mexico. *Atmos. Res.* **2019**, *230*, 104–626. [CrossRef]
15. Mousavi, A.; Sowlat, M.; Lovett, C.; Rauber, M.; Szidat, S.; Boffi, R.; Borgini, A.; De Marco, C.; Ruprecht, A.; Sioutas, C. Source apportionment of black carbon (BC) from fossil fuel and biomass burning in metropolitan Milan, Italy. *Atmos. Environ.* **2019**, *203*, 252–261. [CrossRef]
16. Bond, T.C.; Doherty, S.J.; Fahey, D.W.; Forster, P.M.; Berntsen, T.; DeAngelo, B.J.; Flanner, M.G.; Ghan, S.; Kärcher, B.; Koch, D.; et al. Bounding the role of black carbon in the climate system: A scientific assessment. *JGA Atmos.* **2013**, *118*, 5380–5552. [CrossRef]
17. World Health Organization (WHO). *WHO Global Air Quality Guidelines: Particulate Matter (PM2.5 and PM10), Ozone, Nitrogen Dioxide, Sulfur Dioxide and Carbon Monoxide*; World Health Organization: Geneva, Switzerland, 2021.
18. Manoj, M.; Satheesh, S.; Moorthy, K.; Gogoi, M.; Babu, S. Decreasing trend in black carbon aerosols over the Indian region. *Geophys. Res. Lett.* **2019**, *46*, 2903–2910. [CrossRef]
19. Bessagnet, B.; Allemand, N. *Review on Black Carbon (BC) and Polycyclic Aromatic Hydrocarbons (PAHs) Emission Reductions Induced by PM Emission Abatement Techniques*; TFTEI Techno-Scientific Secretariat: Paris, France, 2020; p. 156.
20. Kunder, R.D.; Schneidemesser, E.; Kuik, F.; Quedenau, J.; Weatherhead, E.C.; Schmale, J. Long-term monitoring of black carbon across Germany. *Atmos. Environ.* **2018**, *185*, 41–52.

21. Zioła, N.; Błaszczak, B.; Klejnowski, K. Long-Term eBC Measurements with the Use of MAAP in the Polluted Urban Atmosphere (Poland). *Atmosphere* **2021**, *12*, 808. [CrossRef]
22. Diapouli, E.; Kalogridis, A.-C.; Markantonaki, C.; Vratolis, S.; Fetfatzis, P.; Colombi, C.; Eleftheriadis, K. Annual Variability of Black Carbon Concentrations Originating from Biomass and Fossil Fuel Combustion for the Suburban Aerosol in Athens, Greece. *Atmosphere* **2017**, *8*, 234. [CrossRef]
23. Helin, A.; Niemi, J.V.; Virkkula, A.; Pirjola, L.; Teinilä, K.; Backman, J.; Aurela, M.; Saarikoski, S.; Rönkkö, T.; Asmi, E.; et al. Characteristics and source apportionment of black carbon in the Helsinki metropolitan area, Finland. *Atmos. Environ.* **2018**, *190*, 87–98. [CrossRef]
24. Beekmann, M.; Prévôt, A.S.H.; Drewnick, F.; Sciare, J.; Pandis, S.N.; Denier van der Gon, H.A.C.; Crippa, M.; Freutel, F.; Poulain, L.; Ghersi, V.; et al. In Situ, satellite measurement and model evidence on the dominant regional contribution to fine particulate matter levels in the Paris megacity. *Atmos. Chem. Phys.* **2015**, *15*, 9577–9591. [CrossRef]
25. Becerril-Valle, M.; Coz, E.; Prévôt, A.S.H.; Mocnik, G.; Pandis, S.N.; Sánchez de la Campa, A.M.; Alastuey, A.; Díaz, E.; Pérez, R.M.; Artíñano, B. Characterization of atmospheric black carbon and co-pollutants in urban and rural areas of Spain. *Atmos. Environ.* **2017**, *169*, 36–53. [CrossRef]
26. Kucbel, M.; Corsaro, A.; Švédová, B.; Raclavská, H.; Raclavský, K.; Juchelková, D. Temporal and seasonal variations of black carbon in a highly polluted European city: Apportionment of potential sources and the effect of meteorological conditions. *J. Environ. Manag.* **2017**, *203*, 1178–1189. [CrossRef] [PubMed]
27. Bernardoni, V.; Ferrero, L.; Bolzacchini, E.; Forello, A.C.; Gregorič, A.; Massabò, D.; Močnik, G.; Prati, P.; Rigler, M.; Santagostini, L.; et al. Determination of Aethalometer multiple-scattering enhancement parameters and impact on source apportionment during the winter 2017/18 EMEP/ACTRIS/COLOSSAL campaign in Milan. *Atmos. Meas. Tech.* **2021**, *14*, 2919–2940. [CrossRef]
28. Krol, M.; Houweling, S.; Bregman, B.; van den Broek, M.; Segers, A.; van Velthoven, P.; Peters, W.; Dentener, F.; Bergamaschi, P. The two-way nested global chemistry-transport zoom model TM5: Algorithm and applications. *Atmos. Chem. Phys.* **2005**, *5*, 417–432. [CrossRef]
29. Gelaro, R.; McCarty, W.; Suárez, M.J.; Todling, R.; Molod, A.; Takacs, L.; Randles, C.; Darmenov, A.; Bosilovich, M.G.; Reichle, R.; et al. The Modern-Era Retrospective Analysis for Research and Applications, Version 2 (MERRA-2). *J. Clim.* **2017**, *30*, 5419–5454. [CrossRef]
30. Schaap, M.; Van Der Gon, H.A.C.D.; Dentener, F.J.; Visschedijk, A.J.H.; Van Loon, M.; ten Brink, H.M.; Putaud, J.P.; Guillaume, B.; Liousse, C.; Builtjes, P.J.H. Anthropogenic black carbon and fine aerosol distribution over Europe. *J. Geophys. Res.* **2004**, *109*, D18207. [CrossRef]
31. Tsyro, S.; Simpson, D.; Tarrason, L.; Klimont, Z.; Kupiainen, K.; Pio, C.; Yttri, K. Modeling of elemental carbon over Europe. *J. Geophys. Res.* **2007**, *112*, D23S19. [CrossRef]
32. Ramanathan, V.; Carmichael, G. Global and regional climate changes due to black carbon. *Nat. Geosci.* **2008**, *1*, 221–227. [CrossRef]
33. Vignati, E.; Karl, M.; Krol, M.; Wilson, J.; Stier, P.; Cavalli, F. Sources of uncertainties in modelling black carbon at the global scale. *Atmos. Chem. Phys.* **2010**, *10*, 2595–2611. [CrossRef]
34. Prank, M.; Sofiev, M.; Tsyro, S.; Hendriks, C.; Semeena, V.; Vazhappilly Francis, X.; Butler, T.; Denier van der Gon, H.; Friedrich, R.; Hendricks, J.; et al. Evaluation of the performance of four chemical transport models in predicting the aerosol chemical composition in Europe in 2005. *Atmos. Chem. Phys.* **2016**, *16*, 6041–6070. [CrossRef]
35. Mircea, M.; Bessagnet, B.; D'Isidoro, M.; Pirovano, G.; Aksoyoglu, S.; Ciarelli, G.; Tsyro, S.; Manders, A.; Bieser, J.; Stern, R. EURODELTA III exercise: An evaluation of air quality models' capacity to reproduce the carbonaceous aerosol. *Atmos. Environ. X* **2019**, *2*, 100018. [CrossRef]
36. Kuik, F.; Lauer, A.; Beukes, J.P.; Van Zyl, P.G.; Josipovic, M.; Vakkari, V.; Laakso, L.; Feig, G.T. The anthropogenic contribution to atmospheric black carbon concentrations in southern Africa: A WRF-Chem modeling study. *Atmos. Chem. Phys.* **2015**, *15*, 8809–8830. [CrossRef]
37. Morino, Y.; Nagashima, T.; Sugata, S.; Sato, K.; Tanabe, K.; Noguchi, T.; Takami, A.; Tanimoto, H.; Ohara, T. Verification of Chemical Transport Models for $PM_{2.5}$ Chemical Composition Using Simultaneous Measurement Data over Japan. *Aerosol Air Qual. Res.* **2015**, *15*, 2009–2023. [CrossRef]
38. Permadi, D.A.; Kim Oanh, N.T.; Vautard, R. Integrated emission inventory and modeling to assess distribution of particulate matter mass and black carbon composition in Southeast Asia. *Atmos. Chem. Phys.* **2018**, *18*, 2725–2747. [CrossRef]
39. Hristova, E.; Veleva, B. Estimation of black carbon concentration in fine particulate matter in urban area. In Proceedings of the 20th International Multidisciplinary Scientific GeoConference SGEM 2020, Albena, Bulgaria, 16–25 August 2020; Volume 20, pp. 415–422.
40. Hristova, E.; Veleva, B.; Naydenova, S.; Gonsalvesh-Musakova, L. Air particulate matter and black carbon concentrations during winter time at two Bulgarian urban sites. In Proceedings of the 21st International Multidisciplinary Scientific GeoConference SGEM 2021, Albena, Bulgaria, 14–22 August 2021. in press.
41. Copernicus Atmosphere Monitoring Service. Available online: https://atmosphere.copernicus.eu/ (accessed on 18 October 2021).
42. National Statistical Institute. Available online: https://www.nsi.bg/en/content/6710/population-towns-and-sex (accessed on 1 June 2021).

43. Hak, C.; Sivertsen, B. *Mission Report. Burgas, 1–12 March 2010—Screening Study*; NILU OR, Scientific Reports—OR 39/2010; Norwegian Institute for Air Research: Kjeller, Norway, 2010; ISBN 978-82-425-2238-2.
44. National State of the Environment Report. Available online: http://eea.government.bg/en/output/soe-report/index.html (accessed on 10 October 2021).
45. EMEP Centre on Emission Inventories and Projections. Data Viewer—Reported Emissions Data. Available online: https://www.ceip.at/data-viewer (accessed on 10 October 2021).
46. Program for Improving the Quality of the Atmospheric Air on the Territory of Sofia Municipality for the Period 2021–2026. Available online: https://www.sofia.bg/en/programa-kav (accessed on 5 October 2021).
47. Program for Improving the Quality of the Atmospheric Air on the Territory of Burgas Municipality for the Period 2021–2026. Available online: https://www.burgas.bg/bg/programi/ (accessed on 5 October 2021).
48. Dimitrova, R.; Velizarova, M. Assessment of the Contribution of Different Particulate Matter Sources on Pollution in Sofia City. *Atmosphere* **2021**, *12*, 423. [CrossRef]
49. Manohar, M.; Atanacio, A.; Button, D.; Cohen, D. MABI—A multi-wavelength absorption black carbon instrument for the measurement of fine light absorbing carbon particles. *Atmos. Pollut. Res.* **2021**, *12*, 133–140. [CrossRef]
50. Cohen, D.D. *Summary of Light Absorbing Carbon and Visibility Measurements and Terms*; ANSTO/External Report ER-790; Australian Nuclear Science and Technology Organisation: Sydney, Australia, 2020; ISBN 1-921268-32-8.
51. Leskinen, A.; Ruuskanen, A.; Kolmonen, P.; Zhao, Y.; Fang, D.; Wang, Q.; Gu, C.; Jokiniemi, J.; Hirvonen, M.R.; Lehtinen, K.E.; et al. The Contribution of Black Carbon and Non-BC Absorbers to the Aerosol Absorption Coefficient in Nanjing, China. *Aerosol Air Qual. Res.* **2020**, *20*, 590–605. [CrossRef]
52. Alas, H.D.C.; Müller, T.; Weinhold, K.; Pfeifer, S.; Glojek, K.; Gregorič, A.; Močnik, G.; Drinovec, L.; Costabile, F.; Ristorini, M.; et al. Performance of microAethalometers: Real-world Field Intercomparisons from Multiple Mobile Measurement Campaigns in Different Atmospheric Environments. *Aerosol Air Qual. Res.* **2020**, *20*, 2640–2653. [CrossRef]
53. Bond, T.C.; Anderson, T.L.; Campbell, D. Calibration and intercomparison of filter-based measurements of visible light absorption by aerosols. *Aerosol Sci. Technol.* **1999**, *30*, 582–600. [CrossRef]
54. Ogren, J.A.; Wendell, J.; Andrews, E.; Sheridan, P.J. Continuous light absorption photometer for long-term studies. *Atmos. Meas. Tech.* **2017**, *10*, 4805–4818. [CrossRef]
55. Weingartnera, E.; Saatho, H.; Schnaiter, M.; Streit, N.; Bitnar, B.; Baltensperger, U. Absorption of light by soot particles: Determination of the absorption coeffcient by means of aethalometers. *Aerosol Sci.* **2003**, *34*, 1445–1463. [CrossRef]
56. Virkkula, A.; Ahlquist, N.; Covert, D.; Arnott, W.; Sheridan, P.J.; Quinn, P.; Coffman, D. Modification, Calibration and a Field Test of an Instrument for Measuring Light Absorption by Particles. *Aerosol Sci. Technol.* **2005**, *39*, 68–83. [CrossRef]
57. Virkkula, A.; Mäkelä, T.; Hillamo, R.; Yli-Tuomi, T.; Hirsikko, A.; Hämeri, K.; Koponen, I. A Simple Procedure for Correcting Loading Effects of Aethalometer Data. *J. Air Waste Manag. Assoc.* **2007**, *57*, 1214–1222. [CrossRef] [PubMed]
58. Drinovec, L.; Močnik, G.; Zotter, P.; Prévôt, A.S.H.; Ruckstuhl, C.; Coz, E.; Rupakheti, M.; Sciare, J.; Müller, T.; Wiedensohler, A.; et al. The "dual-spot" Aethalometer: An improved measurement of aerosol black carbon with real-time loading compensation. *Atmos. Meas. Tech.* **2015**, *8*, 1965–1979. [CrossRef]
59. Drinovec, L.; Gregorič, A.; Zotter, P.; Wolf, R.; Bruns, E.; Prévôt, A.S.H.; Petit, J.-E.; Favez, O.; Sciare, J.; Arnold, I.J.; et al. The filter-loading effect by ambient aerosols in filter absorption photometers depends on the coating of the sampled particles. *Atmos. Meas. Tech.* **2017**, *10*, 1043–1059. [CrossRef]
60. World Meteorological Organization (WMO); Global Atmosphere Watch (GAW). *WMO/GAW Aerosol Measurement Procedures, Guidelines and Recommendations*, 2nd ed.; GAW Report No. 227; World Meteorological Organization: Geneva, Switzerland, 2016; ISBN 978-92-63-11177-7.
61. Arnott, W.; Hamasha, K.; Moosmüller, H.; Sheridan, P.; Ogren, J. Towards Aerosol Light-Absorption Measurements with a 7-Wavelength Nethalometer: Evaluation with a Photoacoustic Instrument and 3-Wavelength Nephelometer. *Aerosol Sci. Technol.* **2005**, *39*, 17–29. [CrossRef]
62. Segura, S.; Estellés, V.; Titos, G.; Lyamani, H.; Utrillas, M.P.; Zotter, P.; Prévôt, A.S.H.; Močnik, G.; Alados-Arboledas, L.; Martínez-Lozano, J.A. Determination and analysis of in situ spectral aerosol optical properties by a multi-instrumental approach. *Atmos. Meas. Tech.* **2014**, *7*, 2373–2387. [CrossRef]
63. Marécal, V.; Peuch, V.-H.; Andersson, C.; Andersson, S.; Arteta, J.; Beekmann, M.; Benedictow, A.; Bergström, R.; Bessagnet, B.; Cansado, A.; et al. A regional air quality forecasting system over Europe: The MACC-II daily ensemble production. *Geosci. Model Dev.* **2015**, *8*, 2777–2813. [CrossRef]
64. Rémy, S.; Kipling, Z.; Flemming, J.; Boucher, O.; Nabat, P.; Michou, M.; Bozzo, A.; Ades, M.; Huijnen, V.; Benedetti, A.; et al. Description and evaluation of the tropospheric aerosol scheme in the European Centre for Medium-Range Weather Forecasts (ECMWF) Integrated Forecasting System (IFS-AER, cycle 45R1). *Geosci. Model Dev.* **2019**, *12*, 4627–4659. [CrossRef]
65. Granier, C.; Darras, S.; Denier van der Gon, H.A.C.; Doubalova, J.; Elguindi, N.; Galle, B.; Gauss, M.; Guevara, M.; Jalkanen, J.-P.; Kuenen, J.; et al. *The Copernicus Atmosphere Monitoring Service Global and Regional Emissions (April 2019 Version)*; Copernicus Atmosphere Monitoring Service (CAMS) Report, 2019; Copernicus Atmosphere Monitoring Service: Reading, UK, 2019. [CrossRef]
66. Kuenen, J.; Dellaert, S.; Visschedijk, A.; Jalkanen, J.-P.; Super, I.; Denier van der Gon, H. *Copernicus Atmosphere Monitoring Service regional Emissions Version 4.2 (CAMS-REG-v4.2)*; Copernicus Atmosphere Monitoring Service: Reading, UK, 2021. [CrossRef]

67. Copernicus Atmosphere Monitoring Service (CAMS). *Regional Production, Updated Documentation Covering All Regional Operational Systems and the ENSEMBLE: Following U2 Upgrade, February 2020*; CAMS—ECMWF Report, Isuued by METEO-FRANCE/G. Collin, CAMS50_2018SC2_D2.0.2-U2_Models_documentation_202003_v2; Copernicus Atmosphere Monitoring Service: Reading, UK, 2020.
68. CAMS European Air Quality Forecasts, ENSEMBLE Data; Copernicus Atmosphere Monitoring Service (CAMS), Atmosphere Data Store (ADS). Available online: https://ads.atmosphere.copernicus.eu/cdsapp#!/dataset/cams-europe-air-quality-forecasts?tab=overview (accessed on 11 October 2021).
69. Stein, A.F.; Draxler, R.R.; Rolph, G.D.; Stunder, B.J.B.; Cohen, M.D.; Ngan, F. NOAA's HYSPLIT atmospheric transport and dispersion modeling system. *Bull. Am. Meteor. Soc.* **2015**, *96*, 2059–2077. [CrossRef]
70. Rolph, G.; Stein, A.; Stunder, B. Real-time Environmental Applications and Display sYstem: READY. Environ. *Model. Softw.* **2017**, *95*, 210–228. [CrossRef]
71. Kalnay, E.; Kanamitsu, M.; Kistler, R.; Collins, W.; Deaven, D.; Gandin, L.; Iredell, M.; Saha, S.; White, G.; Woollen, J.; et al. The NCEP/NCAR 40-Year Reanalysis Project. *Bull. Am. Meteorol. Soc.* **1996**, *77*, 437–472. [CrossRef]
72. National Oceanic and Atmospheric Administration, Earth System Research Laboratories (NOAA/ESRL). Available online: https://psl.noaa.gov/data/composites/day/ (accessed on 15 October 2021).
73. Emissions of Atmospheric Compounds and Compilation of Ancillary Data (ECCAD). Available online: https://eccad3.sedoo.fr/#CAMS-REG-AP (accessed on 15 October 2021).
74. National Institute of Meteorology and Hydrology of Bulgaria. *Monthly Hydrometeorological Bulletin*; National Institute of Meteorology and Hydrology of Bulgaria: Sofia, Bulgaria, 2021; ISSN 1314-894X. Available online: http://www.meteo.bg/ (accessed on 15 October 2021).
75. Fire Information for Resource Management System Service. Available online: https://firms.modaps.eosdis.nasa.gov (accessed on 5 October 2021).

atmosphere

MDPI

Article

Organic and Elemental Carbon in the Urban Background in an Eastern Mediterranean City

Tareq Hussein [1,2,*], Xinyang Li [2], Zaid Bakri [1,3], Andres Alastuey [4], Sharif Arar [5], Afnan Al-Hunaiti [5], Mar Viana [4] and Tuukka Petäjä [2]

1 Department of Physics, School of Science, The University of Jordan, Amman 11942, Jordan; zbakri@mtu.edu or zyd8190198@ju.edu.jo
2 Institute for Atmospheric and Earth System Research (INAR/Physics), University of Helsinki, FI-00014 Helsinki, Finland; xinyang.li@helsinki.fi (X.L.); tuukka.petaja@helsinki.fi (T.P.)
3 Physics Department and Atmospheric Sciences Program, Michigan Technological University, Houghton, MI 49931, USA
4 Institute of Environmental Assessment and Water Research (IDAEA-CSIC), 08034 Barcelona, Spain; andres.alastuey@idaea.csic.es (A.A.); mar.viana@idaea.csic.es (M.V.)
5 Department of Chemistry, School of Science, The University of Jordan, Amman 11942, Jordan; s.arar@ju.edu.jo (S.A.); a.alhunaiti@ju.edu.jo (A.A.-H.)
* Correspondence: tareq.hussein@helsinki.fi or t.hussein@ju.edu.jo

Abstract: The Mediterranean region is an important area for air pollution as it is the crossroads between three continents; therefore, the concentrations of atmospheric aerosol particles are influenced by emissions from Africa, Asia, and Europe. Here we concentrate on an eleven-month time series of the ambient concentration of organic carbon (OC) and elemental carbon (EC) between May 2018–March 2019 in Amman, Jordan. Such a dataset is unique in Jordan. The results show that the OC and EC annual mean concentrations in $PM_{2.5}$ samples were 5.9 ± 2.8 μg m^{-3} and 1.7 ± 1.1 μg m^{-3}, respectively. It was found that the majority of OC and EC concentrations were within the fine particle fraction ($PM_{2.5}$). During sand and dust storm (SDS) episodes OC and EC concentrations were higher than the annual means; the mean values during these periods were about 9.6 ± 3.5 μg m^{-3} and 2.5 ± 1.2 μg m^{-3} in the $PM_{2.5}$ samples. Based on this, the SDS episodes were identified to be responsible for an increased carbonaceous aerosol content as well as $PM_{2.5}$ and PM_{10} content, which may have direct implications on human health. This study encourages us to perform more extensive measurements during a longer time period and to include an advanced chemical and physical characterization for urban aerosols in the urban atmosphere of Amman, which can be representative of other urban areas in the region.

Keywords: urban air quality; PM_{10}; $PM_{2.5}$; OC; EC

Citation: Hussein, T.; Li, X.; Bakri, Z.; Alastuey, A.; Arar, S.; Al-Hunaiti, A.; Viana, M.; Petäjä, T. Organic and Elemental Carbon in the Urban Background in an Eastern Mediterranean City. *Atmosphere* **2022**, *13*, 197. https://doi.org/10.3390/atmos13020197

Academic Editors: Elena Hristova, Manousos Ioannis Manousakas, Anikó Angyal and Maria Gini

Received: 2 December 2021
Accepted: 24 January 2022
Published: 26 January 2022

Publisher's Note: MDPI stays neutral with regard to jurisdictional claims in published maps and institutional affiliations.

1. Introduction

Carbonaceous aerosols found in particulate matter (PM) are mainly in the form of elemental carbon (EC) and organic carbon (OC) [1–8]. These species of aerosols are of worldwide interest due to their vague origins and complicated source apportionment process [9–11]. On the one hand, incomplete combustion processes and wildfires are the major sources of EC [12,13]. On the other hand, OC sources potentially originate from processes that involve chemical reactions of hydrocarbons [14]. There is a contrast between the effects of OC and EC on the climate; EC is involved in the global warming effect due to its strong light-absorption property [15,16], whereas OC is responsible for cooling the atmosphere mainly because it reflects solar radiation [17]; however, some recent publications have reported that some OC (a newly emerged phrase: brown carbon) can significantly absorb light in the region of 300–400 nm and could hinder and oppose the general cooling action [18–21]. In terms of health effects, it has been suggested that increased mortality rates and respiratory diseases are related to OC and EC content exposure [22–32].

In urban areas, OC and EC originate from many sources, both local and regional. They can be transported over long distances, reaching thousands of kilometers away from their source [33–42]. For instance, OC comprises thousands of individual molecules that can be directly emitted as primary emissions or can be formed in the atmosphere from semi-volatile and gaseous precursors over the course of minutes to days. EC is directly emitted from combustion processes, such as mobile sources or biomass burning.

The Mediterranean basin, including the Eastern Mediterranean region, is considered a climate change hotspot due to warming tendencies and decreased precipitation processes [43]. The Eastern Mediterranean region, especially Jordan, is impacted by anthropogenic emissions as well as natural sources (e.g., sand and dust storm episodes (SDS)), which are found to affect PM concentrations in the region [44]. Since there is a lack of information on OC and EC concentrations and ratios in the region, there is a need to monitor and explore their aerosol concentrations, sources, and compositions in this region. Chemical analysis of ambient PM enables the identification of aerosol sources and addressing the relative contributions of different processes. In this study, we aim at characterizing PM_{10} and $PM_{2.5}$ with respect to OC and EC during May 2018–March 2019 in Amman, which is a typical city in the region. This study is important to the Middle East and North Africa region (MENA), especially Jordan, where the measurements took place, as it provides background information for urban aerosol chemical composition. This can be utilized to explore aerosol impacts on climate and health in follow-up analyses.

2. Materials and Methods

2.1. Aerosol Measurement

The measurement campaign took place during May 2018–March 2019 on the rooftop (about 20 m above the ground) of the Department of Physics at the campus of the University of Jordan (32°0129′ N, 35°8738′ E) (Figure S1). This was classified as an urban background in the northern part of Amman, Jordan. The surroundings are a mixture of residential areas and road networks [44].

The aerosol measurement instrumentation included two high-volume samplers (model CAV-A/mb, MCV, S.A., Barcelona, Spain) for PM_{10} and $PM_{2.5}$. The cascade head (model PM1025-CAV, MCV, S.A., Barcelona, Spain) was equipped with a filter (Pallflex, PAL-LXQ250ETDS0150, TISSUQUARTZ 2500 QAT-UP, Merek, New Jersy, USA), which has a diameter of 15 cm. The flow rate was 30 m^3 h^{-1} and the sampler automatically recorded the overall mean temperature and pressure during the sampling session.

The PM_{10} and $PM_{2.5}$ sampling was performed for 24 h every 6 days. We obtained 51 and 48 valid samples for PM_{10} and $PM_{2.5}$, respectively. We also collected several blank samples, which were needed as an accuracy control of the sampling [45].

2.2. Gravimetric and OC/EC Chemical Analysis

Before performing the chemical analyses (including organic carbon (OC) and elemental carbon (EC)), the PM_{10} and $PM_{2.5}$ mass concentrations were determined by gravimetric analysis, which was performed according to the EN1234-1. Accordingly, the particulate matter concentration can be calculated from the filter's weights (difference between post-weight (m_{post}) and pre-weight (m_{pre})) divided by the sampling flow rate (Q (30 m^3 h^{-1})) and sampling period (Δt = 24 h).

A quarter of each sampled filter was taken to the OC and EC analysis according to the EUSAAR2 protocol employing a Sunset Laboratory Dual-Optical Carbonaceous Analyzer [46–48]. The uncertainty in our analysis was approximately 0.2, 0.1, 0.3 μg m^{-3}, respectively, for OC, EC, and TC.

2.3. Ambient Conditions and Air Mass Trajectories Measurement

In addition to the aerosol measurement, the ambient conditions (T, P, RH, wind speed, and direction) were monitored with 5 min resolution (Table S1, Figures S2, S3, and S5–S8) by using a weather station (WH-1080, Clas Ohlson: Art. no. 36-3242). We also calculated the air mass back-trajectories (Hybrid Single-Particle Lagrangian Integrated Trajectory (HYSPLIT) [49,50]. The trajectories were calculated for the previous four days on an hourly basis at arrival heights 100, 500, and 1500 m above ground level.

3. Results and Discussion

3.1. An Overview of PM Concentrations

Throughout the measurement period, the 24 h PM_{10} was within the range 20–190 μg m^{-3} (average 64 ± 39 μg m^{-3}) and the $PM_{2.5}$ was 15–190 μg m^{-3} (average 47 ± 32 μg m^{-3})—see Table S2 in the Supplementary Material. On average, approximately 80% of the PM_{10} was within the $PM_{2.5}$ fraction. According to the Jordanian standards of ambient air quality (JS-1140/2006: annual mean PM_{10} and $PM_{2.5}$ must not exceed 70 μg m^{-3} and 15 μg m^{-3} and 24 h mean must not exceed 120 μg m^{-3} and 65 μg m^{-3}, respectively), the observed overall mean PM_{10} was below its annual limit value but the $PM_{2.5}$ was three times higher than its limit value. Compared to the World Health Organization (WHO) old air quality guidelines [51] (i.e., before 2021) for PM_{10} (annual and 24 h must not exceed 20 μg m^{-3} and 50 μg m^{-3}, respectively) and $PM_{2.5}$ (annual and 24 h must not exceed 10 μg m^{-3} and 25 μg m^{-3}, respectively), the observed annual concentrations here exceeded the annual guideline. By all means, the reported values here exceeded the new WHO air quality guidelines [52] (i.e., after 2021), which was updated to be tighter than the old guidelines for PM_{10} (annual and 24 h must not exceed 15 μg m^{-3} and 45 μg m^{-3}, respectively) and $PM_{2.5}$ (annual and 24 h must not exceed 5 μg m^{-3} and 15 μg m^{-3}, respectively). In general, the reported PM_{10} concentrations in Jordan were higher than the concentrations reported by the WHO (2018) in urban, suburban, and residential sites in countries around the Mediterranean Sea in 2016, especially Turkey (52 ± 18 μg m^{-3}), Italy (25 ± 6 μg m^{-3}), Greece (52 ± 18 μg m^{-3}), Cyprus (37 ± 6 μg m^{-3}), and Malta (38 ± 8 μg m^{-3}). These concentrations were lower than concentrations observed in the following regions: Kuwait (130 ± 35 μg m^{-3}), Palestine (90 μg m^{-3}), Egypt (249–284 μg m^{-3}), and the United Arab Emirates (122–153 μg m^{-3}).

3.2. Organic and Elemental Carbon Concentrations

The TC and OC concentrations followed a rather similar temporal variation as that for the PM concentrations (Figures 1 and 2). The OC concentrations observed in the $PM_{2.5}$ samples were in the range 1.2–17.1 μg m^{-3} (annual mean 5.9 ± 2.8 μg m^{-3}), the EC concentrations were 0.45–6.1 μg m^{-3} (annual mean of 1.7 ± 1.1 μg m^{-3}), and the TC concentrations were 1.7–23.2 μg m^{-3} (annual mean 7.6 ± 3.5 μg m^{-3}). As for the PM_{10}, they were 2.2–17.5 μg m^{-3} (annual mean of 6.5 ± 3.0 μg m^{-3}), 0.5–5.5 μg m^{-3} (annual mean of 1.9 ± 1.1 μg m^{-3}), and 2.7–22.4 μg m^{-3} (annual mean 8.4 ± 3.8 μg m^{-3}), respectively, for the OC, EC, and TC concentrations. On average, the $PM_{2.5}$ contained about 14%, 4%, and 18% OC, EC, and TC (Figure 3). As for the PM_{10}, it was about 12%, 4%, 15%, respectively. As expected for an urban background, the TC was dominated by OC for both the $PM_{2.5}$ and PM_{10} (Figure 4); this indicates a dominating fraction of organic emissions from anthropogenic activities rather than natural emissions, as also reported elsewhere [53–60].

According to $PM_{2.5}$ observations elsewhere in the region (mostly outside of Jordan; see Table 1), the OC and EC concentrations reported herein are comparable. A greater interest might be the extremely high OC concentrations in Beijing (29.1 μg m^{-3}) and Tehran (15.35 ± 6.05 μg m^{-3}) with corresponding $PM_{2.5}$ concentrations of 115 μg m^{-3} and 41.2 μg m^{-3}, respectively [61,62]. As for PM_{10} observations elsewhere (Table 2), the OC concentrations reported here are generally lower than those reported in other regions in the world.

The PM_{10} records in Table 2 show a clear contrast with the OC concentrations. A relatively high OC_{10} concentration record was observed in Lahore, with a value of 63 μg m^{-3}, while a very low concentration was observed in Spain, with a value of 4 μg m^{-3}. For both $PM_{2.5}$ and PM_{10} the OC to EC ratio was generally higher in the summertime than that in the wintertime (Figure 5). The OC/EC ratio in the $PM_{2.5}$ was approximately 5.6 ± 1.5

during May–August and was approximately 2.8 ± 0.9 during November–March. As for the PM_{10}, the ratio was approximately 4.9 ± 1.2 and 3.1 ± 1.4, respectively.

During the measurement period, the ratio $PM_{2.5}/PM_{10}$ was close to one on some days. This was basically due to the domination of fine aerosols, i.e., the absence of sand and dust storm (SDS) episodes. On these occasions, the $OC_{2.5}$ was 4.0–17.1 μg m^{-3} and the OC_{10} was 2.4–15.9 μg m^{-3}. The corresponding $EC_{2.5}$ was 0.7–6.1 μg m^{-3} and EC_{10} was 0.9–5.5 μg m^{-3}.

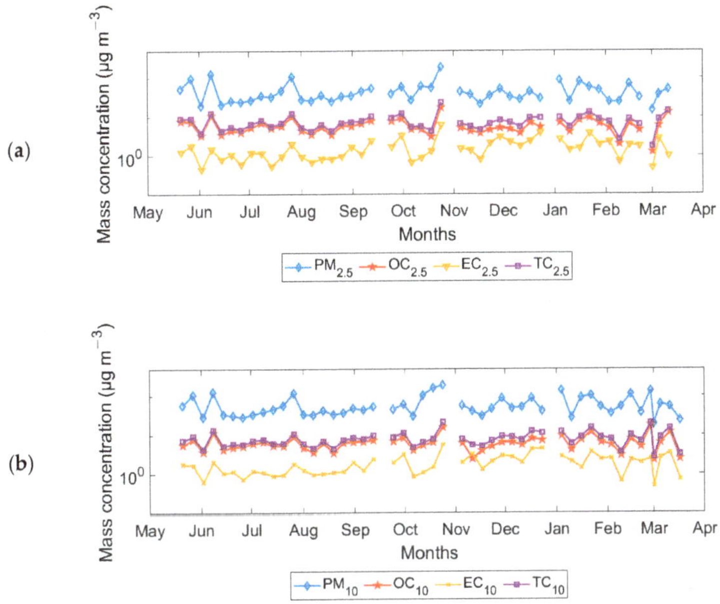

Figure 1. PM and corresponding OC, EC, and TC concentrations: (**a**) fine fraction; (**b**) coarse fraction.

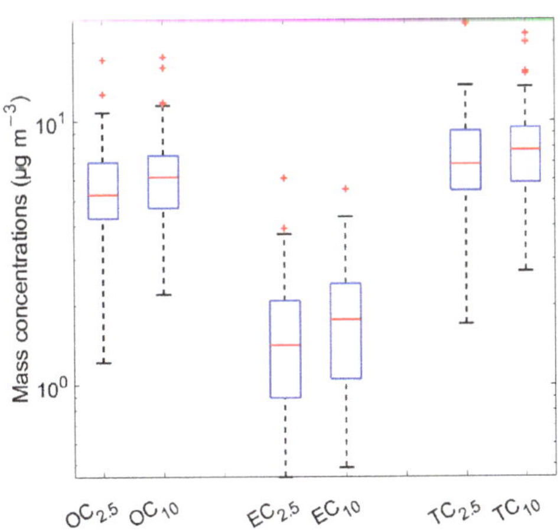

Figure 2. Boxplot of OC, EC, and TC within the $PM_{2.5}$ and PM_{10}. The box plot represents 25th and 75th percentiles and median (red line), the bars represent the 5th and 95th percentiles, and the plus symbols indicate outlayers.

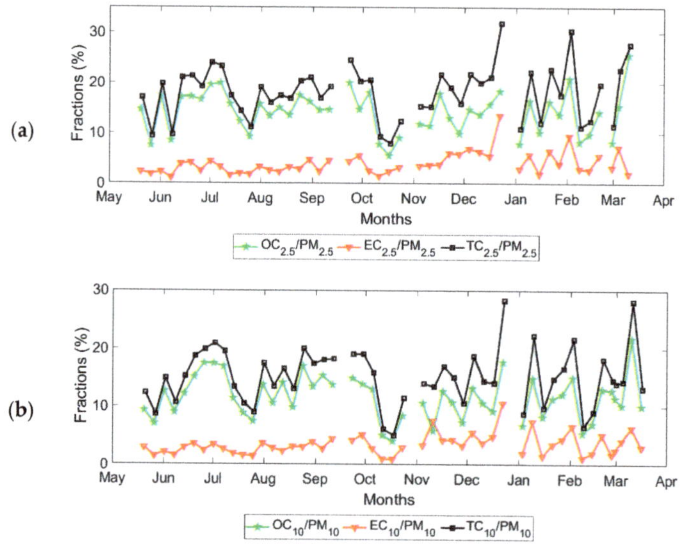

Figure 3. OC, EC, and TC contents (mass concentrations, %) in the corresponding PM concentrations: (**a**) fine fraction; (**b**) coarse fraction.

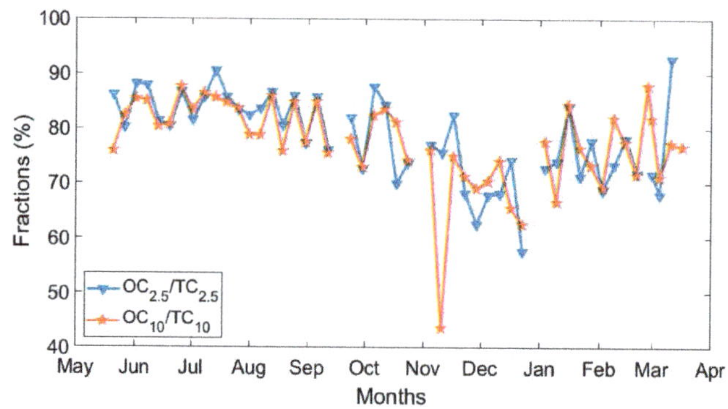

Figure 4. The ratio OC/TC in the $PM_{2.5}$ and PM_{10} concentrations.

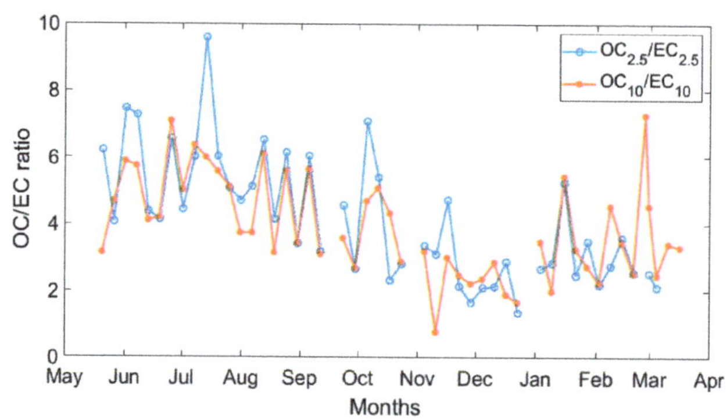

Figure 5. Variation of OC and EC ratio in $PM_{2.5}$ and PM_{10} during the measurement period.

Table 1. OC and EC concentrations in $PM_{2.5}$ reported in Asia, North America, Middle East, and Europe.

Location	Year	Background	$PM_{2.5}$	$OC_{2.5}$	$OC_{2.5}/PM_{2.5}$	$EC_{2.5}$	$EC_{2.5}/PM_{2.5}$	OC/EC	References
Amman, Jordan	2018–2019	Urban and residence	47 ± 32	5.9 ± 2.8	12.5%	1.65 ± 1.06	3.5%	3.6	This study
Beijing, China	2000	Urban and residence	127	29.1	22.9%	10.1	7.95%	2.88	He et al. [61]
Chegongzhuang, China	2000	Urban and residence	115	21.5	18.7%	8.7	7.6%	2.47	He et al. [61]
Seoul, South Korea	1997	Urban	-	2.97	-	0.32	-	-	Kim et al. [62]
Nagoya, Japan	2003–2019	Residence	-	3.3	-	0.7	-	-	Yamagami et al. [63]
Seattle, USA	1996–1999	Urban	8.9 ± 7.5	2.2 (modeled)	24.7%	0.852 (modeled)	9.6%	-	Maykut et al., [64]
Riyadh, Saudi Arabia	2012	Urban	-	4.7 ± 4.4	-	2.1 ± 2.5	-	-	Bian et al. [65]
Athens, Greece	2003	Urban	-	6.8	-	2.2	-	-	Grivas et al. [66]
Tehran, Iran	2013–2014	Urban	41.19	15.35 ± 6.05	37.3%	2.25 ± 0.65	5.5%	6.82 ± 2.30	Arfaeinia et al. [67]
Kuwait	2004–2005	Residence	30.8 ± 16.6	3.4 ± 1.4	11.0%	1.9 ± 0.9	6.2%	1.8	Brown et al. [68]
Amman, Jordan	2007	Residence and commerce	40 ± 9	6.7 ± 0.5	16.8%	2.6 ± 0.8	6.5%	2.8 ± 0.7	von Schneidemesser et al. [3]
Eilat, Israel	2007	Residence	21 ± 4	3.3 ± 0.6	15.7%	0.82 ± 0.1	4.0%	4.1 ± 0.9	von Schneidemesser et al. [3]
East Jerusalem, Palestine	2007	Residence and commerce	27 ± 10	5.6 ± 1.4	20.7%	2.2 ± 0.5	8.1%	2.6 ± 0.7	von Schneidemesser et al. [3]
Beirut, Lebanon	2011	Urban	21.9	5.6	25.6%	1.8	8.22%	-	Waked et al. [69]
Warsaw, Poland	2016	Urban	18.8 ± 11.9	5.56	29.6%	1.47	7.8%	3.7	Juda-Rezler et al. [70]
Barcelona, Spain	2004	Urban	16.4–17.7	3–4 (summer)	17–25%	1–2 (summer)	6–12.5%	-	Viana et al. [71]
Apulia region, Italy	2015	Costal rural	11 ± 6	3.5 ± 2.8	31.8%	0.35 ± 0.18	3.2%	-	Siciliano et al. [72]
Italy	2012–2013	Veneto Province	-	5.5	-	1.3	-	4.54	Khan et al. [73]

Table 2. OC and EC concentrations in PM10 reported in Asia, North America, Middle East, and Europe.

Location	Year	Background	PM_{10}	OC_{10}	OC_{10}/PM_{10}	EC_{10}	EC_{10}/PM_{10}	OC/EC	References
Amman, Jordan	2018–2019	Urban and residence	64 ± 39	6.5 ± 3.04	10.2%	1.9 ± 1.07	3.0%	3.4	This study
Taiyuan, China	2001–2002	Urban	146.36	25.89 (summer)	17.7%	6.82 (summer)	4.7%	-	Tian et al. [74]
Seoul, South Korea	1994	Urban	-	11.1	-	8.39	-	-	Kim et al. [66]
Indo-Gangetic Plain, India	2015–2016	Residence	283 ± 61	74.2 ± 14 (Night)	26.2%	-	-	-	Arif et al. [75]
Indo-Gangetic Plain, India	2015–2016	Residence	167 ± 45	44.3 ± 8.9 (Day)	26.5%	-	-	-	Arif et al. [75]
Mira Loma, USA	2001	Urban plume	-	15.91 ± 6.81	-	1.56 ± 0.56	-	-	Salmon et al. [76]
Lahore, Pakistan	2010	Urban	406.2	63	15.5%	21	5.2%	3.9 ± 1.6	Alam et al. [77]
Thessaloniki, Greece	2012	Urban	51.1 ± 14	11.3 ± 5.0	22.1%	6.56 ± 2.14	12.8%	1.96 ± 1.16	Samara et al. [78]
Barcelona, Spain	2004	Urban	29.5 ± 8.5	4 (summer)	13.6%	1 (summer)	3.4%	-	Viana et al. [71]
Budapest, Hungary	2002	Near-city	54	11	20.4%	3.6	6.7%	-	Salma et al. [79]
Apulia region, Italy	2015	Coastal rural	23 ± 14	5 ± 4	21.7%	0.41 ± 0.19	1.8%	11.3	Siciliano et al. [72]
Prague, Czech Republic	-	Suburb and	33 ± 23	5.5	16.7%	0.74	2.2%	8 ± 3.4	Vodička et al. [57]
		Downtown	37 ± 22	4.8	13.0%	0.8	2.2%	5.8 ± 3.3	Vodička et al. [57]

3.3. Changes during Sand and Dust Storms (SDS)

As per our previous analysis [44], the SDS episodes were classified into three categories based on their origin: S (Sahara), SL (Saharan and Levant), SA (Sahara and Arabia), and SLA (Sahara, Levant, and Arabia); kindly see more information in the Supplementary Materials. During SDS episodes, the mean $OC_{2.5}$ and $EC_{2.5}$ were 9.6 ± 3.5 μg m^{-3} and 2.5 ± 1.2 μg m^{-3}, respectively (Figure 6). During non-dust episodes, the $OC_{2.5}$ and $EC_{2.5}$ were 5.4 ± 1.8 μg m^{-3} and 1.7 ± 0.9 μg m^{-3}, respectively (Figure 7). As for PM_{10}, the OC_{10} and EC_{10} were 8.0 ± 3.5 μg m^{-3} and 2.4 ± 1.3 μg m^{-3} during dust episodes and 5.2 ± 2.0 μg m^{-3} and 2.4 ± 1.3 μg m^{-3} during non-dust episodes, respectively.

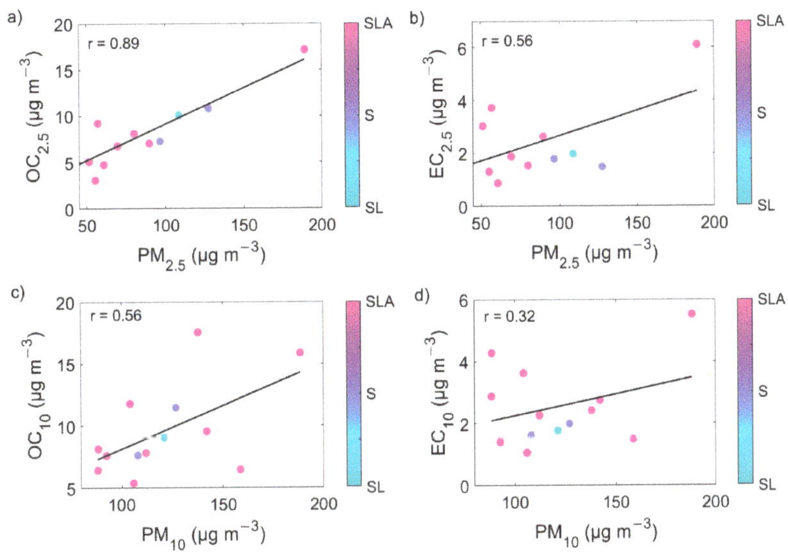

Figure 6. OC and EC versus the corresponding PM concentrations on days (13 in total) with sand and dust storm (SDS) episodes: (**a**) the OC content (here called $OC_{2.5}$) versus its corresponding $PM_{2.5}$, (**b**) the EC versus (here called $EC_{2.5}$) within its corresponding $PM_{2.5}$, (**c**) the OC content (here called OC_{10}) versus its corresponding PM_{10}, and (**d**) the EC versus (here called EC_{10}) within its corresponding PM_{10}.

It was apparent that the concentrations of OC were greatly influenced by the type of aerosols (i.e., coarse dust particles) but the EC was not affected as much. The $OC_{2.5}/PM_{2.5}$ during the SDS episodes carried the highest correlation coefficient of 0.89, which was expected since OC mostly exists in accumulation mode (0.1–1 μm) particles in $PM_{2.5}$ [80]; it also confirmed the dominance of $OC_{2.5}$ in the total carbon content. By comparing both the EC_x/PM_x in the corresponding $PM_{2.5}$ and PM_{10}, we observed that, during SDS episodes, the EC and PM in the fine particle size range were more correlated (r = 0.556, Figure 6b), whereas they were much less correlated (r = 0.32, Figure 6d) in the coarse particle size range. On the other hand, the r values for EC and PM on both particle size ranges were very close on non-SDS days (Figure 7b,d). This result suggests that, during SDS episodes, aerosols have a high potential of transporting OC, which dramatically increased the concentration of OC to the measurement site from other urban centers in the region. As previously mentioned, the regular atmospheric EC concentrations remain at around 2–3 μg m^{-3}.

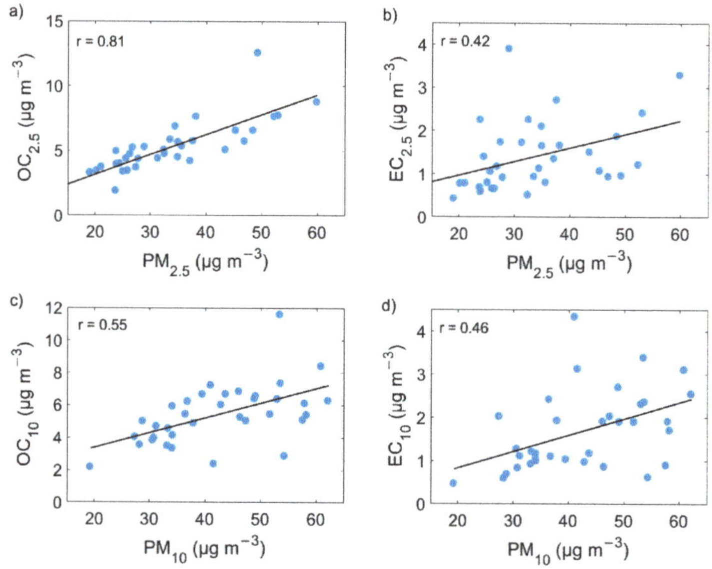

Figure 7. OC and EC versus the corresponding PM concentrations on days (39 in total) without sand and dust storm (SDS) episodes: (**a**) the OC content (here called $OC_{2.5}$) versus its corresponding $PM_{2.5}$, (**b**) the EC versus (here called $EC_{2.5}$) within its corresponding $PM_{2.5}$, (**c**) the OC content (here called OC_{10}) versus its corresponding PM_{10}, and (**d**) the EC versus (here called EC_{10}) within its corresponding PM_{10}.

4. Conclusions

General monitoring requirements for air quality and assessment include ambient air $PM_{2.5}$ and PM_{10}, in addition to some gaseous pollutants; however, there is a lack of monitoring of OC and EC concentrations in many parts of the world. Unfortunately, there has been a lack of information regarding OC, EC, and TC in the Eastern Mediterranean region. In this study, we characterized the PM_{10} and $PM_{2.5}$ with respect to OC/EC during an eleven-month time series (May 2018–March 2019) in the urban atmosphere of Amman, Jordan, which is a typical Eastern Mediterranean city.

The OC found in the $PM_{2.5}$ fraction (i.e., $OC_{2.5}$) was within the range of 1.2–17.1 μg m^{-3}. The corresponding $EC_{2.5}$ was within the range of 0.6–6.1 μg m^{-3}. As for the PM_{10} fraction, the OC_{10} and EC_{10} were within the range of 2.2–17.5 μg m^{-3} and 0.5–5.5 μg m^{-3}, respectively. In percentages, about 14% and 3.9% of the $PM_{2.5}$ were OC and EC, respectively. In the PM10, 11.6% and 3.5% were OC and EC, respectively. These results indicate the domination of anthropogenic activities' emissions over natural sources' emissions.

Sand and dust storm (SDS) episodes were observed during the measurement campaign. During SDS episodes, the mean $OC_{2.5}$ was approximately 9.6 ± 3.5 μg m^{-3}, which is much higher than the annual mean (i.e., 5.9 ± 2.8 μg m^{-3}). Similarly, the $EC_{2.5}$ approached a mean concentration of 2.5 ± 1.2 μg m^{-3} during the SDS episodes, while it had an annual mean of 1.7 ± 1.1 μg m^{-3}. This is evidence of the SDS episodes' role in introducing particulate phase pollutants other than the coarse mode dust particles to the measurement site, or more generally, to the whole region.

This study indicated that a large fraction of carbonaceous aerosol mass most likely originates from anthropogenic activities rather than natural sources. Future studies with detailed source apportionment tools [81,82] are needed to verify this result. Accurate classification of the origins of this type of aerosols can be useful in regulating the involved activities.

Supplementary Materials: The following are available online at https://www.mdpi.com/article/10.3390/atmos13020197/s1, Figure S1:Maps showing (a) the Mediterranean Sea region with Jordan highlighted in red, (b) Jordan with highlights on the geographical locations of main cities, (c) road network and the campus of the University of Jordan (red shaded area) inside Amman, and (d) details of the campus of the University of Jordan with the sampling location (red shaded area) at the middle of the campus, Figure S2:Time series of weather conditions during 1 May 2018—19 March 2019 presented as hourly, daily, and monthly means for (a) ambient temperature, (b) relative humidity, (c) absolute pressure, and (d) wind speed. (e) The rainfall was presented as hourly cumulative precipitation, Figure S3: Back trajectories (96 h) crossing maps at arrival heights (a) 100 m and (b) 1500 m. The arrival location was the campus of the University of Jordan, Amman, Jordan. These maps were generated from the hourly trajectories during 1 May 2018—31 March 2019, Figure S4: Time series of PM10 and PM2.5 concentrations with markups for sand and dust episodes (SDS) and clean air periods (i.e. PM10 concentrations <70 μg/m^3), Figure S5: Back trajectories (96 h) crossing maps during S-type SDS-episodes (indicated on Figure S4) at arrival heights (a) 100 m, (b) 500 m, and (c) 1500 m. The arrival location was the campus of the University of Jordan, Amman, Jordan. These maps were generated from the hourly back trajectories during the sampling dates (+ following day), Figure S6: Back trajectories (96 h) crossing maps during SL-type SDS-episodes (indicated on Figure S4) at arrival heights (a) 100 m, (b) 500 m, and (c) 1500 m. The arrival location was the campus of the University of Jordan, Amman, Jordan. These maps were generated from the hourly back trajectories during the sampling dates (+ following day), Figure S7: Back trajectories (96 h) crossing maps during SLA-type SDS-episodes (indicated on Figure S4) at arrival heights (a) 100 m, (b) 500 m, and (c) 1500 m. The arrival location was the campus of the University of Jordan, Amman, Jordan. These maps were generated from the hourly back trajectories during the sampling dates (+ following day), Figure S8: Back trajectories (96 h) crossing maps during low PM10 concentrations (indicated on Figure S4) at arrival heights (a) 100 m, (b) 500 m, and (c) 1500 m. The arrival location was the campus of the University of Jordan, Amman, Jordan. These maps were generated from the hourly back trajectories during the sampling dates (+ following day), Table S1: Particulate matter concentrations and overall average temperature and pressure according to the sampling schedule. The aerosol sampler reported the 24-h mean temperature (T) and pressure (P) during sampling sessions, Table S2: PM2.5 and PM10 concentrations (μg m^{-3}) and corresponding OC and EC concentrations (μg m^{-3}), Table S3: Sand and Dust Storm (SDS) episodes according to type and observation during the sampling period. The type of SDS is denoted as: Saharan (S); Saharan and Levant (SL); Saharan, Arabian, and Levant (SAL); Saharan, Arabian, Levant, and Ahvaz (SALA). The source region was verified according to the back trajectories analysis for crossing maps on the sampling day (+ following day). The date here indicates the start of the sampling day.

Author Contributions: Conceptualization, T.H. and M.V.; methodology, T.H., A.A.-H., S.A., A.A., and M.V.; validation, T.H., M.V., A.A., and Z.B.; formal analysis, T.H., Z.B., and X.L.; investigation, T.H., A.A.-H., T.P., S.A., and M.V.; resources, T.H., T.P., and M.V.; data curation, X.L., and Z.B.; writing—original draft preparation, T.H., M.V., A.A.-H., Z.B., and X.L.; writing—review and editing, T.H., X.L., Z.B., A.A., S.A., A.A.-H., M.V., and T.P.; visualization, X.L., Z.B., and T.H.; supervision, T.H., S.A., A.A.-H., and M.V.; project administration, T.H. and M.V.; funding acquisition, T.H. All authors have read and agreed to the published version of the manuscript.

Funding: This research was funded by the Deanship of Academic Research (DAR, project number 2015) at the University of Jordan. Academy of Finland Center of Excellence (project No. 272041). eCOST action (inDUST, project number CA16202) support via the Short Term Scientific Mission (STSM) mobility grant.

Institutional Review Board Statement: Not applicable.

Informed Consent Statement: Not applicable.

Data Availability Statement: Data will be available upon request.

Acknowledgments: This research was part of a close collaboration between the University of Jordan, the Institute of Environmental Assessment and Water Research (IDAEA-CSIC), and the Institute for Atmospheric and Earth System Research (INAR/Physics, University of Helsinki). We acknowledge the financial support provided by the Deanship of Academic Research (DAR, project No. 2015) at the University of Jordan, Academy of Finland Center of Excellence (project No. 272041), ERA-

PLANET, trans-national project SMURBS (grant agreement n. 689443 funded under the EU Horizon 2020 Framework Programme and Academy of Finland via the Center of Excellence in Atmospheric sciences and NanoBioMass (project number 1307537). The eCOST action (inDUST, project number CA16202) is acknowledged for supporting this research via the Short Term Scientific Mission (STSM) mobility grant. The work was supported by the Spanish Ministry of Science and Innovation (Project CEX2018-000794-S) and by AGAUR (project 2017 SGR41). This manuscript was written and completed during the sabbatical leave of the first author (Tareq Hussein) that was spent at the University of Helsinki and supported by the University of Jordan during 2019.

Conflicts of Interest: The authors declare no conflict of interest.

References

1. Geron, C. Carbonaceous aerosol over a Pinus taeda forest in Central North Carolina, USA. *Atmos. Environ.* **2009**, *43*, 959–969. [CrossRef]
2. Lin, P.; Hu, M.; Deng, Z.; Slanina, J.; Han, S.; Kondo, Y.; Takegawa, N.; Miyazaki, Y.; Zhao, Y.; Sugimoto, N. Seasonal and diurnal variations of organic carbon in $PM_{2.5}$ in Beijing and the estimation of secondary organic carbon. *J. Geophys. Res.* **2009**, *114*, D00G11. [CrossRef]
3. von Schneidemesser, E.; Zhou, J.; Stone, E.A.; Schauer, J.J.; Qasrawi, R.; Abdeen, Z.; Shpund, J.; Vanger, A.; Sharf, G.; Moise, T.; et al. Seasonal and spatial trends in the sources of fine particle organic carbon in Israel, Jordan, and Palestine. *Atmos. Environ.* **2010**, *44*, 3669–3678. [CrossRef]
4. Carrico, C.M.; Bergin, M.H.; Shrestha, A.B.; Dibb, J.E.; Gomes, L.; Harris, J.M. The importance of carbon and mineral dust to seasonal aerosol properties in the Nepal Himalaya. *Atmos. Environ.* **2003**, *37*, 2811–2824. [CrossRef]
5. Schauer, J.J.; Rogge, W.F.; Hildemann, L.M.; Mazurek, M.A.; Cass, G.R.; Simoneit, B.R.T. Source apportionment of airborne particulate matter using organic compounds as tracers. *Atmos. Environ.* **1996**, *30*, 3837–3855. [CrossRef]
6. Sillanpää, M.; Frey, A.; Hillamo, R.; Pennanen, A.S.; Salonen, R.O. Organic, elemental and inorganic carbon in particulate matter of six urban environments in Europe. *Atmos. Chem. Phys.* **2005**, *5*, 2869–2879. [CrossRef]
7. Stone, E.A.; Snyder, D.C.; Sheesley, R.J.; Sullivan, A.P.; Weber, R.J.; Schauer, J.J. Source apportionment of fine organic aerosol in Mexico City during the MILAGRO experiment 2006. *Atmos. Chem. Phys.* **2008**, *8*, 1249–1259. [CrossRef]
8. Kanakidou, M.; Seinfeld, J.H.; Pandis, S.N.; Barnes, I.; Dentener, F.J.; Facchini, M.C.; van Dingenen, R.; Ervens, B.; Nenes, A.; Nielsen, C.J.; et al. Organic aerosol and global climate modelling: A review. *Atmos. Chem. Phys.* **2005**, *5*, 1053–1123. [CrossRef]
9. Bond, T.C.; Doherty, S.J.; Fahey, D.W.; Forster, P.M.; Berntsen, T.; Deangelo, B.J.; Flanner, M.G.; Ghan, S.; Kärcher, B.; Koch, D.; et al. Bounding the role of black carbon in the climate system: A scientific assessment. *J. Geophys. Res. Atmos.* **2013**, *118*, 5380–5552. [CrossRef]
10. Ramanathan, V.; Carmichael, G. Global and regional climate changes due to black carbon. *Nat. Geosci.* **2008**, *1*, 221–227. [CrossRef]
11. Robinson, A.L.; Donahue, N.M.; Shrivastava, M.K.; Weitkamp, E.A.; Sage, A.M.; Grieshop, A.P.; Lane, T.E.; Pierce, J.R.; Pandis, S.N. Rethinking organic aerosols: Semivolatile emissions and photochemical aging. *Science* **2007**, *315*, 1259–1262. [CrossRef] [PubMed]
12. McMurry, P.; Shepherd, M.; Vickery, J. *Particulate Matter Science for Policy Makers: A Narsto Assessment*; Cambridge University Press: New York, NY, USA, 2004.
13. Hu, Y.; Odman, M.T.; Russell, A.G. Top-down analysis of the elemental carbon emissions inventory in the United States by inverse modeling using Community Multiscale Air Quality model with decoupled direct method (CMAQ-DDM). *J. Geophys. Res. Atmos.* **2009**, *114*, 11987. [CrossRef]
14. Zhu, J.J.; Chen, Y.C.; Shie, R.H.; Liu, Z.S.; Hsu, C.Y. Predicting carbonaceous aerosols and identifying their source contribution with advanced approaches. *Chemosphere* **2021**, *266*, 128966. [CrossRef] [PubMed]
15. Jacobson, M.Z. Strong radiative heating due to the mixing state of black carbon in atmospheric aerosols. *Nature* **2001**, *409*, 695–697. [CrossRef] [PubMed]
16. Menon, S.; Hansen, J.; Nazarenko, L.; Luo, Y. Climate effects of black carbon aerosols in China and India. *Science* **2002**, *297*, 2250–2253. [CrossRef]
17. Turpin, B.J.; Huntzicker, J.J. Identification of secondary organic aerosol episodes and quantitation of primary and secondary organic aerosol concentrations during SCAQS. *Atmos. Environ.* **1995**, *29*, 3527–3544. [CrossRef]
18. Feng, Y.; Ramanathan, V.; Kotamarthi, V.R. Brown carbon: A significant atmospheric absorber of solar radiation? *Atmos. Chem. Phys.* **2013**, *13*, 8607–8621. [CrossRef]
19. Laskin, A.; Laskin, J.; Nizkorodov, S.A. Chemistry of Atmospheric Brown Carbon. *Chem. Rev.* **2015**, *115*, 4335–4382. [CrossRef]
20. Pandey, A.; Hsu, A.; Tiwari, S.; Pervez, S.; Chakrabarty, R.K. Light absorption by organic aerosol emissions rivals that of black carbon from residential biomass fuels in South Asia. *Environ. Sci. Technol. Lett.* **2020**, *7*, 266–272. [CrossRef]
21. Pani, S.K.; Lin, N.-H.; Griffith, S.M.; Chantara, S.; Lee, C.-T.; Thepnuan, D.; Tsai, Y.I. Brown carbon light absorption over an urban environment in northern peninsular Southeast Asia. *Environ. Pollut.* **2021**, *276*, 116735. [CrossRef]
22. Chuang, K.J.; Chan, C.C.; Su, T.C.; Lee, C.T.; Tang, C.S. The effect of urban air pollution on inflammation, oxidative stress, coagulation, and autonomic dysfunction in young adults. *Am. J. Respir. Crit. Care Med.* **2007**, *176*, 370–376. [CrossRef] [PubMed]

23. Ostro, B.; Feng, W.Y.; Broadwin, R.; Green, S.; Lipsett, M. The effects of components of fine particulate air pollution on mortality in California: Results from CALFINE. *Environ. Health Perspect.* **2007**, *115*, 13–19. [CrossRef] [PubMed]
24. Peng, R.D.; Bell, M.L.; Geyh, A.S.; McDermott, A.; Zeger, S.L.; Samet, J.M.; Dominici, F. Emergency admissions for cardiovascular and respiratory diseases and the chemical composition of fine particle air pollution. *Environ. Health Perspect.* **2009**, *117*, 957–963. [CrossRef] [PubMed]
25. Zhou, J.; Ito, K.; Lall, R.; Lippmann, M.; Thurston, G. Time-series analysis of mortality effects of fine particulate matter components in Detroit and seattle. *Environ. Health Perspect.* **2011**, *119*, 461–466. [CrossRef]
26. Wang, Y.C.; Lin, Y.K. Mortality associated with particulate concentration and Asian dust storms in Metropolitan Taipei. *Atmos. Environ.* **2015**, *117*, 32–40. [CrossRef]
27. Pope, C.A.; Dockery, D.W. Health Effects of Fine Particulate Air Pollution: Lines that Connect. *J. Air Waste Manag. Assoc.* **2006**, *56*, 709–742. [CrossRef]
28. Pope III, C.A. Lung Cancer, Cardiopulmonary Mortality, and Long-term Exposure to Fine Particulate Air Pollution. *JAMA* **2002**, *287*, 1132–1141. [CrossRef]
29. Hoek, G.; Krishnan, R.M.; Beelen, R.; Peters, A.; Ostro, B.; Brunekreef, B.; Kaufman, J.D. Long-term air pollution exposure and cardio- respiratory mortality: A review. *J. Environ. Health* **2013**, *12*, 43. [CrossRef]
30. Engling, G.; Gelencsér, A. Atmospheric brown clouds: From local air pollution to climate change. *Elements* **2010**, *6*, 223–228. [CrossRef]
31. Park, R.J.; Jacob, D.J.; Chin, M.; Martin, R.V. Sources of carbonaceous aerosols over the United States and implications for natural visibility. *J. Geophys. Res. Atmos.* **2003**, *108*, 12. [CrossRef]
32. Mauderly, J.L.; Chow, J.C. Health effects of organic aerosols. *Inhal. Toxicol.* **2008**, *20*, 257–288. [CrossRef] [PubMed]
33. Small, I.; van der Meer, J.; Upshur, R.E. Acting on an environmental health disaster: The case of the Aral Sea. *Environ. Health Perspect.* **2001**, *109*, 547–549. [CrossRef] [PubMed]
34. Menéndez, I.; Díaz-Hernández, J.L.; Mangas, J.; Alonso, I.; Sánchez-Soto, P.J. Airborne dust accumulation and soil development in the North-East sector of Gran Canaria (Canary Islands, Spain). *J. Arid Environ.* **2007**, *71*, 57–81. [CrossRef]
35. McTainsh, G.; Strong, C. The role of aeolian dust in ecosystems. *Geomorphology* **2007**, *89*, 39–54. [CrossRef]
36. Goudie, A.S. Dust storms: Recent developments. *J. Environ. Manag.* **2009**, *90*, 89–94. [CrossRef]
37. Karanasiou, A.; Moreno, N.; Moreno, T.; Viana, M.; de Leeuw, F.; Querol, X. Health effects from Sahara dust episodes in Europe: Literature review and research gaps. *Environ. Int.* **2012**, *47*, 107–114. [CrossRef]
38. Rezazadeh, M.; Irannejad, P.; Shao, Y. Climatology of the Middle East dust events. *Aeolian Res.* **2013**, *10*, 103–109. [CrossRef]
39. Almasi, A.; Mousavi, A.R.; Bakhshi, S.; Namdari, F. Dust storms and environmental health impacts. *J. Middle East Appl. Sci. Technol.* **2014**, *8*, 353–356.
40. Goudie, A.S. Desert dust and human health disorders. *Environ. Int.* **2014**, *63*, 101–113. [CrossRef]
41. Díaz, J.; Linares, C.; Carmona, R.; Russo, A.; Ortiz, C.; Salvador, P.; Trigo, R.M. Saharan dust intrusions in Spain: Health impacts and associated synoptic conditions. *Environ. Res.* **2017**, *156*, 455–467. [CrossRef]
42. Middleton, N.J. Desert dust hazards: A global review. *Aeolian Res.* **2017**, *24*, 53–63. [CrossRef]
43. Contini, D.; Vecchi, R.; Viana, M. Carbonaceous aerosols in the atmosphere. *Atmosphere* **2018**, *9*, 181. [CrossRef]
44. Hussein, T.; Li, X.; Al-Dulaimi, Q.; Daour, S.; Atashi, N.; Viana, M.; Alastuey, A.; Sogacheva, L.; Arar, S.; Al-Hunaiti, A.; et al. Particulate matter concentrations in a middle eastern city – an insight to sand and dust storm episodes. *Aerosol Air Qual. Res.* **2020**, *20*, 2780–2792. [CrossRef]
45. Querol, X.; Alastuey, A.; Rodrıguez, S.; Viana, M.M.; Artıñano, B.; Salvador, P.; Mantilla, E.; do Santos, S.G.; Patier, R.F.; de La Rosa, J.; et al. Levels of particulate matter in rural, urban and industrial sites in Spain. *Sci. Total Environ.* **2004**, *334*, 359–376. [CrossRef] [PubMed]
46. Birch, M.E.; Cary, R.A. Elemental Carbon-Based Method for Monitoring Occupational Exposures to Particulate Diesel Exhaust. *Aerosol Sci. Technol.* **1996**, *25*, 221–241. [CrossRef]
47. Cavalli, F.; Viana, M.; Yttri, K.E.; Genberg, J.; Putaud, J.-P. Toward a standardised thermal-optical protocol for measuring atmospheric organic and elemental carbon: The EUSAAR protocol. *Atmos. Meas. Technol.* **2010**, *3*, 79–89. [CrossRef]
48. Viana, M.; Maenhaut, W.; Chi, X.; Querol, X.; Alastuey, A. Comparative chemical mass closure of fine and coarse aerosols at two sites in south and west Europe: Implications for EU air pollution policies. *Atmos. Environ.* **2007**, *41*, 315–326. [CrossRef]
49. Draxler, R. *Personal Communication (Memo)*; Hess, G.D., Ed.; Air Resources Laboratory: Silver Spring, MD, USA; Bureau of Meteorology Research Centre: Melbourne, Australia, 1997.
50. Stein, A.F.; Draxler, R.R.; Rolph, G.D.; Stunder, B.J.B.; Cohen, M.D.; Ngan, F. NOAA's HYSPLIT Atmospheric Transport and Dispersion Modeling System. *Bull. Am. Meteorol. Soc.* **2015**, *96*, 2059–2077. [CrossRef]
51. Ambient (Outdoor) Air Quality Database 2018 by Country and City. Available online: https://www.who.int/airpollution/data/aap_air_quality_database_2018_v14.xlsx?ua=1,ap_air_quality_database_2018_v14.xlsx (accessed on 20 March 2020).
52. WHO Global Air Quality Guidelines. Available online: https://apps.who.int/iris/bitstream/handle/10665/345329/9789240034228-eng.pdf?sequence=1andisAllowed=y (accessed on 28 November 2021).
53. Millet, D.B. Atmospheric volatile organic compound measurements during the Pittsburgh Air Quality Study: Results, interpretation, and quantification of primary and secondary contributions. *J. Geophys. Res.* **2005**, *110*, D07S07. [CrossRef]

54. Saarikoski, S.; Timonen, H.; Saarnio, K.; Aurela, M.; Järvi, L.; Keronen, P.; Kerminen, V.-M.; Hillamo, R. Sources of organic carbon in fine particulate matter in northern European urban air. *Atmos. Chem. Phys.* **2008**, *8*, 6281–6295. [CrossRef]
55. Genberg, J.; Hyder, M.; Stenström, K.; Bergström, R.; Simpson, D.; Fors, E.O.; Jönsson, J.Å.; Swietlicki, E. Source apportionment of carbonaceous aerosol in southern Sweden. *Atmos. Chem. Phys.* **2011**, *11*, 11387–11400. [CrossRef]
56. Hu, W.W.; Hu, M.; Deng, Z.Q.; Xiao, R.; Kondo, Y.; Takegawa, N.; Zhao, Y.J.; Guo, S.; Zhang, Y.H. The characteristics and origins of carbonaceous aerosol at a rural site of PRD in summer of 2006. *Atmos. Chem. Phys.* **2012**, *12*, 1811–1822. [CrossRef]
57. Vodička, P.; Schwarz, J.; Ždímal, V. Analysis of one year's OC/EC data at a Prague suburban site with 2-h time resolution. *Atmos. Environ.* **2013**, *77*, 865–872. [CrossRef]
58. Heal, M.R.; Hammonds, M.D. Insights into the Composition and Sources of Rural, Urban and Roadside Carbonaceous PM_{10}. *Environ. Sci. Technol.* **2014**, *48*, 8995–9003. [CrossRef] [PubMed]
59. Huang, X.H.H.; Bian, Q.J.; Louie, P.K.K.; Yu, J.Z. Contributions of vehicular carbonaceous aerosols to $PM_{2.5}$ in a roadside environment in Hong Kong. *Atmos. Chem. Phys.* **2014**, *14*, 9279–9293. [CrossRef]
60. Huang, X.H.; Bian, Q.; Ng, W.M.; Louie, P.K.; Yu, J.Z. Characterization of PM2.5 Major Components and Source Investigation in Suburban Hong Kong: A One Year Monitoring Study. *Aerosol Air Qual. Res.* **2014**, *14*, 237–250. [CrossRef]
61. He, K.; Yang, F.; Ma, Y.; Zhang, Q.; Yao, X.; Chan, C.K.; Cadle, S.; Chan, T.; Mulawa, P. The characteristics of $PM_{2.5}$ in Beijing, China. *Atmos. Environ.* **2001**, *35*, 4959–4970. [CrossRef]
62. Kim, Y.P.; Moon, K.C.; Lee, J.H.; Baik, N.J. Concentrations of carbonaceous species in particles at Seoul and Cheju in Korea. *Atmos. Environ.* **1999**, *33*, 2751–2758. [CrossRef]
63. Yamagami, M.; Ikemori, F.; Nakashima, H.; Hisatsune, K.; Ueda, K.; Wakamatsu, S.; Osada, K. Trends in $PM_{2.5}$ Concentration in Nagoya, Japan, from 2003 to 2018 and Impacts of $PM_{2.5}$ Countermeasures. *Atmosphere* **2021**, *12*, 590. [CrossRef]
64. Maykut, N.N.; Lewtas, J.; Kim, E.; Larson, T.V. Source apportionment of $PM_{2.\ 5}$ at an urban IMPROVE site in Seattle, Washington. *Environ. Sci. Technol.* **2003**, *37*, 5135–5142. [CrossRef]
65. Bian, Q.; Alharbi, B.; Shareef, M.M.; Husain, T.; Pasha, M.J.; Atwood, S.A.; Kreidenweis, S.M. Sources of $PM_{2.5}$ carbonaceous aerosol in Riyadh, Saudi Arabia. *Atmos. Chem. Phys.* **2018**, *18*, 3969–3985. [CrossRef]
66. Grivas, G.; Cheristanidis, S.; Chaloulakou, A. Elemental and organic carbon in the urban environment of Athens. Seasonal and diurnal variations and estimates of secondary organic carbon. *Sci. Total Environ.* **2012**, *414*, 535–545. [CrossRef] [PubMed]
67. Arfaeinia, H.; Hashemi, S.E.; Alamolhoda, A.A.; Kermani, M. Evaluation of organic carbon, elemental carbon, and water soluble organic carbon concentration in $PM_{2.5}$ in the ambient air of Sina Hospital district, Tehran, Iran. *J. Adv. Environ. Health Res.* **2016**, *4*, 95–101. [CrossRef]
68. Brown, K.W.; Bouhamra, W.; Lamoureux, D.P.; Evans, J.S.; Koutrakis, P. Characterization of particulate matter for three sites in Kuwait. *J. Air Waste Manag. Assoc.* **2008**, *58*, 994–1003. [CrossRef] [PubMed]
69. Waked, A.; Afif, C.; Brioude, J.; Formenti, P.; Chevaillier, S.; Haddad, I.E.; Seigneur, C. Composition and source apportionment of organic aerosol in Beirut, Lebanon, during winter 2012. *Aerosol Sci. Technol.* **2013**, *47*, 1258–1266. [CrossRef]
70. Juda-Rezler, K.; Reizer, M.; Maciejewska, K.; Błaszczak, B.; Klejnowski, K. Characterization of atmospheric $PM_{2.5}$ sources at a Central European urban background site. *Sci. Total Environ.* **2020**, *713*, 136729. [CrossRef]
71. Viana, M.; Chi, X.; Maenhaut, W.; Querol, X.; Alastuey, A.; Mikuška, P.; Večeřa, Z. Organic and elemental carbon concentrations in carbonaceous aerosols during summer and winter sampling campaigns in Barcelona, Spain. *Atmos. Environ.* **2006**, *40*, 2180–2193. [CrossRef]
72. Siciliano, T.; Siciliano, M.; Malitesta, C.; Proto, A.; Cucciniello, R.; Giove, A.; Genga, A. Carbonaceous PM_{10} and $PM_{2.5}$ and secondary organic aerosol in a coastal rural site near Brindisi (Southern Italy). *Environ. Sci. Pollut. Res.* **2018**, *25*, 23929–23945. [CrossRef]
73. Khan, M.B.; Masiol, M.; Formenton, G.; Di Gilio, A.; de Gennaro, G.; Agostinelli, C.; Pavoni, B. Carbonaceous $PM_{2.5}$ and secondary organic aerosol across the Veneto region (NE Italy). *Sci. Total Environ.* **2016**, *542*, 172–181. [CrossRef]
74. Tian, Y.Z.; Xiao, Z.M.; Han, B.; Shi, G.L.; Wang, W.; Hao, H.Z.; Zhu, T. Seasonal study of primary and secondary sources of carbonaceous species in PM_{10} from five northern Chinese cities. *Aerosol Air Qual. Res.* **2013**, *13*, 148–161. [CrossRef]
75. Arif, M.; Kumar, R.; Kumar, R.; Zusman, E.; Singh, R.P.; Gupta, A. Assessment of indoor & outdoor black carbon emissions in rural areas of Indo-Gangetic Plain: Seasonal characteristics, source apportionment and radiative forcing. *Atmos. Environ.* **2018**, *191*, 227–240. [CrossRef]
76. Salmon, L.G.; Mayo, P.R.; Cass, G.R.; Seinfeld, J.H. *Determination of Elemental Carbon and Organic Carbon Concentrations During the Southern California Children's Health Study, 1999–2001*; Final report preparaed for the California Air Resources Board and the California Environmental Protection Agency, Research Division, University of California: Berkeley, CA, USA, 2004.
77. Alam, K.; Mukhtar, A.; Shahid, I.; Blaschke, T.; Majid, H.; Rahman, S.; Rahman, N. Source apportionment and characterization of particulate matter (PM_{10}) in urban environment of Lahore. *Aerosol Air Qual. Res.* **2014**, *14*, 1851–1861. [CrossRef]
78. Samara, C.; Voutsa, D.; Kouras, A.; Eleftheriadis, K.; Maggos, T.; Saraga, D.; Petrakakis, M. Organic and elemental carbon associated to PM_{10} and $PM_{2.5}$ at urban sites of northern Greece. *Environ. Sci. Pollut. Res.* **2014**, *21*, 1769–1785. [CrossRef] [PubMed]
79. Salma, I.; Chi, X.; Maenhaut, W. Elemental and organic carbon in urban canyon and background environments in Budapest, Hungary. *Atmos. Environ.* **2004**, *38*, 27–36. [CrossRef]

80. Shrivastava, M.; Cappa, C.D.; Fan, J.; Goldstein, A.H.; Guenther, A.B.; Jimenez, J.L.; Kuang, C.; Laskin, A.; Martin, S.T.; Ng, N.L.; et al. Recent advances in understanding secondary organic aerosol: Implications for global climate forcing. *Rev. Geophys.* **2017**, *55*, 509–559. [CrossRef]
81. Manousakas, M.I.; Florou, K.; Pandis, S.N. Source Apportionment of Fine Organic and Inorganic Atmospheric Aerosol in an Urban Background Area in Greece. *Atmosphere* **2020**, *11*, 330. [CrossRef]
82. Chen, G.; Sosedova, Y.; Canonaco, F.; Fröhlich, R.; Tobler, A.; Vlachou, A.; Daellenbach, K.R.; Bozzetti, C.; Hueglin, C.; Graf, P.; et al. Time-dependent source apportionment of submicron organic aerosol for a rural site in an alpine valley using a rolling positive matrix factorisation (PMF) window. *Atmos. Chem. Phys.* **2021**, *21*, 15081–15101. [CrossRef]

atmosphere

MDPI

Article

Impact of Vehicle Fleet Modernization on the Traffic-Originated Air Pollution in an Urban Area—A Case Study

Piotr Holnicki [1,*], Zbigniew Nahorski [1,2] and Andrzej Kałuszko [1]

1 Systems Research Institute, Polish Academy of Sciences, 01-447 Warsaw, Poland; nahorski@ibspan.waw.pl or Z.Nahorski@wit.edu.pl (Z.N.); kaluszko@ibspan.waw.pl (A.K.)
2 Warsaw School of Information Technology (WIT), 01-447 Warsaw, Poland
* Correspondence: holnicki@ibspan.waw.pl

Abstract: The main subject of this paper is an analysis of the influence of changes in the air pollution caused by road traffic, due to its modernization, on the air quality in Warsaw conurbation, Poland. Using the Calpuff model, simulations of the yearly averaged concentrations of NOx, CO, PM_{10}, and $PM_{2.5}$ were performed, together with an assessment of the population exposure to individual pollutions. Source apportionment analysis indicates that traffic is the main source of NOx and CO concentrations in the city atmosphere. Utilizing the Euro norms emission standards, a scenario of vehicle emission abatement is formulated based on the assumed general vehicle fleet modernization and transition to Euro 6 emission standards. Computer simulations show a reduction in NOx concentrations attributed to emission mitigation of passenger cars, trucks and vans, and public transport buses, respectively. On the other hand, improving air quality in terms of CO concentrations depends almost exclusively on gasoline vehicle modernization. The implementation of the considered scenario causes an adequate reduction in the population exposure and related health effects. In particular, implementation of the scenario discussed results in a 47% reduction (compared with the baseline value) in the attributable yearly deaths related to NOx pollution. In spite of a substantial contribution of vehicle traffic to the overall PM pollution, modernization of the fuel combustion causes only minor final effects because the dominant share of PM pollution in Warsaw originates from the municipal sector and the transboundary inflow.

Keywords: urban air quality; road traffic emission; emission abatement; Euro norm limits; population exposure

Citation: Holnicki, P.; Nahorski, Z.; Kałuszko, A. Impact of Vehicle Fleet Modernization on the Traffic-Originated Air Pollution in an Urban Area—A Case Study. *Atmosphere* **2021**, *12*, 1581. https://doi.org/10.3390/atmos12121581

Academic Editor: Antonio Donateo

Received: 18 October 2021
Accepted: 26 November 2021
Published: 27 November 2021

1. Introduction

Air pollution remains the main environmental health risk in Europe, especially in urban conurbations, where many inhabitants suffer from poor air quality. The source apportionment analysis reveals [1,2] that in many cities, the transportation system is the dominating emission category in urban air quality degradation. Road traffic contributes to negative health effects mainly by primary nitrogen oxide (NOx) pollution, but also via the primary and secondary components of particulate matter. In many cities, the European air quality standards established by the Ambient Air Quality Directive [3,4] or Air Quality Guidelines [5] are regularly exceeded. While PM pollutions originate from various categories of sources (with a minor contribution of traffic to the primary emission), NO_2 concentrations almost exclusively originate from road traffic. Moreover, NO_2 pollution in urban conurbations is mainly due to emissions from diesel vehicles [6,7]. Traffic is also an important source of the resuspended, coarse fractions of PM_{10}. Hence, recently published European studies assessed the effects of traffic policies, intended to reduce concentrations of the above species in large cities.

Numerous earlier urban scale studies addressed emissions of the road transport pollutants, where potential traffic-related scenarios are discussed to abate NOx or CO pollutions.

Examples are CO pollution in Copenhagen [8]; CO pollution in Luxembourg [9]; integrated NO_2 and CO analysis for Turin [10]; integrated analysis including human exposure and emission abatement techniques for London [11] or [12]. Some recent studies [7,13,14] consider various NOx emission abatement scenarios, based on the implementation of Euro 6 emission standards in urban traffic for gasoline and diesel vehicles. Scenario analysis was applied to assess the effects of traffic policies, intended to reduce NO_2 concentrations in eight European cities [15]. Very important and useful in such an approach is the *Urban* NO_2 *Atlas* report [6], where the traffic emission characteristics for 30 European cities are presented, broken down into diesel and gasoline vehicles and 6 Euro emission standards.

Characteristics of the transportation system in Warsaw given in [6] were utilized by Holnicki et al. [16], who discussed the complex improvement of air quality by vehicle fleets and municipal sector modernizations. Aside from traffic, the municipal sector is another dominant emission category contributing to air quality degradation in Warsaw [16]. This study investigates the possible improvement of the environmental quality in the Warsaw conurbation by reducing traffic-related pollution as a result of implementing low-emission technologies (represented by the Euro norms) in vehicle fleets operating in Warsaw. As a result of emission abatement scenarios, improvements in air quality and the related public health indexes are assessed.

Warsaw suffers from high concentrations of air pollutants that characterize the urban environment, which is also quite common for other European conurbations. Among them are particulate matter, sulfur- and nitrogen oxides, carbon monoxide, and polycyclic aromatic hydrocarbons. Traffic-induced emission is one of the dominating categories of air pollution. The number of vehicles registered in the conurbation has steadily increased, particularly in the last decade [17,18]. The traffic-originated emissions are mainly responsible for concentrations of NO_x, CO, and PM_{10}, the last mainly via the resuspended coarse fraction [16,19]. In particular, high NO_x and PM_{10} annual mean concentrations during the last decade exceed the WHO limits [7,20]. Road traffic also contributes to the overall concentrations of $PM_{2.5}$ and the highly toxic benzo[a]pyrene (B[a]P), but to a lesser extent, because the dominant share of these pollutants is attributed to the municipal sector, mainly due to coal combustion.

As shown below, the final environmental effect of the traffic emission abatement obviously depends on the number of driving vehicles, but also on the vehicle fleet structure and technology applied for emission reduction (cf. Tables S1–S3). The above factors are included in the analysis of the emission limitation scenario, which is presented in the next section. To fully assess environmental benefits arising from the implementation of the discussed solutions, the reference data for the year 2012 are chosen, prior to the baseline state.

2. Methods

Dispersion of atmospheric pollutants in Warsaw conurbation was assessed using the Gaussian regional model CALPUFF, along with the CALMET meteorological preprocessor. Spatial maps of the annual average concentrations of the major air pollutants were obtained in order to determine areas where the permissible concentration levels of individual pollutants were exceeded and to identify the sources responsible for the exceedances.

In order to assess the real impact of the transport sector on air quality and to quantify the advantage of implementing the technology of low-emission vehicles, four categories of sources were considered in the total emission field [16,20]: (a) high energy point sources (24), (b) other industrial point sources (3880), (c) linear sources of the road network (7285), and (d) municipal area sources (6962). The external (transboundary) inflow of pollutants was included via boundary conditions of the forecasting model. The emission data used in computer simulations were adopted from the official database of the Mazovian Voivodship Inspectorate of Environment Protection.

For calculation of the pollutant's concentrations, the Warsaw metropolitan area (about 520 km^2 within administrative boundaries) was digitized with a homogeneous 0.5 km $\times$ 0.5 km grid. For the discretized emission data, the resulting concentrations were calculated in 2248 elementary mesh receptors. Calculations performed for emission and meteorological data in the year 2012 generated the final pollution amount, which was used as a baseline for emission scenario analysis.

The resulting concentration maps show the spatial distributions of the main pollutants. They also indicate which pollutants exceed the limit values and which violations are critical. The linear structure of the CALPUFF model allowed us to indicate the emission categories responsible for standard violations. Moreover, it was possible to quantify the share of an individual emission category in the total concentration at a given receptor point or in a city district (source apportionment). This is important for implementing an emission abatement strategy to improve the city's air quality. Results presented in [16] show that the annual mean concentrations of NOx, PM_{10}, $PM_{2.5}$, and B(a)P exceed in some districts the limit values prescribed by the WHO (30 $\mu g/m^3$, 40 $\mu g/m^3$, 20 $\mu g/m^3$, 1 ng/m^3, respectively).

Subsequently, some scenarios for reducing road traffic emissions were developed, in which implementation of Euro 6 emission standards for the selected categories of vehicles was considered. Construction of NO_2 emission scenarios was based on the results presented in [6], where the NO_2 emission rates per fuel and Euro standards for 30 cities are collected using the national emission data. The yearly emission volume of each category was calculated as a product of yearly kilometers driven and the respective emission rate.

There was a significant difference in NOx emissions rate between diesel and gasoline vehicles, with definite domination of the former ones. This discrepancy was taken into account in the emission scenarios discussed below. In terms of fuel consumption by vehicle fleets in Warsaw, there was a significant prevalence of low emission standards (E0–E3), both for gasoline and diesel vehicles, similar to other Eastern European cities. Among them, the Warsaw transportation system and its emission properties were fully specified by the fleet composition, fuel type used, yearly distance driven, and emission rates connected with the Euro standards, depending on the vehicle production year. These characteristics, formulated for Warsaw vehicle fleet NOx emissions in the report [6] in graphical form, can be presented in the tabular form [16], presented below.

Table 1 presents the annual NO_2 emissions in Warsaw, attributed to each fuel type and Euro standards. These data were used to assess the emission reduction rate connected with the emission policy scenarios for the transportation system, in particular those related to adapting the transportation emission rates to Euro 6 standards. Moreover, the policy of the Warsaw environmental protection authorities regarding public transport was taken into account. Implementation of the hybrid/LNG/CNG technology in the public transportation system was considered. Results presented in Section 3 suggest realistic emission reduction scenarios, which can significantly contribute to the conurbation air quality improvement.

Table 1. Reference NO_2 vehicle emissions in Warsaw per fuel and Euro emission standards.

	Euro Standard	Distance [km $\times 10^6$]	Emis. Rate [g/km]	Emission [kg $\times 10^3$]	Share [%]	Share
GASOLINE	E0	789	2.5	1974	2%	17,000 (19%)
	E1	15,789	0.45	7105	8%	
	E2	24,737	0.25	6184	7%	
	E3	11,579	0.15	1737	2%	
	E4–E6	13,684	0.1	1368	2%	1368
		emission		18,368	21%	18,368

Table 1. *Cont.*

	Euro Standard	Distance [km × 10^6]	Emis. Rate [g/km]	Emission [kg × 10^3]	Share [%]	Share
DIESEL	E0	6316	0.66	4168	5%	52,742 (59%)
	E1	25,789	0.73	18,826	22%	
	E2	24,211	0.8	19,368	22%	
	E3	11,579	0.83	96,11	11%	
	E4–E6	26,316	0.65	17,105	20%	17,105
		emission		69,079	79%	69,847
		TOTAL		87,447	100%	

3. Results—Emission Abatement Scenarios

3.1. The Nitrogen Oxide Pollution

As previously mentioned, road traffic is a dominating source responsible for NOx pollution in urban areas. As seen from Figure 1, the transportation sector in Warsaw contributes 77% to the overall NOx concentration, while among other emission categories, 10% comes from the area sources and 8% from the transboundary inflow.

The emission scenario considered in this study consisted of reducing the nitrogen oxides pollution by achieving ultimately the Euro 6 emission standards for passenger cars, including taxis (39% emission share), and vans and trucks (25% share), as well as the application of low-emission technologies in the urban transport buses (13% share).

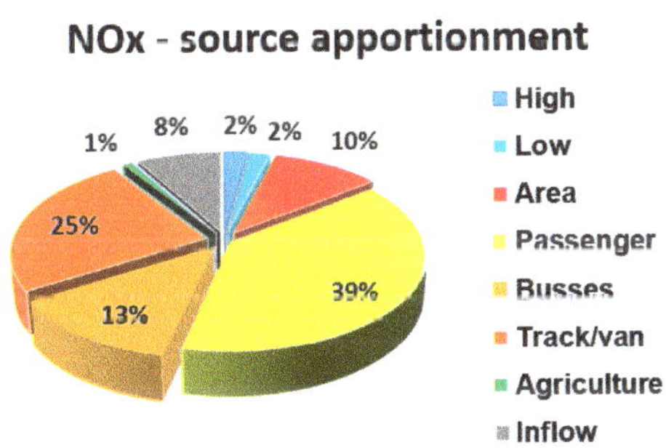

Figure 1. Source apportionment for NOx emissions.

The data listed in Table 1 were used to calculate the annual emission volume attributed to vehicle categories and Euro emission standards. In our study, we assumed Euro 3 emission standards for all pre-E4 vehicles, both diesel and gasoline, as the Euro 3 cars prevail among the older ones. This is a conclusion formulated in [16], that is, distances traveled in Poland by E3 vehicles visibly prevail starting from 2010. The Euro 6 emission standards were adopted as the target to be obtained by both vehicle categories. The share of the total NOx emission from the older (pre-E4) vehicle categories was 19% for gasoline and 59% for diesel vehicles (Table 1). It follows from the Euro emission standards (Tables S1 and S2) that the emission reduction rate, referring to the transition from E3 to E6, was 0.4 for gasoline vehicles and 0.16 for diesel vehicles, which enables determining the target NOx emissions for conversion from pre-E4 to E6 emission standards. Emission rates for E4–E6 vehicles were left unchanged. The above estimations showed that meeting the Euro 6 emission standards for passenger cars in Warsaw would reduce NOx emissions by about 60%, as compared with the baseline level.

Public transport buses have a 13% share of the total NOx emission (Figure 1). The city authorities have launched an initiative to replace the bus fleet with low-emission vehicles, and ultimately public transport will entirely be served by such vehicles (hybrid, LNG/CNG). Reduction in the NOx emission from this sector is estimated at 40–60% by experts from the Voivodship Inspectorate of Environmental Protection. Moreover, transition to Euro 5 emission standards by heavy-duty vehicles (25% share), according to Table S3 can also reduce the respective contribution of trucks and vans by approximately 60%. The emission abatement results in a reduction in the related inhabitants' exposure, calculated as the averaged final concentration weighted by population density. The computed values of the exposure index are shown in all pollution concentration maps presented below.

Bikes are increasingly replacing cars as a means of transportation to move around the city. There are currently about 75,000 bikes in Warsaw [21], including 12% of the public bike-sharing system (Veturillo and scooters). As shown in [22], 80% of cycling traffic in the city refers to transportation, communication, and mobility. Hence, some additional, although a rather slight improvement in air quality, can be obtained, taking into account the growing share of bicycles in urban transportation. Extensive measurements of this movement performed in Warsaw showed that during March–December (as the cycling is negligible during the first two months due to unfavorable Winter conditions), an average of 1.02 million bicycles per month are actively used (Figure 2). We further predict that the growing share of bicycle traffic gradually replaces passenger cars, proportionally to the distance traveled.

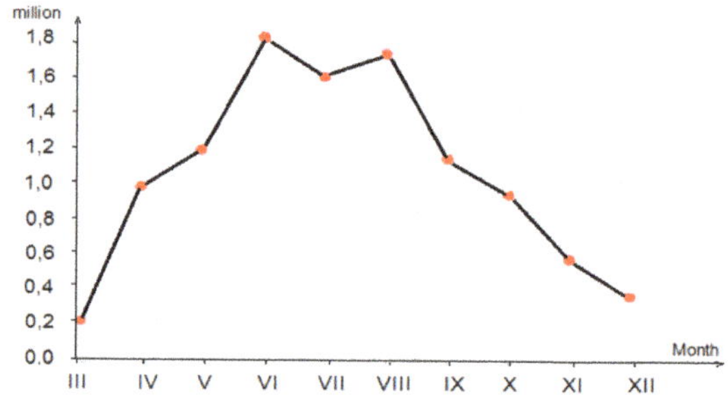

Figure 2. Annual distribution of bicycle traffic in Warsaw (bike/month) (y-axis) in the following months (x-axis) of the year 2019, based on street measurements [21].

Assuming an average distance for daily travel/bike as 10 km (5 km as an average distance between home and destination) results in 300 km monthly and 3000 km yearly travel per bike, as biking is usually too difficult in the first two winter months. Hence, the total yearly distance traveled by bikes can be assessed as 3060 million km, of which 80% (2448 million km) can be attributed to urban transportation. This can be compared with the yearly distance traveled by passenger cars (Table 1), 160,789 million km. Hence, due to a partial replacement of vehicle traffic by cycling, the potential possibility to reduce NO_2 emissions is 1.52%.

A slight improvement in urban air quality is also associated with the growing share of low-emission vehicles (hybrid/electric). In recent years, (2018/19), approximately 4300 [23] passenger cars of this type were registered in Warsaw. A comparison of this number with the total number of passenger cars in the city, estimated at around 1,200,000 vehicles [23], yields the percentage share of low-emission vehicles (and related emission reduction) of about 0.36%. Finally, both of the above effects can lead to about 1.9% abatement of NO_2 emissions in the passenger car category.

3.2. *Carbon Monoxide*

The final distribution of NOx concentration in the city, corresponding to the adopted vehicle fleet emission scenario, is presented in Figure 3. The attached concentration maps show the effect of implementing the discussed scenarios for subsequent categories of vehicle fleets. The effect of cycling (included in Figure 3b) is minor. It reduces the maximum concentration by 0.7 μg/m³ and the population exposure by 0.2 μg/m³, respectively. The comparison of Figure 3a and Figure 3d maps shows that full implementation of Euro 6 emission standards decreases residents' exposure to NOx pollution by almost 50%, while at the same time, the concentration limit value is slightly exceeded in only two receptors.

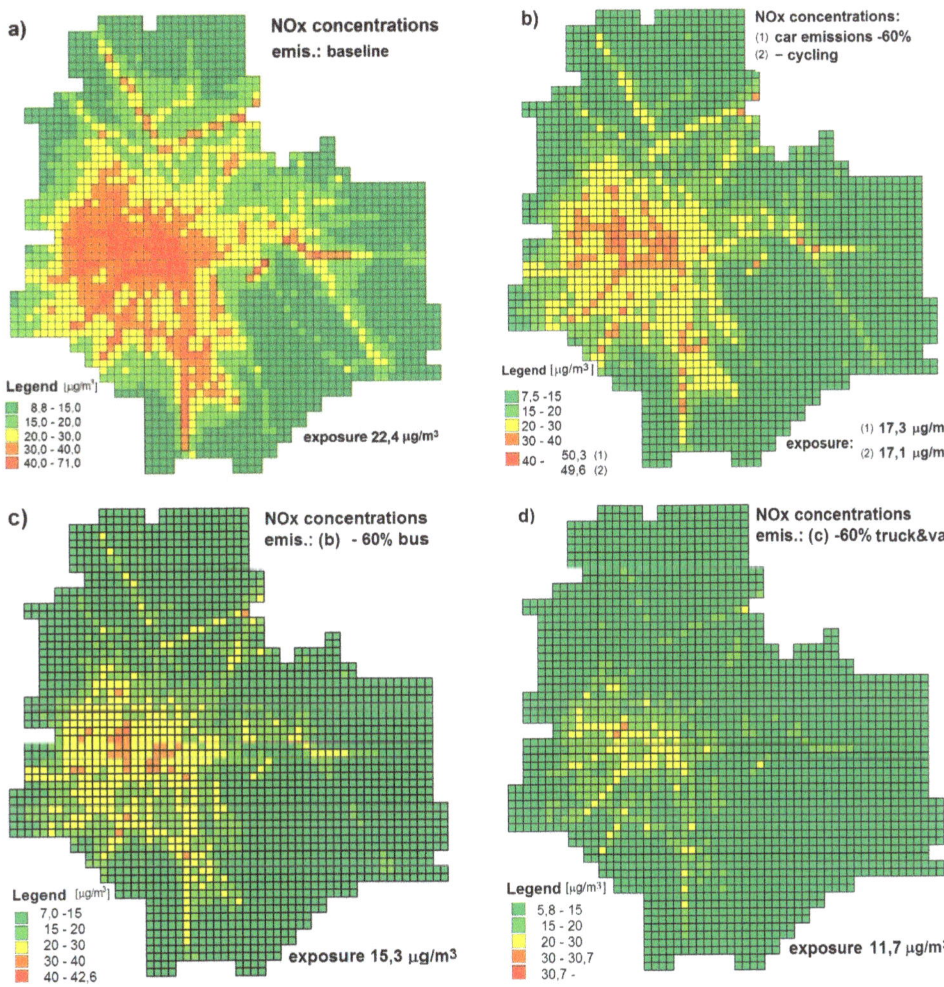

Figure 3. NOx concentrations: (**a**) baseline emission; (**b**) +Euro 6 emission standards in passenger cars (emission −60%, included bikes and hybrid/electric cars); (**c**) +bus emission reduction (−60%); (**d**) +Euro 5 emission standards met by trucks and vans (emission −60%).

As shown in [16], CO concentrations strongly depend on vehicle traffic (share of 52%—traffic, 45%—transboundary inflow, 3%—local area sources), and any method of emission tightening will also help to reduce this very harmful pollution in the urban area. The CO vehicle emission profile fundamentally differs from that for NOx, and gasoline vehicles have a decidedly dominant share in CO emissions. This fact is also reflected in the Euro standards for CO emissions, as can be seen in Tables S1 and S2, where the transition from E3 to E6 emission standards denotes a significant reduction in CO emission for the gasoline vehicles and only a minor change for the diesel vehicles.

Adoption of Euro 6 as the target emission standards, in this case, implies a reduction in the baseline CO emissions by about 57% (cf. Table S1) for all pre-E4 gasoline vehicles. Maps in Figure 4 compare the reference distributions of the annual mean CO concentration and that which relates to the Euro 6 emission standards applied to passenger cars. The maps show a significant reduction in CO pollution, especially in the central districts.

Since CO pollution attributed to line-source emission (52% share) comes almost exclusively from passenger (gasoline) car traffic, the final effect of substituting it by cycling is more evident than in the case of NOx pollution. Figure 4 presents the spatial maps of CO concentrations for baseline emissions (Figure 4a), for implemented Euro 6 emission standards in passenger cars (Figure 4b), and for the additional effect of increased cycling (Figure 4c). A significant improvement in the air quality here is mainly due to the high emission standards applied, but the partial replacement of car traffic by bicycles causes a visible reduction in maximum concentration by 8 μg/m^3 and population exposure by 2.5 μg/m^3. The respective reduction in CO concentration refers to the entire map, including the minimum value.

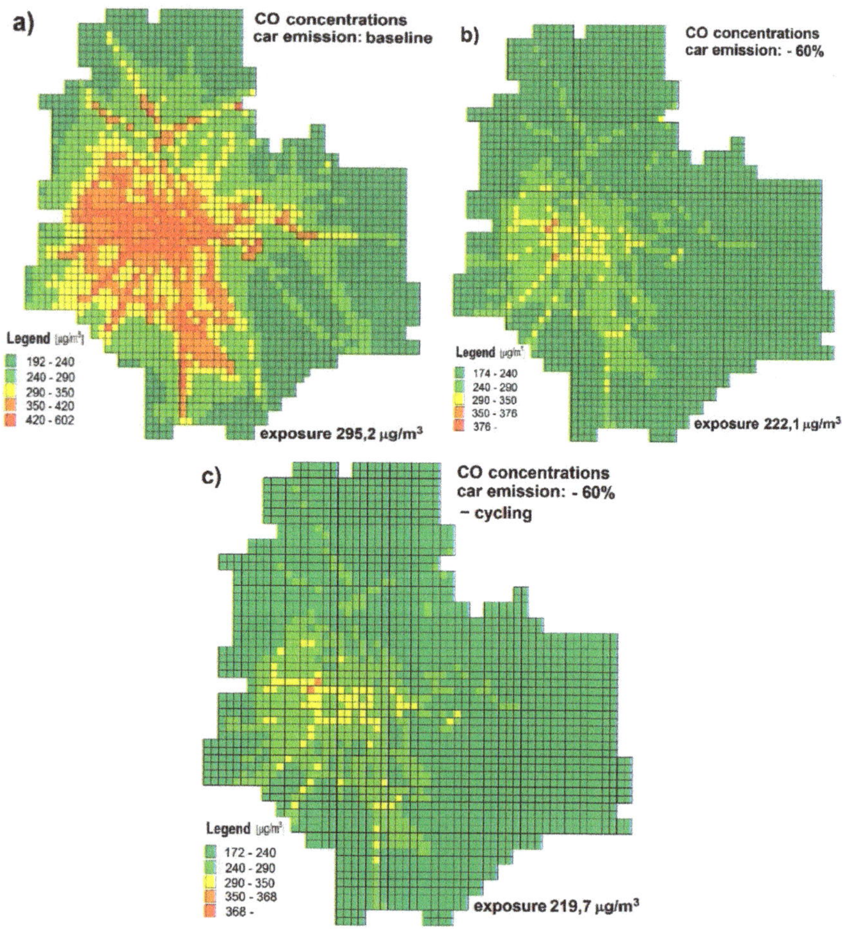

Figure 4. (**a**) The baseline CO concentration; (**b**) +Euro 6 emission standards implemented for passenger car traffic; (**c**) +traffic partially substituted by cycling.

3.3. The Particulate Matter

As seen from [16], road traffic significantly contributes to the particulate matter pollution of an urban area, especially in the case of PM_{10} concentration (33%). However, in this case, implementation of the Euro 6 emission standards will bring only a slight improvement in air quality, for at least two reasons. First of all, PM_{10} pollution emitted by vehicles consists of two components: fraction emitted directly by moving vehicles (primary emission) and resuspended one (secondary emission). As for the rest of the pollution, it comes from all types of sources, including external inflow.

Measurements carried out in Warsaw (Table 2; [24]) revealed that the secondary emission of PM_{10} accounts for about 82% of the total volume. Moreover, the remaining 18% (primary emission) also consists of two components: emissions from fuel combustion and from friction (abrasion of tires, brake linings, etc.). As a result, fuel combustion has only about a 12% share in PM_{10} total emission. Hence, ultimately, implementation of the Euro 6 emission standards in passenger cars and busses can reduce concentration in the city only about 2–3% of PM_{10}, with a minor improvement that can be seen only in the central districts.

Table 2. Emissions of particulate matter from the transportation sector in Warsaw.

Emission		PM_{10} [Mg/y]	PM_{10} [%]	$PM_{2.5}$ [Mg/y]	$PM_{2.5}$ [%]
Primary	Combustion	554	11.5	471	41
	Abrasion	308	6.5	104	9
Re-suspended		3910	81.9	566	50
TOTAL		4772	100	1141	100

Contributions of the primary and resuspended fractions of the final $PM_{2.5}$ emissions are more balanced (compare Table 2), with the share of the fuel combustion being dominant. However, due to the relatively small contribution of vehicle traffic [20] to the overall $PM_{2.5}$ pollution (below 8%), the final air quality improvement from implementing Euro 6 standards in the transportation sector is also minor (below 3%).

Since the use of cycling in urban transportation concerns a possible reduction in passenger car traffic only, the resulting impact on PM concentrations will be also minor. Assuming approximately 50% share of passenger cars in particulate matter emissions, utilizing cycling can reduce an average concentration, as well as the related population exposure, for PM_{10} by about 0.4–0.5%, and for $PM_{2.5}$ by less than 0.1%. In the concentration maps, this will be visible as a slight reduction in the maximum value, compared with the baseline state (A similar effect can be seen in Figure 3b for NOx concentration).

As shown in [16], a significant reduction in PM concentrations, as well as B(a)P pollution, requires a decisive modernization of the municipal sector.

4. Discussion and Conclusions

The paper analyzed emission scenarios of vehicle fleet modernization in Warsaw, aimed at mitigation of the traffic-related air pollution and the related health impact. Similar to other Eastern European conurbations, road traffic in Warsaw has a large share of vehicles meeting only pre-E4 emission standards [6]. This means there is a significant potential for improving air quality through the implementation of low-emission technologies, which was confirmed (especially in the case of NOx concentrations) by the results presented in Section 3.

Regarding the pre-modernization impact of the transport sector on air quality degradation in the city, the total concentrations of NO_x, PM_{10}, $PM_{2.5}$, B(a)P exceed the WHO limit values, while the CO limit is not exceeded. The contribution of individual vehicle categories (diesel or gasoline) varies depending on the type of pollution. Diesel engines definitely have a dominating contribution to NO_x emissions. Gasoline passenger car emissions have the greatest impact on CO concentrations, as shown in Figure 4, where buses and other diesel vehicles have practically no impact. Hence, the modernization of passenger car fleets, particularly concerning diesel vehicles, will improve the urban air quality, mainly due to the significant reduction in NO_x pollution (Figure 3). The development of bicycle transportation that replaces car transportation does not considerably improve the urban air quality. However, the impact of this change is slightly more visible in the case of CO concentration, as a result of the dominant impact of passenger cars on this type of pollution (Figure 4). Nevertheless, bicycling is beneficial for health [25–27]. In the long run, one can also expect a visible improvement in NO_x pollution due to a clearly growing share of hybrid and electric vehicles.

As stated above, vehicles have a considerable share in PM_{10} pollution (33%), but the possible final effects of emission reduction are very small in this case. There are two reasons for this result: (a) low-emission vehicles (also hybrid/electric) reduce approximately 18% of vehicles' primary emissions, compared with 82% of the secondary (resuspended) emissions; (b) only about 11% of the total emission is due to fuel combustion, while the rest comes from friction or resuspension (cf. Table 2). Hence, implementation of any low-emission technology can result in about a 2–3% reduction in final PM_{10} concentration, which is mainly visible in the city center. The basic fractions of $PM_{2.5}$ emissions are more distributed (Table 2), but due to the rather small contribution of cars to the total $PM_{2.5}$ pollution, the final result is also minor. However, a comprehensive effort toward the modernization of road vehicles, together with modernization of HDV diesel vehicles and buses by their replacement with low-emission ones leads to a considerable reduction in the pollution of nitrogen oxides (NO_x), carbon monoxide (CO), and—although to a lesser extent—PMs, B(a)P, SOx, BaP, and C_6H_6. As a result, it undoubtedly contributes to the improvement in urban air quality.

An additional issue to address is the influence of the emission reduction on health effects. As indicated in [20,28], the total NO_x pollution in Warsaw causes a substantial population health risk and is responsible for about 16% of the attributable deaths (non-accidental mortality), 14% of which is assigned directly to the traffic-originated pollution. Modernization of passenger vehicles, including buses, as well as trucks and vans, cause, respectively, a 23%, 35%, and 47% decrease in the inhabitants' exposure to NO_x. Proportionally, this means, respectively, a decrease of 87,133 and 179 attributable yearly deaths (out of the baseline 380). Despite the dominant influence of $PM_{2.5}$ on the assigned mortality rates of residents (82%), the effect of vehicle fleet modernization, in this case, is minor, due to a small share of vehicles (~3%) against a massive impact of area-source pollution and transboundary inflow (share of 45%). A similar remark concerns also other pollutants related to road traffic that have an insignificant share in air pollution, such as B(a)P or SO_2. This is due to the fact that the above pollutions are also emitted by the municipal sector, whose contribution is dominant. It is additionally increased due to a large share of external inflow from Warsaw surroundings. Hence, abatement of the municipal sector emission and transboundary inflow would help more in improving Warsaw air quality and decreasing population health risk. Implementation of the latter abatement policy is, however, more troublesome, also because the abatement of transboundary flow requires national authorities to be involved.

Supplementary Materials: The following are available online at https://www.mdpi.com/article/10.3390/atmos12121581/s1, Table S1. Euro Norms for the gasoline cars. Table S2. Euro Norms for the diesel cars. Table S3. Euro Norms for HDV diesel cars.

Author Contributions: Conceptualization, P.H.; methodology, P.H. and Z.N.; validation, P.H., Z.N. and A.K.; formal analysis, Z.N.; resources, A.K.; data curation, A.K.; writing—original draft preparation, P.H.; writing—review and editing, Z.N. All authors have read and agreed to the published version of the manuscript.

Funding: This research received no external funding.

Institutional Review Board Statement: Not applicable.

Informed Consent Statement: Not applicable.

Data Availability Statement: Not applicable.

Acknowledgments: Research conducted by Z. Nahorski was partially supported by the National Science Center, Poland under the Grant DEC-2018/30/Q/HS4/00764.

Conflicts of Interest: The authors declare no conflict of interest.

References

1. EC. Urban Air Pollution–What Are the Main Sources across the World? 2015. Available online: https://ec.europa.eu/jrc/en/news/what-are-main-sources-urban-air-pollution (accessed on 20 November 2021).
2. WHO. Database on Source Apportionment Studies for Particulate Matter in the Air (PM10 and PM2.5). 2015. Available online: https://www.who.int/quantifying_ehimpacts/global/source_apport/en/ (accessed on 20 November 2021).
3. EC; AAQD. Directive 2008/50/EC of the European Parliament and of the Council of 21 May 2008 on Ambient Air Quality and Cleaner Air for Europe. 2008. Available online: https://eur-lex.europa.eu/eli/dir/2008/50/oj (accessed on 20 November 2021).
4. EEA. Air Quality in Europe—2019 Report. EEA Report, No 10. 2019. Available online: https://www.eea.europa.eu/publications/air-quality-in-europe-2019 (accessed on 20 November 2021).
5. WHO. Ambient (Outdoor) Air Pollution. 2018. Available online: https://www.who.int/news-room/fact-sheets/detail/ambient- (accessed on 20 November 2021).
6. Degraeuwe, B.; Pisoni, E.; Peduzzi, E.; De Meij, A.; Monforti-Ferrario, F.; Bodis, K.; Mascherpa, A.; Astorga-Llorens, M.; Thunis, P.; Vignati, E. *Urban NO_2 Atlas (EUR 29943 EN)*; Publications Office of the European Union: Luxembourg, 2019.
7. EC. Air Quality: Traffic Measures Could Effectively Reduce NO_2 Concentrations by 40% in Cities. 2019. Available online: https://ec.europa.eu/jrc/en/news/air-quality-traffic-measures-could-effectively-reduce-no2-concentrations-40-europe-s-cities (accessed on 20 November 2021).
8. Berkowicz, R.; Winther, M.; Ketzel, M. Traffic pollution modelling and emission data. *Environ. Model. Softw.* **2016**, *21*, 454–460. [CrossRef]
9. Buchholz, S.; Krein, A.; Junk, J.; Heinemann, G.; Hoffmann, L. Simulation of Urban-Scale Air Pollution Patterns in Luxembourg: Contributing Sources and Emission Scenarios. *Environ. Model. Assess.* **2013**, *18*, 271–283. [CrossRef]
10. Calori, G.; Clemente, M.; De Maria, R.; Finardi, S.; Lollobrigida, F.; Tinarelli, G. Air quality integrated modelling in Turin urban area. *Environ. Model. Softw.* **2006**, *21*, 468–476. [CrossRef]
11. Mediavilla-Sahagún, A.; ApSimon, H.M. Urban scale integrated assessment for London: Which emission reduction strategies are more effective in attaining prescribed PM10 air quality standards by 2005? *Environ. Model. Softw.* **2006**, *21*, 501–513. [CrossRef]
12. Rith, M.; Fillone, A.M.; Biona, J.B.M.M. Energy and environmental benefits and policy implications for private passenger vehicles in an emerging metropolis of Southern Asia–A case study of Metro Manila. *Appl. Energy* **2020**, *275*, 115240. [CrossRef] [PubMed]
13. Pisoni, E.; Thunis, P.; Clappier, A. Application of the SHERPA source-receptor relationships, based on the EMEP MSC-W model, for the assessment of air quality policy scenarios. *Atmos. Environ. X* **2019**, *4*, 100047. [CrossRef]
14. Thunis, P.; Degraeuwe, B.; Pisoni, E.; Ferrari, F.; Clappier, A. On the design and assessment of regional air quality plans: The SHERPA approach. *J. Environ. Manag.* **2016**, *183*, 952–958. [CrossRef] [PubMed]
15. Degraeuwe, B.; Thunis, P.; Clappier, A.; Weiss, M.; Lefebvre, W.; Janssen, S.; Vranckx, S. Impact of passenger car NOx emissions on urban NO_2 pollution–Scenario analysis for 8 European cities. *Atmos. Environ.* **2017**, *171*, 330–337. [CrossRef]
16. Holnicki, P.; Kałuszko, A.; Nahorski, Z. Analysis of emission abatement scenario to improve urban air quality. *Arch. Environ. Prot.* **2021**, *47*, 103–114. [CrossRef]
17. Available online: https://smoglab.pl/warszawa-ma-prawie-dwa-razy-wiecej-zarejestrowa-nych-pojazdow-na-km2-niz-krakow-wroclaw-i-berlin (accessed on 20 October 2019). (In Polish)
18. Available online: https://www.auto-swiat.pl/wiadomosci/aktualnosci/czy-w-warszawie-naprawde-jest-wiecej-aut-niz-mieszkancow/4xqb44s (accessed on 20 September 2021). (In Polish)
19. Kiesewetter, G.; Borken-Kleefeld, J.; Schöpp, W.; Heyes, C.; Thunis, P.; Bessagnet, B.; Terrenoire, E.; Gsella, A.; Amann, M. Modelling NO_2 concentrations at the street level in the GAINS integrated assessment model: Projections under current legislation. *Atmos. Chem. Phys.* **2014**, *14*, 813–829. [CrossRef]
20. Holnicki, P.; Tainio, M.; Kałuszko, A.; Nahorski, Z. Burden of mortality and disease attributable to multiple air pollutants in Warsaw, Poland. *Int. J. Environ. Res. Public Health* **2017**, *14*, 1359. [CrossRef] [PubMed]
21. Available online: https://um.warszawa.pl/waw/rowery/-/pomiary-ruchu-rowerowego-2020-1 (accessed on 5 July 2021). (In Polish)
22. Kwiatkowski, M.A.; Szymańska, D. Cycling policy in strategic documents of Polish cities. *Environ. Dev. Sustain.* **2021**, *23*, 10357–10377. [CrossRef]
23. Available online: https://www.portalsamorzadowy.pl/wydarzenia-lokalne/warszawa-spadla-liczba-rejestracji-ale-rosnie-ilosc-samochodow-z-napedem-ekologicznym,252401 (accessed on 8 October 2020). (In Polish)

24. Voivodship Inspectorate of Environment Protection. Report for the Year 2012. (WIOŚ 2012). Environment Quality in Mazovian Voivodship in the Year 2012. Available online: https://keep.eu/partners/6593/Voivodship-Inspectorate-of-EEN/ (accessed on 20 September 2021). (In Polish)
25. Tainio, M.; de Nazelle, A.J.; Götschi, T.; Kahlmeier, S.; Rojas-Rueda, D.; Nieuwenhuijsen, M.J.; de Sá, T.H.; Kelly, P.; Woodcock, J. Can air pollution negate the health benefits of cycling and walking? *Prev. Med.* **2016**, *87*, 233–236. [CrossRef] [PubMed]
26. Zapata-Diomedi, B.; Knibbs, L.D.; Ware, R.S.; Heesch, K.C.; Tainio, M.; Woodcock, J.; Veerman, J.L. A shift from motorised travel to active transport: What are the potential health gains for an Australian city? *PLoS ONE* **2017**, *12*, e0184799. [CrossRef] [PubMed]
27. Giallouros, G.; Kouis, P.; Papatheodorou, S.I.; Woodcock, J.; Tainio, M. The long-term impact of restricting cycling and walking during high air pollution days on all-cause mortality: Health impact Assessment study. *Environ. Int.* **2020**, *140*, 105679. [CrossRef] [PubMed]
28. Tainio, M. Burden of disease caused by local transport in Warsaw, Poland. *J. Transp. Health* **2015**, *2*, 423–433. [CrossRef] [PubMed]

atmosphere

Article

Basic Facts about Numerical Simulations of Atmospheric Composition in the City of Sofia

Vladimir Ivanov * and **Ivelina Georgieva**

National Institute of Geophysics, Geodesy and Geography, Bulgarian Academy of Sciences, Acad. G. Bonchev Str., Bl. 3, 1113 Sofia, Bulgaria; iivanova@geophys.bas.bg
* Correspondence: vivanov@geophys.bas.bg; Tel.: +359-29793328

Abstract: The atmospheric composition in urban areas is one of the primary tasks in air pollution studies. The research aims to provide a statistically reliable assessment of the atmospheric composition climate of the city of Sofia—typical and extreme features of the special/temporal behavior, annual means, seasonal and diurnal variations. For that purpose, extensive numerical simulations of the atmospheric composition fields in Sofia city have been performed. Three models were chosen as modeling tools. We used WRF as a meteorological pre-processor, CMAQ as a chemical transport model, and SMOKE as the emission pre-processor of Models-3 system. We developed the following conclusions. The daily concentration changes of the two essential air pollution species—nitrogen dioxide (NO_2) and fine particle matters (FPRM, particulate matter ($PM_{2.5}$), which has a diameter between 0 and 2.5 micrometers)—have different magnitudes. Second, the emissions relative contributions to the concentration of different species could be different, varying from 0% to above 100%. The contributions of different emission categories to other species surface concentrations have various diurnal courses. Last, the total concentration change (ΔC) is different for each pollutant. The sign of the contributions of some processes is evident. Still, some may have different signs depending on the type of emissions, weather conditions, or topography.

Keywords: atmospheric composition; dynamic and chemical processes; ensemble of numerical simulation; process analysis; contribution of different emission sources

Citation: Ivanov, V.; Georgieva, I. Basic Facts about Numerical Simulations of Atmospheric Composition in the City of Sofia. *Atmosphere* **2021**, *12*, 1450. https://doi.org/10.3390/atmos12111450

Academic Editors: Elena Hristova, Manousos Ioannis Manousakas, Anikó Angyal and Maria Gini

Received: 30 September 2021
Accepted: 28 October 2021
Published: 2 November 2021

1. Introduction

The atmospheric composition in urban areas is one of the primary tasks in air pollution studies. The air pollution climate in urban areas has not been systematically studied yet in Bulgaria, but some air pollution modeling for the city of Sofia had been performed, and air pollution forecast for the city is operationally occurring [1–5]. Recently, extensive studies for long enough simulation periods and reasonable resolution of the atmospheric composition status in Bulgaria have been carried out using up-to-date modeling tools and detailed and reliable input data [6–25]. The next step in studying the atmospheric composition climate is performing simulations on an urban scale. Research works dealing with this topic include domains with different sizes and scales from meters to kilometers [26–34]. The models' simulations in the smallest scales usually deal with sub-urban-sized areas as streets and neighborhoods. The simulations aim to construct an ensemble comprehensive enough to provide a statistically reliable assessment of the atmospheric composition climate of the city of Sofia—typical and extreme features of the special/temporal behavior, annual means, seasonal and diurnal variations, etc. Some evaluations of the contribution of different pollution sources and dynamic and chemical processes to the atmospheric composition of the city of Sofia will be presented in the paper.

Methodology: Extensive numerical simulations of the atmospheric composition fields in Sofia city have been recently performed, and an ensemble, comprehensive enough as to provide a statistically reliable assessment of the atmospheric composition climate, was constructed. The US EPA Models-3 system was chosen as a modeling tool. WRF [35] used as meteorological pre-processor, CMAQ—the Community Multiscale Air Quality System [36,37], being the Chemical Transport Model (CTM) of the system, and SMOKE—the Sparse Matrix Operator Kernel Emissions Modeling System [38]—the emission pre-processor of Models-3 system.

As the NCEP Global Analysis Data with one-degree resolution is used as meteorological background, the system nesting capabilities were applied for downscaling the simulations to a 1 km resolution (Figure 1). Results for the Sofia city domain (D5) and two locations—Orlov most and Bistritsa—are presented in this paper. Orlov most is located in the center of the city and represents a typical urban-polluted site. Bistritsa is located on the semi-mountain outskirts and represents a rural site of the domain.

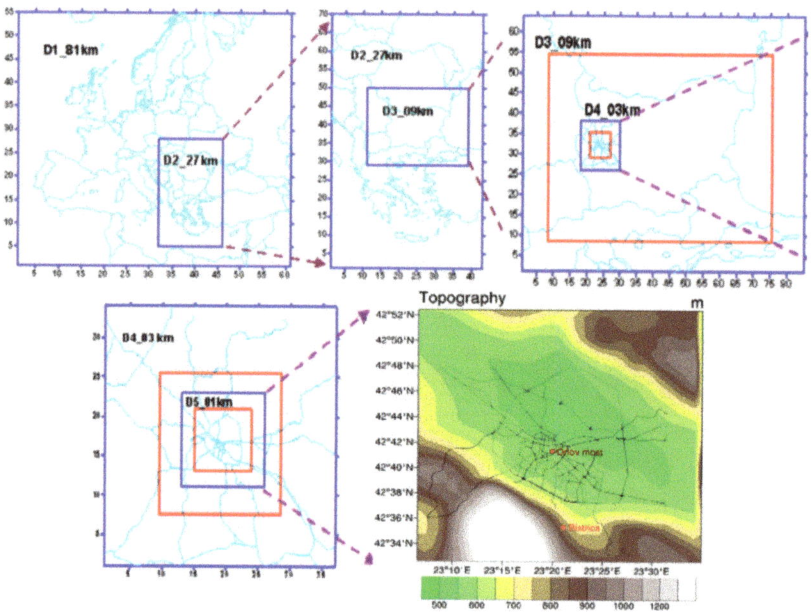

Figure 1. Five computational domains (CMAQ domains are nested in WRF ones)—D1 81 × 81 km (Europe), D2 27 × 27 km (Balkan Peninsula), D3 9 × 9 km (Bulgaria), D4 3 × 3 km (Sofia municipality) and D5 1 × 1 km (Sofia city).

The TNO inventory with resolution 0.25° × 0.125° in 10 SNAP categories for 2010 [39–41] is exploited for the territories outside Bulgaria in the mother CMAQs domain. For the Bulgarian domains, the National inventory for 2010 as provided by Bulgarian Executive Environmental Agency is used. All simulations were performed for a 7-year period from 2008 to 2014 with two-way nesting mod on. Special pre-processing procedures were created for introducing temporal profiles and speciation of the emissions [42].

The study is based on a large number of numerical simulations carried out daily for all periods. Different characteristics of the numerically obtained concentration fields and for five emission scenarios, with all the emissions included and reduced of factor 0.8, including emissions from energetics, non-industrial and industrial combustions, and road transport, will be demonstrated in the present paper. Results concerning the contribution of the different emission categories are demonstrated. The air pollution pattern is also formed as a result of the interaction of different processes; thus, knowing the contribution of each one of these processes for different meteorological conditions and given emission spatial configuration and temporal profiles is useful for understanding the atmospheric composition and air pollutants origin and behavior. To analyze the contribution of different dynamic and chemical processes, the CMAQ "integrated process rate analysis" option was applied. The procedure allows the concentration change for each compound to be presented as a sum of the contribution of each one of the processes, which determines the air pollution concentration.

2. Materials and Methods

The sample size of the 7-year simulation is large and comprehensive enough to allow processing a variety of statistical quantities. We can calculate different statistics for the entire Sofia city domain, as well as for chosen locations—dispersion, the absolute minimum and maximum concentrations, a percent of the averages in the given interval of values, the probability density function, skewness, kurtosis, etc.

We present plots for the main statistical characteristics of the ensemble characteristics of the air pollution, namely seasonal and annual ones for some pollutants NO_2 (nitrogen dioxide) and FPRM (fine particulate matter), averaged for the Sofia city region and in two locations in the city—Orlov most (point situated at the city center) and Bistrica (situated at semi-mountain outskirts). The plots give graphics of mean concentrations, maximum ones for entire simulation (period/ensemble), and curves noted as 10%, 25%, 75%, and 90%, for the 10th, 25th, 75th, and 90th percentile, respectively. The resulting curves suggest that 50% of the cases fall into the 25–75% interval, and 80% of the cases fall into the 10–90% interval. The plotted curves give a good enough idea for the statistical characteristics of the ensemble—dispersion, asymmetry, excess kurtosis, without being shown explicitly. The axes are given in logarithmic scale due to the high absolute minimal concentrations and the impossibility to distinguish the intervals of cases with different concentrations from the absolute ones on one plot. The absolute minimum concentrations are close to zero in most of the cases; thus, they are not shown on the logarithmic plots.

As already stated, the simulated fields ensemble is large enough to allow statistical treatment. In particular, the probability density functions for each of the atmospheric compounds can be calculated, with the respective seasonal and diurnal variations, for each of the points of the simulation grid or averaged over the territory of the city. Knowing the probability density functions means knowing everything about the ensemble. An example of spatial and diurnal variations of the annual ensembles of surface NO_2 and FPRM behaviors ate two locations, the typically urban site "Orlov most" and "Bistritsa", are shown in the paper.

The emission inventory, used in the simulations, includes 10 emission categories (SNAP categories) and allows the evaluation of the contribution of various anthropogenic activities to the overall picture of air pollution in the city of Sofia:

1. SNAP 1 (Combustion in energetics) reduced with factor 0.8;
2. SNAP 2 (Non-industrial combustion plants) reduced with factor 0.8;
3. SNAP 3 (Combustion in manufacturing industry) reduced with factor 0.8;
4. Production processes;
5. Extraction and distribution of fossil fuels;
6. Solvent and other product use;
7. SNAP 7 (Road transport) reduced with factor 0.8;
8. Other mobile sources and machinery;
9. Waste treatment and disposal;
10. Agriculture.

The used SNAP categorization of emissions, reduced by some factor, allows evaluation of the contribution of road transport, energetic, industrial, and non-industrial combustions to the atmospheric composition in the city. The concentrations for each scenario of reduced SNAP's were calculated for each day of the period, and the relative contribution of the emissions for each of the scenarios was calculated in the following way.

If an arbitrary pollution characteristic (concentration, deposition, process contribution, etc.) for a given grid point, or averaged over chosen domain, obtained with all the emissions accounted for, is denoted by ϕ, then ϕ_m is the respective characteristic obtained when the emissions form source category m is reduced by a factor of α. In such a case, the quantity φ_m can be interpreted as the relative (in %) contributions of emission category m to the formation of the characteristic φ:

$$\varphi_m = \frac{1}{1-\alpha}\frac{\phi - \phi_m}{\phi}\cdot 100 \quad (1)$$

More than one selected nomenclature for sources of air pollution (SNAP) category emissions can be reduced by a factor of α, and thus the joint contribution of several or all SNAP categories to the formation of the pollution characteristic ϕ can be evaluated. Obtained relative source contributions can also be averaged for the entire ensemble, thus providing the "climate" of the emission contributions, in particular the "typical" annual and seasonal contributions.

The reason why the emissions from a given category are reduced by a chosen factor and not simply at zero is that by completely removing the emissions from a given category, we can obtain much smaller concentrations, which may change the rate of some nonlinear chemical reactions. Moreover, the significant reduction of the concentrations may change the compound diffusion through the domain boundaries, which is why it is a general practice for the simulations to evaluate the contribution of emissions from a given category to be performed as applied and not removed, but reduced emissions from this category, which is normally performed in such studies. The reduction of 20% is preferred by many authors in atmospheric pollution studies.

Five emission scenarios will be considered here: Simulations with all the emissions included and with the emissions from all the SNAP categories (SNALL), SNAP categories 1 (energetics—SN1), SNAP categories 2 (non-industrial combustions—SN2), SNAP categories 3 (industrial combustions—SN3) and SNAP categories 7 (road transport—SN7) for Sofia reduced by a factor of 0.8. This makes it possible to evaluate the contribution of all the emissions, as well as the emissions from road transport, energetics, industrial and non-industrial combustions to the atmospheric composition in the city. The relative contribution of the emissions for each scenario was calculated for each day of this 7-year period, and then, by averaging over the ensemble, the typical fields of relative contributions of these emissions to the surface concentrations of each of the compounds were calculated for the four seasons and annually. For all the emission categories, the pattern of the contribution fields is complex, which reflects the emission source configuration, the heterogeneity of topography, land use, and meteorological conditions. In order to demonstrate the emission contribution behavior in a simpler and easy to comprehend way, the respective fields can be averaged over some domain, which makes it possible to follow and compare the diurnal behavior of the respective contributions for different species. Graphics of the diurnal evolution of the "typical" relative contribution annually and seasonal emissions of SNAP categories 1, 2, 3, 7 and all the emissions to the surface concentrations of NO_2, FPRM, averaged for the territory of Sofia city and for Orlov most and Bistritsa, are shown in figures.

Atmospheric pollution is a result of the interaction of different dynamic and chemical processes. The consideration of the interaction and contribution of these processes provides a possibility for an explanation of the entire picture of the air pollution in Sofia city. The processes that influence the formation of the air pollution patterns are HADV—horizontal advection, ZADV—vertical advection, HDIF—horizontal diffusion, VDIF—vertical diffusion, EMIS—emissions, DDEP—dry deposition, CLDS—cloud processes, CHEM—chemical processes, and AERO—aerosol processes. In the current section, we present mainly results from the high-performance computing simulations, which evaluate the contribution of different dynamic transportation and transformation processes of air pollutants, which form the air pollution climate in Sofia city. The function of the CMAQ model—"integrated process rate analysis" was used for this task—a specific option that gives an opportunity to estimate the role of each one of the former processes in air pollution formation. In that way, the concentration change ΔC_i of the *i*-th pollutant for a given time interval from t to [t+ Δt], can be present as a sum of the contributions of the different processes that determine the concentration change, i.e., the equation for transport and transformation of pollutants can be written in the form:

$$\Delta c_i^1 = (\Delta c_i^1)_{hdif} + (\Delta c_i^1)_{vdif} + (\Delta c_i^1)_{h\,adv} + (\Delta c_i^1)_{v\,adv} + (\Delta c_i^1)_{drydep} + (\Delta c_i^1)_{emiss} + (\Delta c_i^1)_{chem} + (\Delta c_i^1)_{cloud} + (\Delta c_i^1)_{aero} \quad (2)$$

The figures show the annual and seasonal averaged contributions of the HADV, ZADV, HDIF, VDIF, EMIS, DDEP, CLDS, CHEM, AERO leading to the different air pollutants (NO_2, FPRM) formations, averaged for some locations of the domain, as well as for the entire one—Sofia city. All the plots demonstrated are for the first model layer. The total change of the concentration (ΔC) is also plotted, and we can see its sign—positive or negative and that it has a well pronounced diurnal course. Different contributions with different values dominate the formation of the air pollutants, which is traced on the graphics and which contribution of given processes dominate, in a given time and sign.

3. Results

The Section 3 is separated into three parts and presents characteristics of the numerically obtained concentration fields, the contribution of different emission sources, and process analysis of the atmospheric composition for the entire domain Sofia city and two locations—Orlov most and Bistritsa.

Graphics of the diurnal evolution of the "typical" relative contribution—annual and seasonal emissions—of SNAP categories 1, 2, 3, 7 and the emissions of the surface concentrations of NO_2 and FPRM, averaged for the territory of Sofia city and for Orlov most and Bistritsa, are shown.

The annual and seasonal averaged contributions of the processes leading to the formation of pollutants (NO_2, FPRM), averaged for some locations of the domain as well as for the entire Sofia city, are presented. The total change of the concentration (ΔC) is also plotted, and we can see its sign—positive or negative—has a well pronounced diurnal course. Different contributions with different values dominate the formation of the air pollutants, which is traced on the graphics—where a given process dominates, at a given time and sign. The values of ΔC differ in different seasons and are defined from the superposition of the contributions of the different processes, which are also with a different signs.

3.1. Characteristics of the Numerically Obtained Concentration Fields, Contribution of Different Emission Sources and Different Dynamic and Chemical Processes to the Atmospheric Composition in Sofia

The graphics in Figure 2 show that the average NO_2 concentrations have a well-expressed diurnal course with a maximum in the early hours and a minimum in the afternoon. The average concentrations are mostly in the 80% interval of cases (between the two green curves 10–90%). The average concentrations are asymmetrically located in the different pieces of a number of case intervals during the day as well as in the different seasons. The seasonal course shows that the absolute maximum of NO_2 concentrations are highest in the autumn and the winter, which is probably due to the bigger frequency of the stable atmospheric stratification cases and the impeded vertical turbulent transport of NO_2. The averaged ensemble annual and seasonal contributions of the different sources (Figure 3), leading to the formation of surface NO_2 in Sofia, sheds light from another perspective. The results suggest that the contribution of the different sources varies in each season but with a similar diurnal course. Everywhere during the night, the biggest natural contribution is from emissions of all SNAP categories, followed by one of the road transport (SNALL and SN7), about 30% around midnight during all seasons. The contribution of the other sources in the morning hours has a small peak of around 10%.

Figure 4 shows the annually and seasonally averaged contributions of the processes of NO_2 formation averaged for the entire domain. The main contribution for the NO_2 formation has chemical processes with a positive sign. The horizontal advection is positive in the morning and afternoon and becomes negative at noon. The vertical advection is opposite to the horizontal one. The dry deposition and vertical diffusion have a negative contribution, although with smaller magnitudes. We can outline the positive contribution of the emissions at all hours of the day. The contribution of the other processes is almost zero.

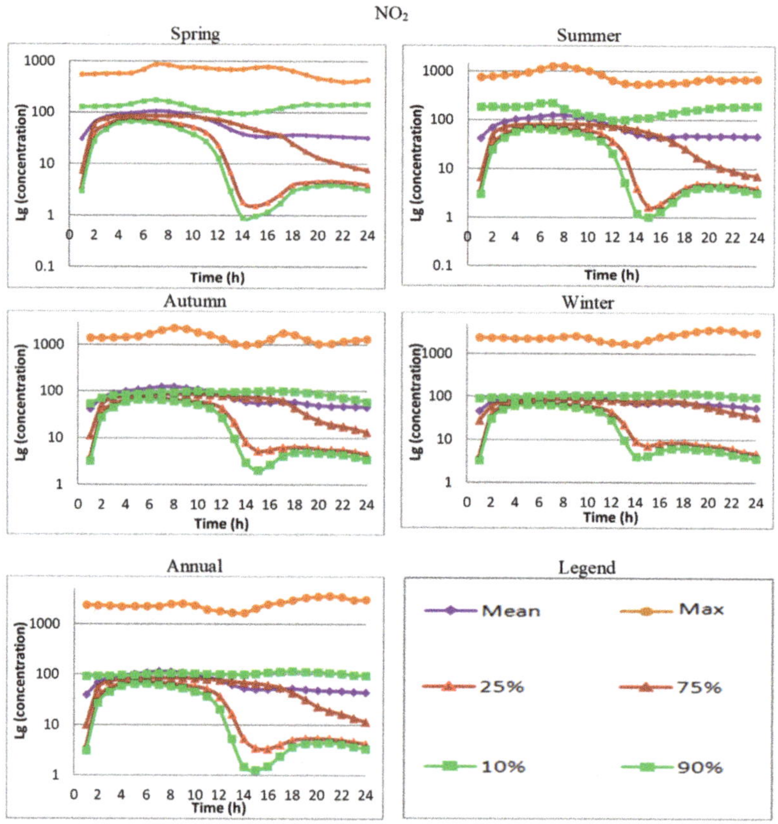

Figure 2. Spatial and diurnal variations of the annual and seasonal ensembles of surface NO_2 ($\mu g/m^3$) concentration behavior in logarithmic scale for Sofia for the period 2008–2014.

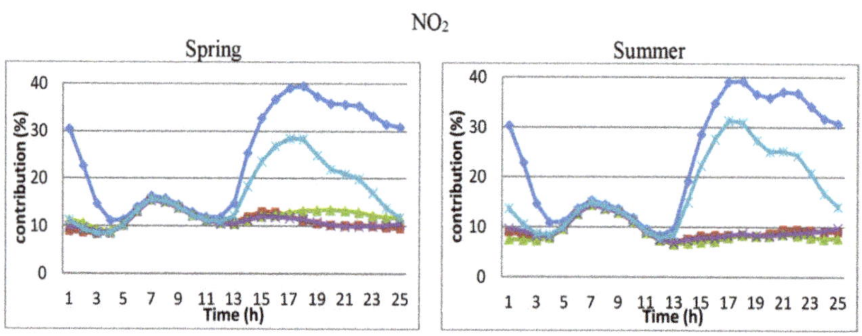

Figure 3. *Cont.*

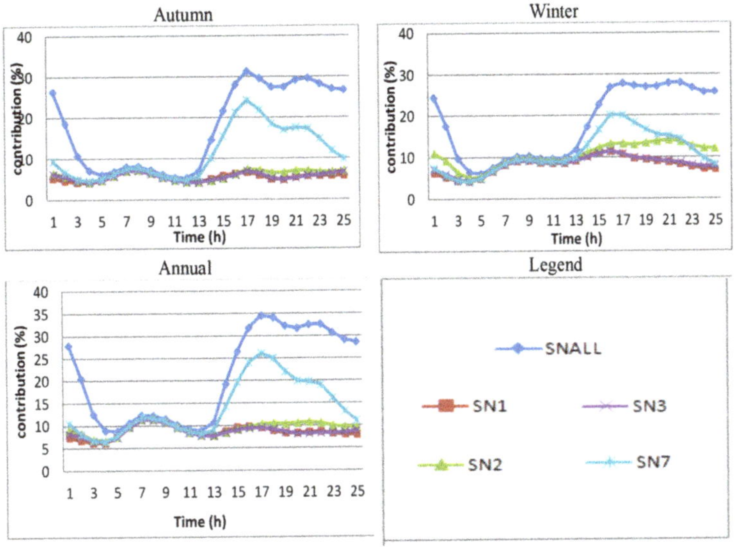

Figure 3. Annual and seasonal averaged relative contribution (%) of all emissions and emissions from different SNAP categories to the formation of NO_2 concentration in Sofia for the period 2008–2014.

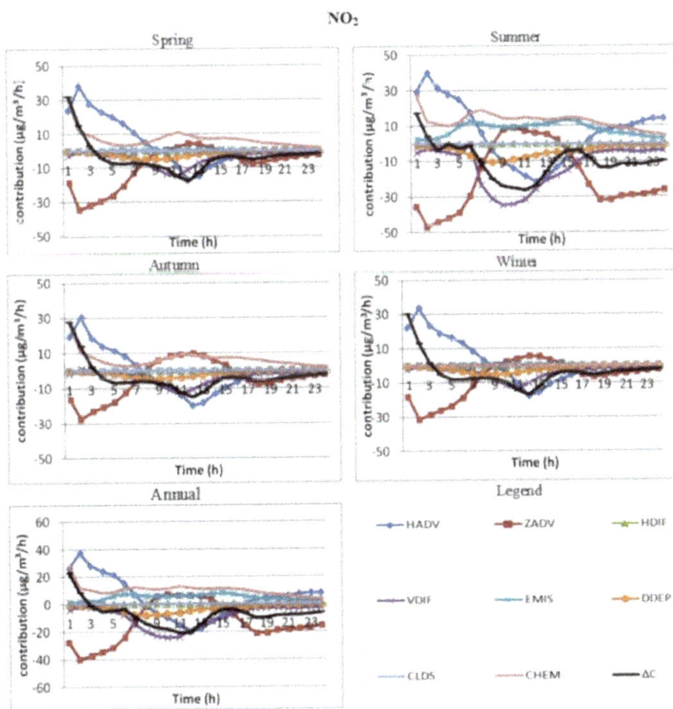

Figure 4. Annual, seasonal and diurnal course of the contribution of the different dynamic and chemical processes of the formation of NO_2 concentrations (μg/m^3/h) averaged for the territory of Sofia for the period 2008–2014.

The plots in Figure 5 suggest that the average FPRM concentrations have a well-expressed diurnal and seasonal course and asymmetrical distribution during the day and seasons for different intervals. The average concentrations fall into the interval containing 80% of cases during the warm months and above that for the cold ones, as during the winter, coinciding with the 90% curve. The absolute maximum and average concentrations are highest in the autumn and winter (stable stratification). There is a well-expressed maximum during the early hours for NO_2 and a minimum around noon. The possible reasons for that maximum are the stable atmosphere and intensive road traffic in the early morning. The results for the averaged by ensemble annual and seasonal contributions of any source leading to the formation of surface FPRM for Sofia city are different (Figure 6), suggesting that the contribution of the different sources varies seasonally but has an equal diurnal course. The dominating natural contribution is one of the sources of all SNAP categories (SNALL), about 40%, followed by one of the road transport (SN7), at about 30%. The contribution of the other sources is about 10% in all seasons. The diurnal distribution shows that the different SNAP categories have maximum contribution in the morning hours and afternoon, and minimum around noon and during the night, which correspond to the concentrations graphics.

The annually and seasonally averaged contributions of processes for FPRM formation averaged for the entire Sofia city domain are shown in Figure 7. The main positive contribution has vertical diffusion for the entire day and vertical advection around noon. The contribution of the vertical advection in the morning and afternoon is negative. The horizontal advection is opposite to the vertical one. The dry deposition has a negative contribution with a maximum around noon. The aerosol processes have a negative sign in all seasons except for winter. The processes are more active during the winter in comparison with other seasons. The winter is outlined with a large positive contribution of emissions and decreases to a negative contribution of vertical diffusion. The other processes have almost zero contributions.

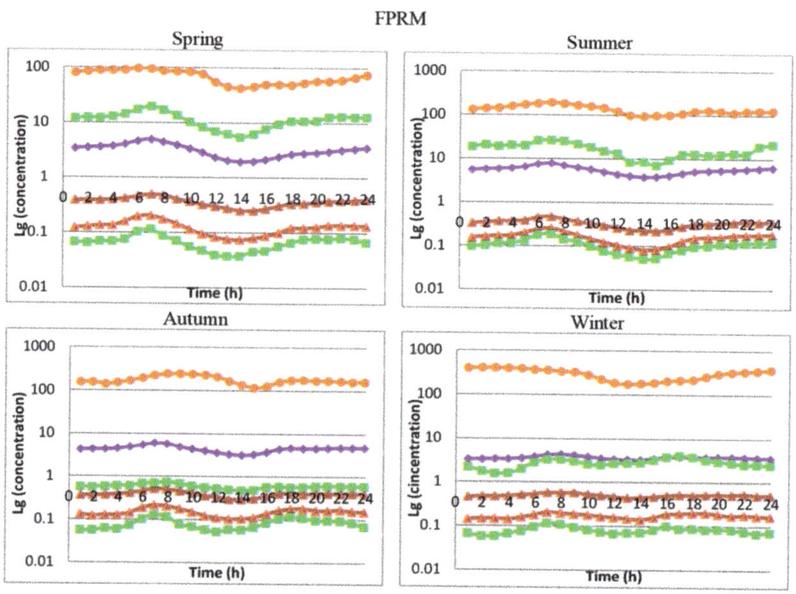

Figure 5. *Cont.*

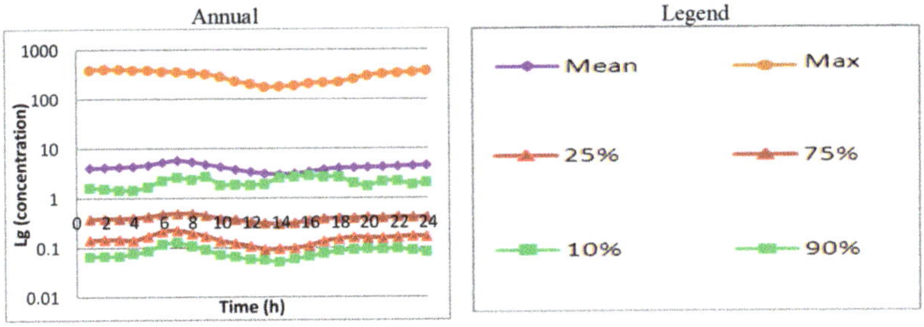

Figure 5. Spatial and diurnal variations of the annual and seasonal ensembles of surface FPRM (μg/m^3) concentrations behavior in logarithmic scale for Sofia for the period 2008–2014.

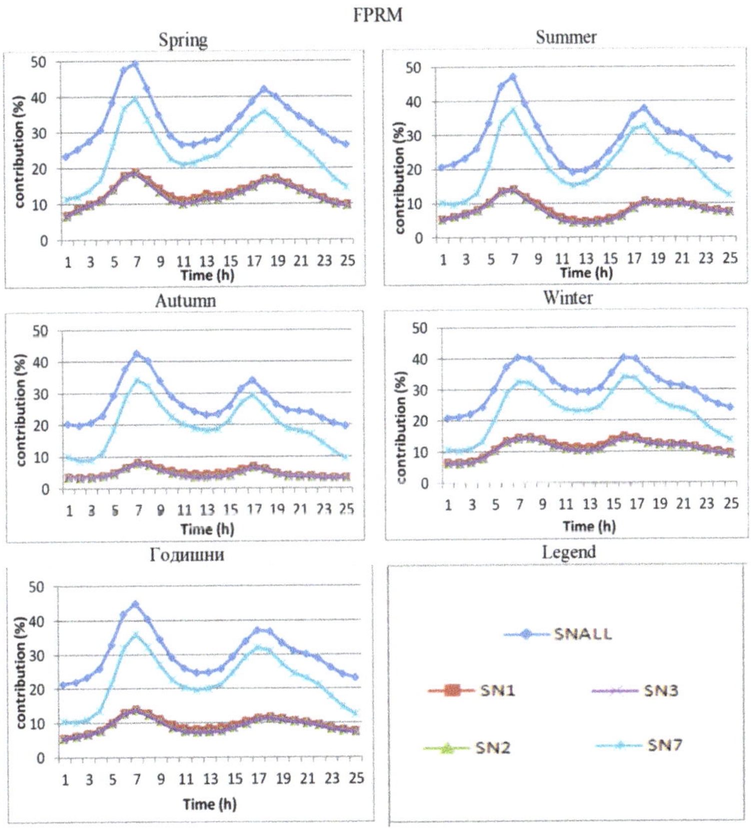

Figure 6. Annual and seasonal averaged relative contribution (%) of all emissions and the emissions from different SNAP categories of the formation of FPRM concentrations for Sofia for the period 2008–2014.

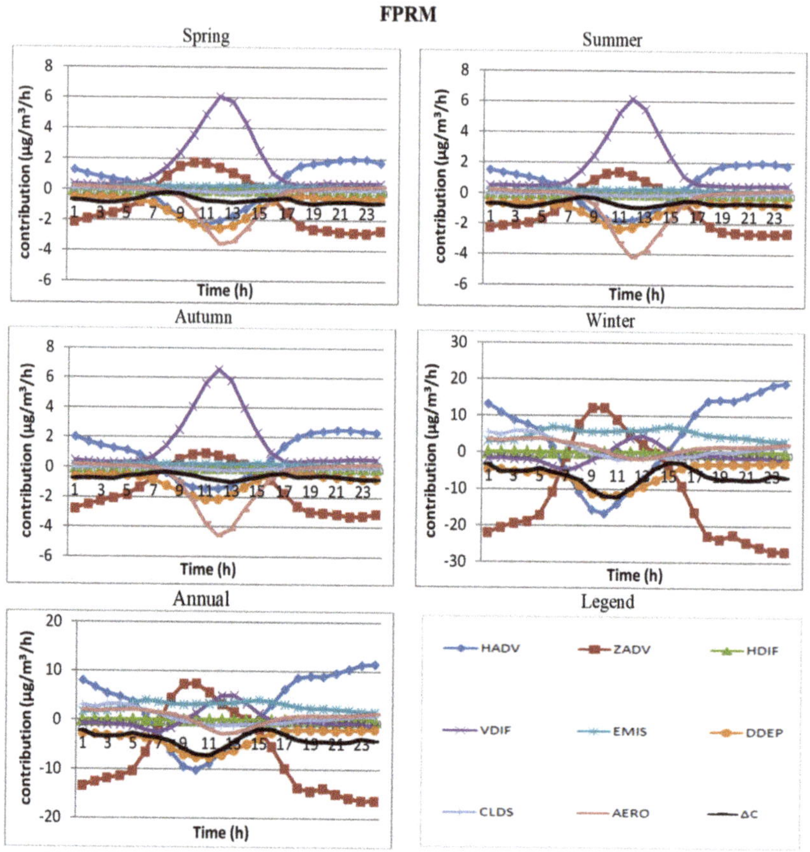

Figure 7. Annual, seasonal and diurnal course of the contribution of the different dynamic and chemical processes for the formation of FPRM concentrations (μg/m^3/h) averaged for the territory of Sofia for the period 2008–2014.

3.2. Characteristics of the Numerically Obtained Concentration Fields, Contribution of Different Emission Sources and Different Dynamic and Chemical Processes to the Atmospheric Composition in Orlov Most

Figure 8 shows the annually and seasonally averaged NO_2 concentrations for the Orlov most station. The results on the plots show that the average concentrations have a well-expressed diurnal course with a maximum in the early morning hours and a minimum in the afternoon. The average concentrations are mostly in the 80% interval of cases (between the two green curves, at 10–90%). The average concentrations are asymmetrically located in the different intervals of a number of cases during the day as well as the seasons. The seasonal course suggests that the absolute maximum NO_2 concentrations are highest in autumn and winter. The last is probably due to the more cases with stable atmospheric stratification and the impeded turbulent exchange of NO_2 in the vertical direction. The averaged contributions from the different sources for NO_2 formation in each season are different but with an equal diurnal course (Figure 9). The main contribution during the night is from the emissions of all SNAP sources (SNALL)—about 80%, followed by traffic (SN7)—about 50%, in spring, summer, and autumn. The second largest contribution in winter is from the industry (SN3). The contribution of all other sources in the morning hours has a peak of about 20%. The contribution of the sources from non-industrial burning (SN2) is the least.

The average annual and seasonal contributions of the processes of NO_2 for Orlov most are shown in Figure 10. The main contribution for NO_2 formation is chemical processes, which have a positive sign, as well as a high positive contribution of emissions in winter. Horizontal advection is positive during the entire day, and vertical advection is opposite. Vertical diffusion also has a negative contribution, although with smaller values and a maximum around noon. The contribution of the other processes is at almost zero magnitude.

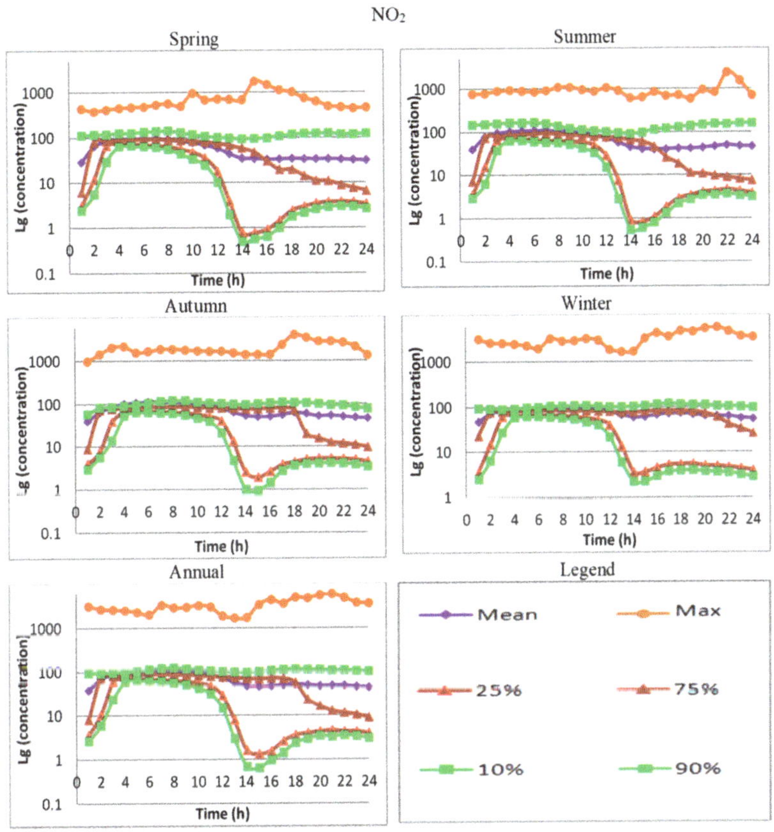

Figure 8. Spatial and diurnal variations of the annual and seasonal ensembles of surface NO_2 ($\mu g/m^3$) concentrations behavior in logarithmic scale for Orlov most for the period 2008–2014.

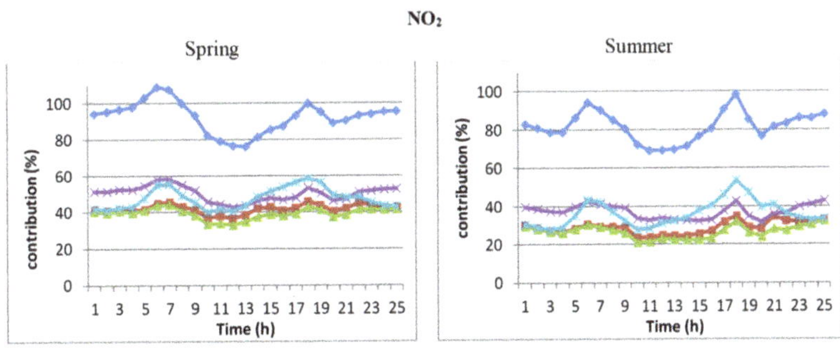

Figure 9. *Cont.*

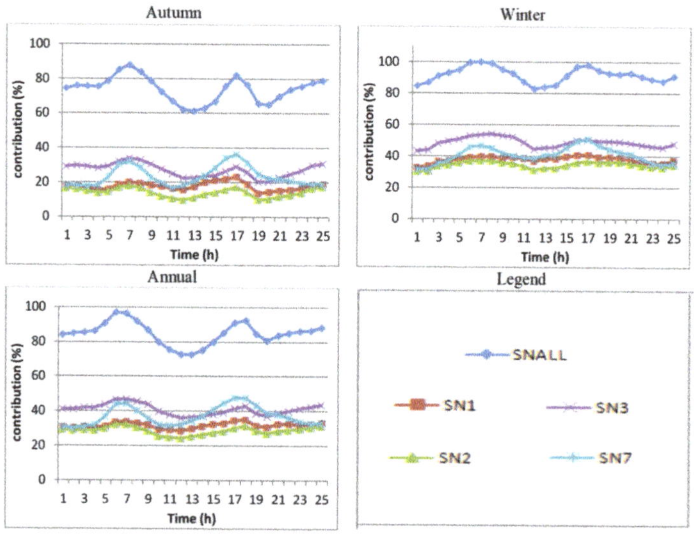

Figure 9. Annual and Seasonal averaged relative contribution (%) of all emissions and the emissions from different SNAP categories of the formation of NO_2 concentrations for Orlov most for the period 2008–2014.

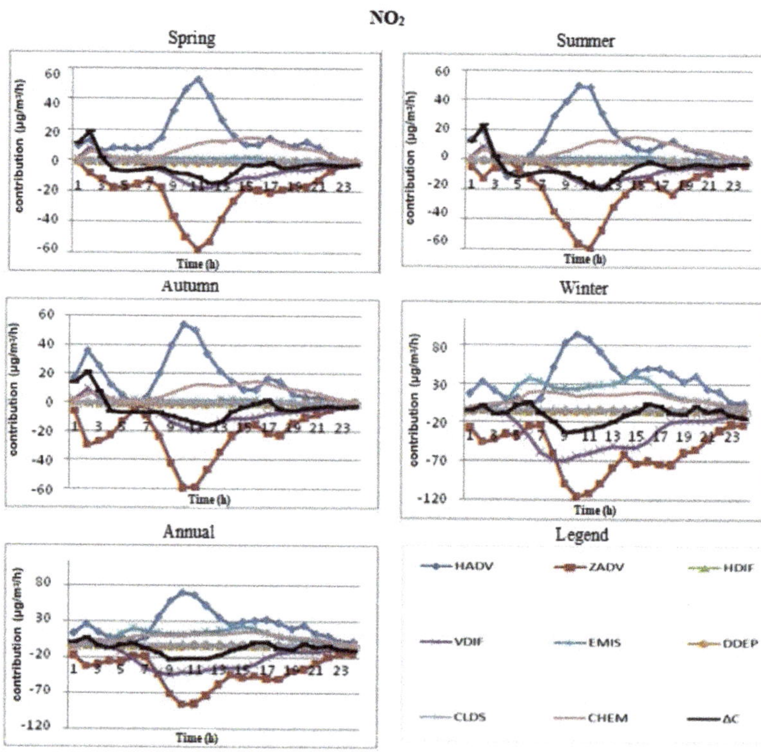

Figure 10. Annual, seasonal and diurnal course of the contribution of the different dynamic and chemical processes of the formation of NO_2 concentrations ($\mu g/m^3/h$) averaged for Orlov most for the period 2008–2014.

The annually and seasonally averaged concentrations of the FPRM for the Orlov most station are shown in Figure 11. The curves show that the average concentrations have well-expressed diurnal and seasonal courses and asymmetric location in different intervals of numbers of cases during the day and seasons. The average concentrations fall into the 80% cases interval during the warm months and the above interval for the cold months. The winter average concentrations coincide with the 90% curve. The absolute maximum and average concentrations are highest for autumn and winter (stable stratification). The early morning hours are characterized by well-expressed maximum and minimum around noon. Probable reasons for this maximum are the stable atmosphere and intensive traffic early morning, which probably play a role in the results for the ensemble annual and seasonal average contributions of different sources for the formation of FPRM (Figure 12). The contribution of the different types of sources varies among the seasons but with the same diurnal course. The main contribution is from sources of all SNAPs (SNALL)—about 70%—followed by the one of the road transport (SN7)—about 60%. The contribution from the other sources is about 20% in all seasons. The distribution during the course of the day has maximum contributions in the morning and afternoon and a minimum around noon and during the night.

Figure 13 shows the same characteristics as in Figure 10, but for the FPRM formation. The main positive contribution to the FPRM formation is vertical in the afternoon with vertical advection for all hours of the day. The phase of horizontal advection is opposite the vertical, which shows a negative contribution. The dry deposition is negative with a maximum around noon. The aerosol processes during all seasons except for winter have a negative contribution. The processes are more intense during the winter in comparison with the other seasons, which is outlined with a large positive contribution of emissions and which decreases to negative for vertical diffusion. The contribution of the other processes is almost zero.

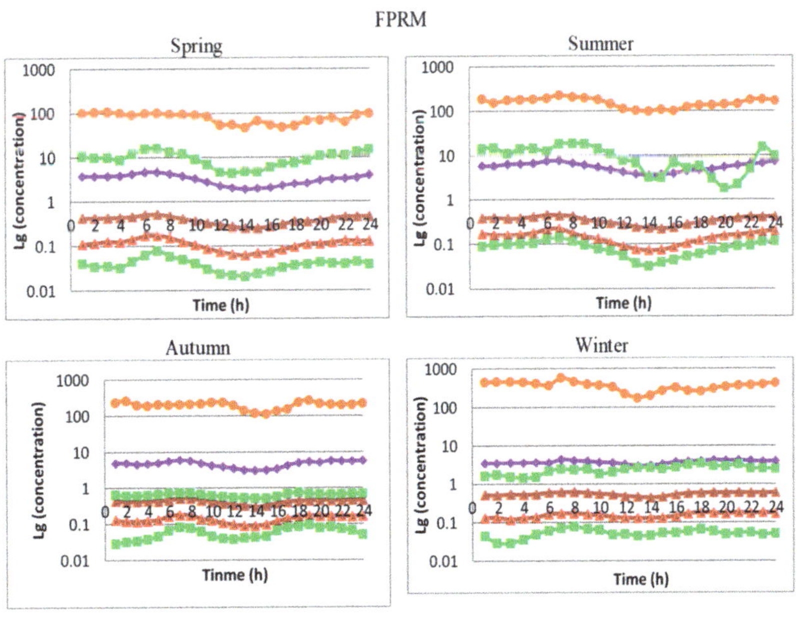

Figure 11. *Cont.*

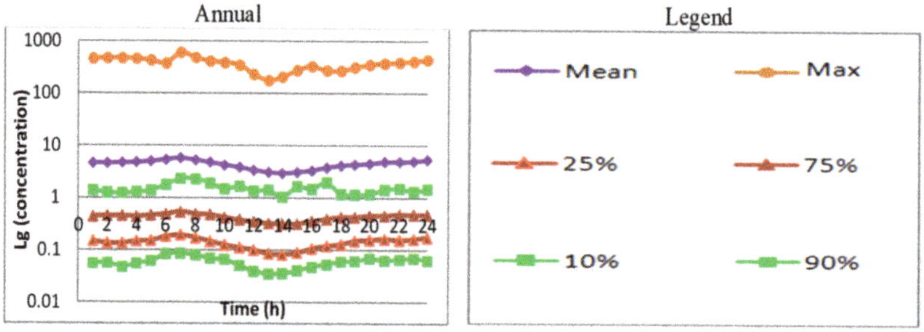

Figure 11. Spatial and diurnal variations of the annual and seasonal ensembles of surface FPRM ($\mu g/m^3$) concentrations behavior in logarithmic scale for Orlov most for the period 2008–2014.

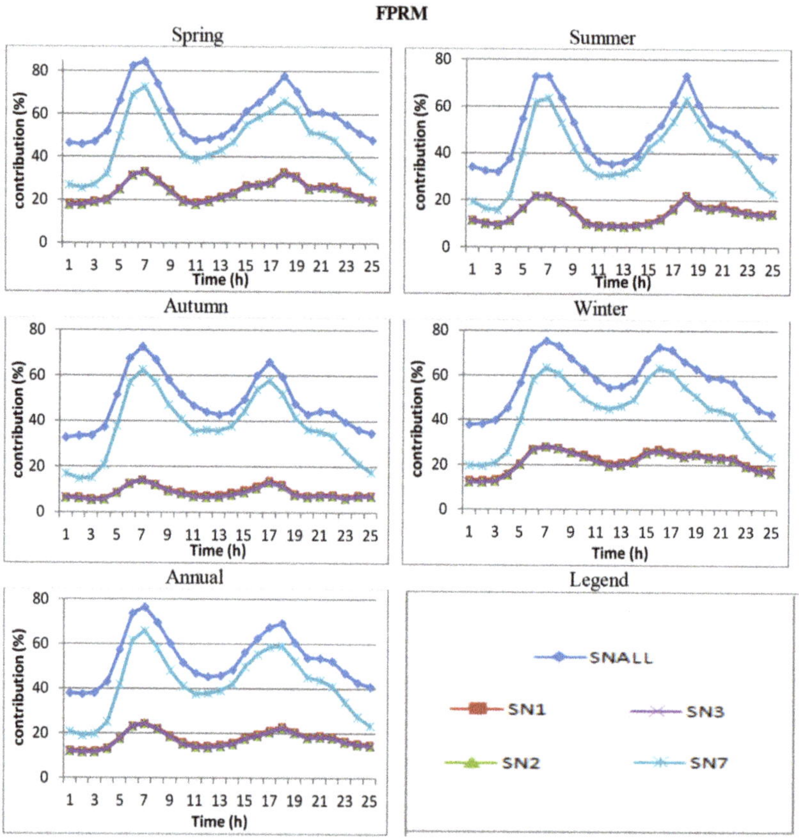

Figure 12. Annual and seasonal averaged relative contribution (%) of all emissions and the emissions from different SNAP categories of the formation of FPRM concentrations for Orlov most for the period 2008–2014.

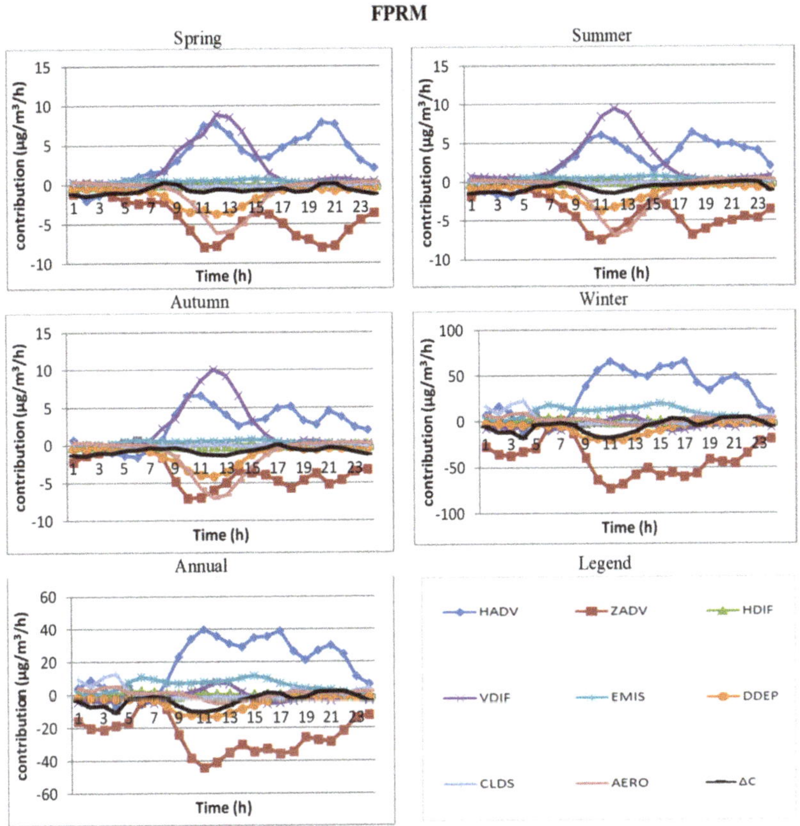

Figure 13. Annual, seasonal and diurnal course of the contribution of the different dynamic and chemical processes to the formation of FPRM concentrations (μg/m^3/h) averaged for Orlov most for the period 2008–2014.

3.3. Characteristics of the Numerically Obtained Concentration Fields, Contribution of Different Emission Sources and Different Dynamic and Chemical Processes to the Atmospheric Composition in Bistritsa

Figure 14 shows the annually and seasonally averaged NO_2 concentrations for the Bistritsa station. The graphics show that the average concentrations have well-expressed diurnal course with a maximum in the early morning hours and a minimum in the afternoon. The average concentration curves follow the 75% and fall into the 80% interval of cases. The locations of the average concentrations are asymmetric for different pieces of the number of cases interval during the day and for the seasons. The seasonal course suggests that the maximum nitrogen dioxide concentrations are highest in autumn and winter. The ensemble-averaged contributions for surface NO_2 formation at Bistritsa point (Figure 15) differ for each type of source and in each season, but their diurnal course is almost equal, although not for Orlov most. The contributions are positive, and the largest one in the afternoon and during the night is SNALL, followed by road transport (SN7). The contribution of the other sources in the morning hours peaks at about 20%. The contributions of the sources from energetics (SN1) and non-industrial burning (SN2) reach 15% in the afternoon during the spring and the winter.

The averaged annual and seasonal contributions of the processes of NO_2 averaged for Bistritsa are shown in Figure 16. The main contribution for NO_2 formation is chemical processes, which have a positive sign. Horizontal advection is positive in morning hours and negative at noon, with a peak for all seasons, and vertical advection is opposite. Vertical diffusion also has a negative contribution, although with smaller values and a maximum around noon. The contribution of the other processes is of almost zero magnitude.

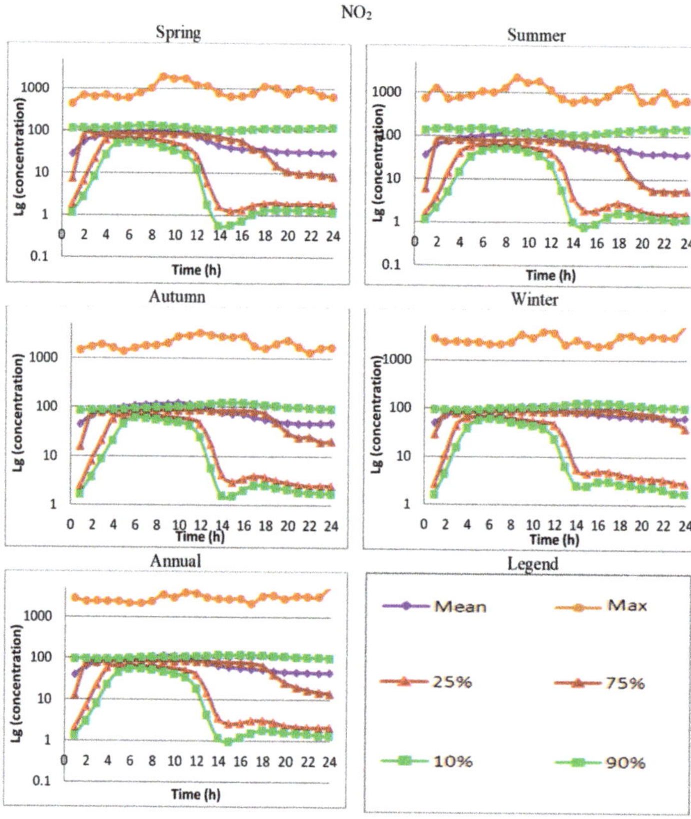

Figure 14. Spatial and diurnal variations of the annual and seasonal ensembles of surface NO_2 ($\mu g/m^3$) concentrations behavior in logarithmic scale for Bistritsa for the period 2008–2014.

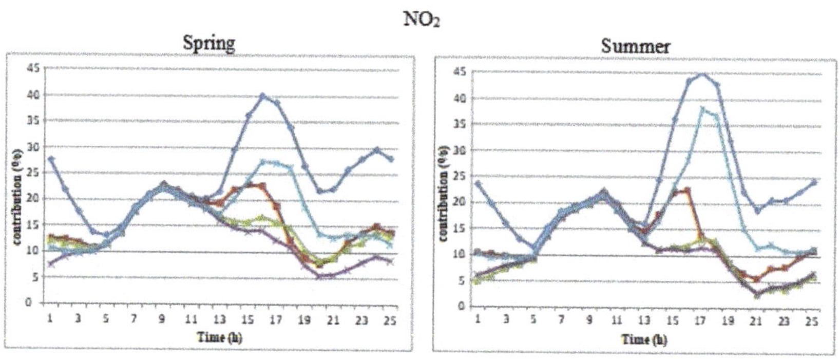

Figure 15. *Cont.*

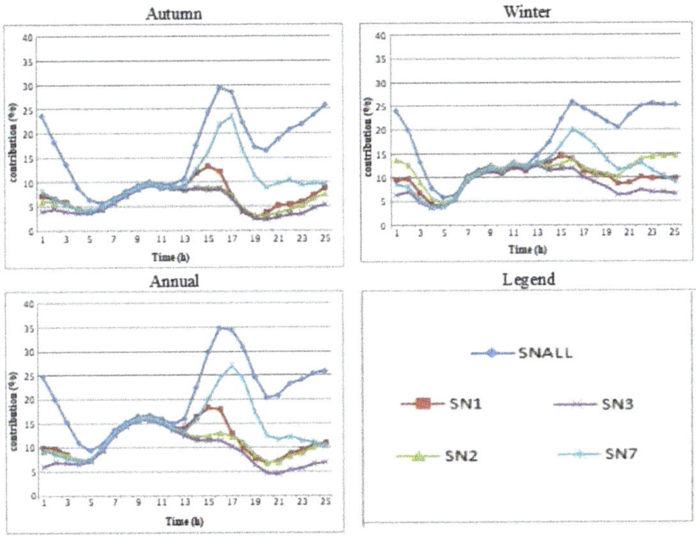

Figure 15. Annual and seasonal averaged relative contribution (%) of all the emissions and the emissions from different SNAP categories of the formation of NO_2 concentrations for Bistritsa for the period 2008–2014.

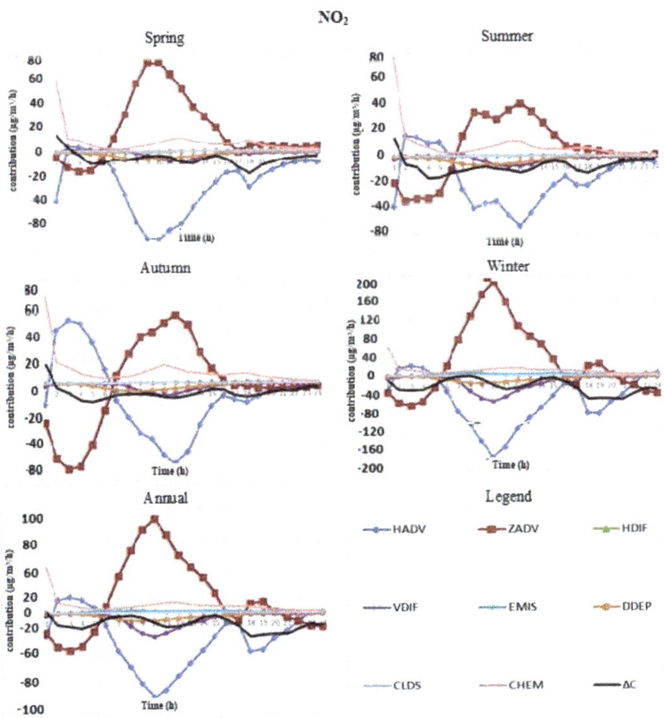

Figure 16. Annual, seasonal and diurnal course of the contribution of the different dynamic and chemical processes of the formation of NO_2 concentrations ($\mu g/m^3/h$) averaged for Bistritsa for the period 2008–2014.

The annually and seasonally averaged concentrations of the FPRM for the Bistritsa station are shown in Figure 17. The plots show that the average concentrations have a well-expressed diurnal and seasonal course and are asymmetrically located in different intervals of numbers of cases during the day and seasons. The average concentrations fall into the 80% cases interval during the warm months and above it for the cold months. The winter average concentrations coincide with the 90% curve. The absolute maximum and average concentrations are highest for autumn and winter. The early morning hours are characterized by a well-expressed maximum for NO_2, with a minimum around noon, which may be due to the stable atmosphere and intensive traffic in the early morning, which probably influence the patterns of the contributions of different sources for surface FPRM formation around Bistritsa (Figure 18). The averaged contributions are different for each season. The main contribution is naturally from sources of all SNAP categories (SNALL)—about 40%—followed by road transport (SN7)—about 30%. In the afternoon, the sources from combustion in the production and transformation of energy (SN1) have a maximum of about 20%. The contribution from other sources is about 10% for all seasons. The daily distribution suggests that SNAPs have a maximum in the morning and afternoon and a minimum around noon and night.

Figure 19 shows the same characteristics as in Figure 16, but for FPRM formation. The main positive contribution to the FPRM formation is vertical diffusion for all hours of the day and vertical advection around noon. The phase of horizontal advection is opposite to the vertical one, which has a negative contribution. The dry deposition is negative with a maximum around noon. The processes are more intense during the winter in comparison with other seasons. The contribution of the other processes is almost zero.

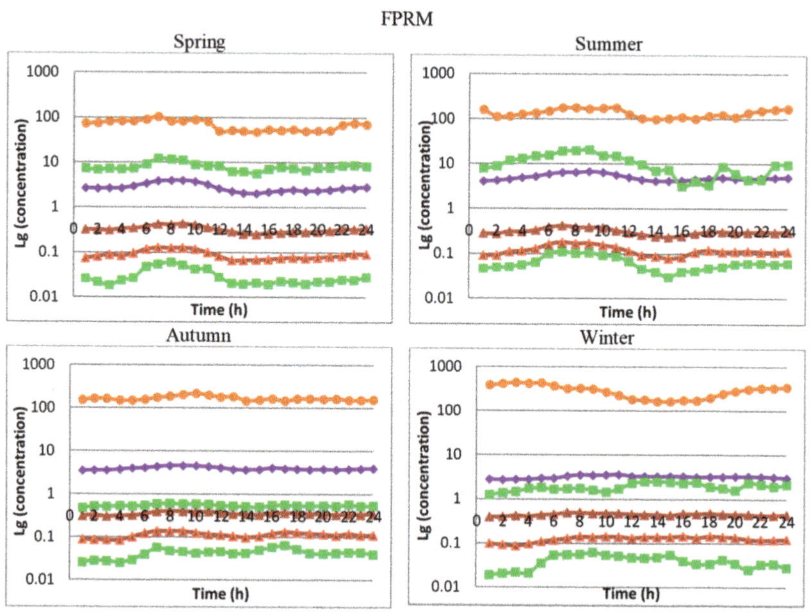

Figure 17. *Cont.*

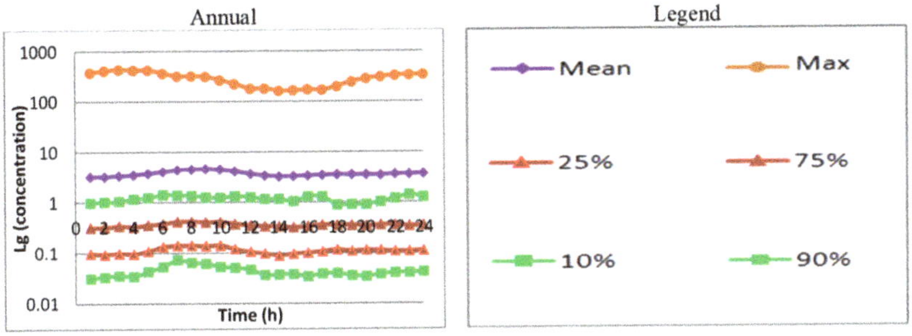

Figure 17. Spatial and diurnal variations of the annual and seasonal ensembles of surface FPRM (μg/m^3) concentrations behavior in logarithmic scale for Bistritsa for the period 2008–2014.

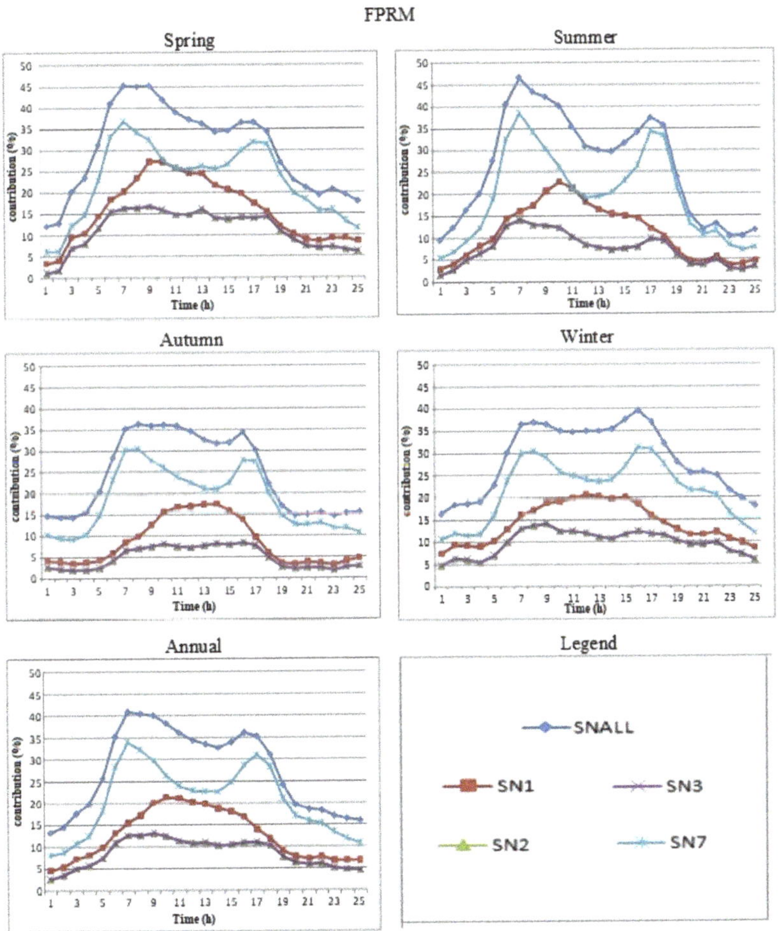

Figure 18. Annual and seasonal averaged relative contribution (%) of all the emissions and the emissions from different SNAP categories of the formation of FPRM concentrations for Bistritsa for the period 2008–2014.

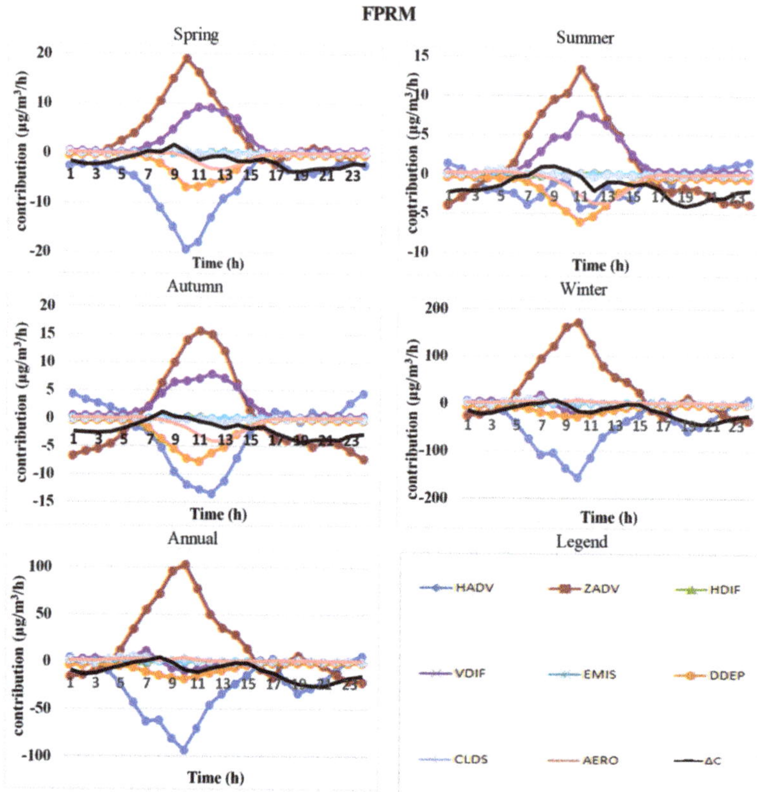

Figure 19. Annual, seasonal and diurnal course of the contribution of the different dynamic and chemical processes of the formation of FPRM concentrations (μg/m^3/h) averaged for Bistritsa for the period 2008–2014.

4. Conclusions

Statistical processing was performed, and the probability density function was calculated for each of the atmospheric compounds with the corresponding seasonal and diurnal fluctuations for each of the points of the grid or the territory of the city. Two points were selected: a mountain type "Bistritsa" and a typical city "Orlov most." Spatial and diurnal variations of the annual ensemble of NO_2 and FPRM are considered. The minimum and maximum curves and the 25%, 75%, 10%, and 90% probability curves are shown. The curves show the imaginary concentrations for which the probability of each simulation is less than 25%, 75%, 10%, and 90%, respectively.

The concentrations of NO_2, FPRM of Bistritsa are of the same order as Orlov Most. The NO_2 and FPRM concentrations around noon are at a local minimum and are larger during morning hours due to the combination of factors, including traffic and atmospheric stability. The NO_2 concentrations are at a minimum around noon, probably because of the more intense turbulent mixing and the slope effect. The FPRM concentrations do not have significant diurnal variations. The ensemble behavior of NO_2 and FPRM is significantly asymmetric for both selected sites.

For all emission categories, the pattern of the contribution fields is complex, which reflects the emission source configuration and the heterogeneity of topography, land use, and meteorological conditions. Plots of this kind can give a good qualitative impression of the spatial complexity of emission contributions. In order to demonstrate emission contribution behavior in a simpler and easier way, the respective fields can be averaged over some domain, which makes it possible to follow and compare the diurnal behavior of the respective contributions of different species. The results presented in the paper are a first glance at the atmospheric composition status in urban areas; thus, few decisive conclusions can be made at this stage of the study. Different emissions, relative to contribution of the concentration of different species, can be different, varying from 80% to above 100%. The contributions of different emission categories to different species surface concentrations have different diurnal courses. For all of the pollutants, the contribution of SNALL is dominant, but this contribution of emissions is less than 100%, which means that part of the concentrations is formed from sources outside the Sofia city, due to transport into the domain. The contribution of all SN7 (road transport) to NO_2 surface concentrations is positive and reaches about 50% around the busiest traffic roads. The SN7 emissions have dominant contributions to the NO_2 and FPRM surface concentrations. The relative contribution of SNALL and SN7 to the formation of FPRM, as well as for NO_2 at the busiest traffic place, have two maximums in the diurnal course.

The estimation of the contribution of emissions from different source categories is valuable information that can be useful for the definition of measures for improving the air quality in Sofia by reducing emissions. Moreover, knowing the diurnal course of the contributions for a specific time period can suggest an optimal emission temporal regime in order to mitigate air pollution for the given episode.

The results produced by the CMAQ "integrated process rate analysis" demonstrate the complex behavior and interaction of the different processes. Further analysis of these processes, their spatial, diurnal, and seasonal variability, and interaction can be helpful for an explanation of the overall picture and origin of the pollution in the considered region. For the entire domain of Sofia, and for each of the selected items, the total concentration change (ΔC), leading to a change in a concentration, is determined mainly by a small number of dominating processes that have large values and may have opposite signs and phases. The total concentration change (ΔC) is different for each pollutant. The sign of the contributions of some of the processes is obvious, but some may have different signs, depending on the type of emissions as well as weather conditions and topography. In general, it can be concluded that the contributions of different processes have different behaviors and interact in a complex way.

The "integrated process rate analysis" is not often applied in urban air quality simulations; thus, in the case of the present study, it enriches the entire picture of the atmospheric composition climate of the city of Sofia. This ensemble treatment of the contribution of different processes, however, does not give an easy answer to how the processes interact and how exactly they form the air composition. It will probably be more fruitful to consider the process contribution for a given episode together with the specific meteorological conditions for that episode.

The obtained results and the corresponding conclusions made in this paper are in good agreement with the general idea of how urban atmospheric composition is formed; thus, they do not add general knowledge on the subject. The research and applied contribution of the paper are that it presents quantitative estimations specific to the city of Sofia. Such an extensive and comprehensive study for the city of Sofia had not been made before.

The models and the entire procedure used in the present paper can, of course, be applied to other cities. Such studies have been made for many cities. A crucial point, however, is the emission inventory, and the activities for preparing it are specific for each city, depending on the raw data available.

Author Contributions: Conceptualization, I.G. and V.I.; methodology, I.G.; software, I.G.; validation, I.G. and V.I.; formal analysis, V.I.; investigation, V.I.; resources, I.G.; data curation, V.I.; writing—original draft preparation, V.I.; writing—review and editing, V.I. and I.G.; visualization, I.G.; supervision, V.I.; project administration, I.G.; funding acquisition, V.I. All authors have read and agreed to the published version of the manuscript.

Funding: This work has been carried out in the framework of the National Science Program "Environmental Protection and Reduction of Risks of Adverse Events and Natural Disasters", approved by the Resolution of the Council of Ministers no. 577/17.08.2018 and supported by the Ministry of Education and Science (MES) of Bulgaria (agreement no. D01-363/17.12.2020) and has been partially supported by the National Center for High-performance and Distributed Computing (NCHDC), part of National Roadmap of RIs under grant no. D01-387/18.12.2020.

Institutional Review Board Statement: Not applicable.

Informed Consent Statement: Not applicable.

Data Availability Statement: The input data—large scale meteorological background and the emission inventories are described in the paper. The output from the computer simulations, which, after processing and generalization is the basis of the results, reported in the paper, is, unfortunately, still not publicly available. This is due to the requirements of the projects, which financially support the present study.

Acknowledgments: Deep gratitude to the organizations and institutes (TNO, NCEP/NCAR, ECA&D, Unidata, MPI-M and all others), which provide free of charge software and data. Without their innovative data services and tools, this study would not be possible.

Conflicts of Interest: The authors declare no conflict of interest.

References

1. Syrakov, D.; Etropolska, I.; Prodanova, M.; Slavov, K.; Ganev, K.; Miloshev, N.; Ljubenov, T. Downscaling of Bulgarian Chemical Weather Forecast from Bulgaria Region to Sofia City, American Institute of Physics. *Conf. Proc.* **2013**, *1561*, 120–132.
2. Dimitrova, R.; Velizarova, M. Assessment of the Contribution of Different Particulate Matter Sources on Pollution in Sofia City. *Atmosphere* **2021**, *12*, 423. [CrossRef]
3. Georgieva, I.; Gadzhev, G.; Ganev, K.; Miloshev, N. Analysis of The Contribution of Different Processes (Chemical and Dynamical) Which form the Atmospheric Composition in Sofia. In Proceedings of the 19th International Conference on Harmonisation within Atmospheric Dispersion Modelling for Regulatory Purposes, Bruges, Belgium, 3–6 June 2019.
4. Gadzhev, G.; Georgieva, I.; Ganev, K.; Miloshev, N. Contribution of Different Emission Sources to the Atmospheric Composition Formation in City of Sofia. *Int. J. Environ. Pollut.* **2018**, *64*, 47–57. [CrossRef]
5. Gadzhev, G.; Georgieva, I.; Ganev, K.; Ivanov, V.; Miloshev, N.; Chervenkov, H.; Syrakov, D. Climate Applications in a Virtual Research Environment Platform. *Scalable Comput. Pract. Exp.* **2018**, *19*, 107–118. [CrossRef]
6. Gadzhev, G.; Ganev, K.; Mukhtarov, P. HPC Simulations of the Atmospheric Composition Bulgaria's Climate (on the Example of Coarse Particulate Matter Pollution). In *Proceedings of the International Conference on High Performance Computing, Borovets, Bulgaria, 2–6 September 2019*; Studies in Computational Intelligence; Springer: Cham, Switzerland, 2021; pp. 221–233.
7. Gadzhev, G.; Ganev, K.; Mukhtarov, P. Statistical Moments of the Vertical Distribution of Air Pollution Over Bulgaria. In *Proceedings of the 12th International Conference on Large-Scale Scientific Computing, Sozopol, Bulgaria, 10–14 June 2019*; Lirkov, I., Margenov, S., Eds.; Lecture Notes in Computer Science (including subseries Lecture Notes in Artificial Intelligence and Lecture Notes in Bioinformatics); Springer: Cham, Switzerland, 2020; pp. 213–219.
8. Gadzhev, G.; Ganev, K. Vertical Structure of Some Pollutant Over Bulgaria-Ozone and Nitrogen Dioxide. *SGEM* **2018**, *18*, 449–454. [CrossRef]
9. Gadzhev, G.; Ganev, K.; Miloshev, N. Numerical Study of the Atmospheric Composition Climate of Bulgaria-Validation of the Computer Simulation Results. *Int. J. Environ. Pollut.* **2015**, *57*, 189–201. [CrossRef]
10. Gadzhev, G.; Ganev, K.; Miloshev, N.; Syrakov, D.; Prodanova, M. HPC Simulations of the Fine Particulate Matter Climate of Bulgaria. In *Proceedings of the 8th International Conference on Numerical Methods and Applications, Borovets, Bulgaria, 20–24 August 2014*; Lecture Notes in Computer Science (Including Subseries Lecture Notes in Artificial Intelligence and Lecture Notes in Bioinformatics); Springer: Cham, Switzerland, 2015; pp. 178–186. [CrossRef]
11. Gadzhev, G.; Ganev, K.; Miloshev, N.; Syrakov, D.; Prodanova, M. Analysis of the Processes Which form the Air Pollution Pattern over Bulgaria. In *Proceedings of the 9th International Conference on Large-Scale Scientific Computations, Sozopol, Bulgaria, 3–7 June 2013*; Lecture Notes in Computer Science (Including Subseries Lecture Notes in Artificial Intelligence and Lecture Notes in Bioinformatics); Springer: Berlin/Heidelberg, Germany, 2014; pp. 390–396. [CrossRef]
12. Gadzhev, G.; Ganev, K.; Miloshev, N.; Syrakov, D.; Prodanova, M. Some Basic Facts about the Atmospheric Composition in Bulgaria-Grid Computing Simulations. In *Proceeding of the 9th International Conference on Large-Scale Scientific Computations, Sozopol, Bulgaria, 3–7 June 2013*; Lecture Notes in Computer Science (Including Subseries Lecture Notes in Artificial Intelligence and Lecture Notes in Bioinformatics); Springer: Berlin/Heidelberg, Germany, 2014; pp. 484–490. [CrossRef]
13. Gadzhev, G.; Ganev, K.; Prodanova, M.; Syrakov, D.; Atanasov, E.; Miloshev, N. Multi-Scale Atmospheric Composition Modelling for Bulgaria. In *Air Pollution Modeling and Its Application XXII*; NATO Science for Peace and Security Series C: Environmental Security; Springer: Dordrecht, The Netherlands, 2013; Volume 137, pp. 381–385.

14. Gadzhev, G.; Ganev, K.; Prodanova, M.; Syrakov, D.E.; Miloshev, N.; Georgiev, G. Some Numerically Studies of the Atmospheric Composition Climate of Bulgaria. In Proceeding of the AIP Conference Proceedings, Proceedings of the 5th International Conference for Promoting the Application of Mathematics in Technical and Natural Sciences, Albena, Bulgaria, 24–29 June 2013; pp. 100–111. [CrossRef]
15. Brandiyska, A.; Ganev, K.; Syrakov, D.; Prodanova, M.; Miloshev, N.; Gadzhev, G. Bulgarian Emergency Response System for Release of Hazardous Pollutants-Brief Description and Some Examples. *Int. J. Environ. Pollut.* **2012**, *50*, 3–11. [CrossRef]
16. Gadzhev, G.; Ganev, K.; Syrakov, D.; Miloshev, N.; Prodanova, M. Contribution of Biogenic Emissions to the Atmospheric Composition of the Balkan Region and Bulgaria. *Int. J. Environ. Pollut.* **2012**, *50*, 130–139. [CrossRef]
17. Gadzhev, G.; Syrakov, D.; Ganev, K.; Brandiyska, A.; Miloshev, N.; Georgiev, G.; Prodanova, M. Atmospheric Composition of the Balkan Region and Bulgaria. Study of the Contribution of Biogenic Emissions. In Proceeding of the AIP Conference Proceedings, Proceedings of the 3th International Conference on Application of Mathematics in Technical and Natural Sciences, Albena, Bulgaria, 20–25 June 2011; pp. 200–209.
18. Todorova, A.; Gadzhev, G.; Jordanov, G.; Syrakov, D.; Ganev, K.; Miloshev, N.; Prodanova, M. Numerical Study of Some High PM10 Level Episodes. *Int. J. Environ. Pollut.* **2011**, *46*, 69–82. [CrossRef]
19. Ganev, K.; Syrakov, D.; Todorova, A.; Gadzhev, G.; Miloshev, N.; Prodanova, M. Study of regional dilution and transformation processes of the air pollution from road transport. *Int. J. Environ. Pollut.* **2011**, *44*, 62–70. [CrossRef]
20. Todorova, A.D.; Ganev, K.G.; Syrakov, D.E.; Prodanova, M.; Georgiev, G.J.; Miloshev, N.G.; Gadjhev, G.K. Bulgarian Emergency Response System for Release of Hazardous Pollutants-Design and First Tests. In *Air Pollution Modeling and Its Application XXI*; NATO Science for Peace and Security Series C: Environmental Security; Steyn, D., Trini Castelli, S., Eds.; Springer: Dordrecht, The Netherlands, 2011; pp. 263–268. [CrossRef]
21. Gadzhev, G.; Jordanov, G.; Ganev, K.; Prodanova, M.; Syrakov, D.; Miloshev, N. Atmospheric Composition Studies for the Balkan Region. In *Proceedings of the 7th International Conference on Numerical Methods and Applications, Borovets, Bulgaria, 20–24 August 2010*; Lecture Notes in Computer Science (Including Subseries Lecture Notes in Artificial Intelligence and Lecture Notes in Bioinformatics); Springer: Berlin/Heidelberg, Germany, 2011; pp. 150–157.
22. Gadzhev, G.; Ganev, K.; Jordanov, G.; Miloshev, N.; Todorova, A.; Syrakov, D.; Prodanova, M. Transport and Transformation of Air Pollution from Road and Ship Transport-Joint Analysis of Regional Scale Impacts and Interactions. In Proceedings of the International Conference on Transport, Atmosphere and Climate, Aachen, Germany, Maastricht, The Netherlands, 22–25 June 2009; pp. 33–37.
23. Ganev, K.; Syrakov, D.; Gadzhev, G.; Prodanova, M.; Jordanov, G.; Miloshev, N.; Todorova, A. Joint Analysis of Regional Scale Transport and Transformation of Air Pollution from Road and Ship Transport. In *Proceedings of the 7th International Conference on Large-Scale Scientific Computations, Sozopol, Bulgaria, 4–8 June 2009*; Lecture Notes in Computer Science (Including Subseries Lecture Notes in Artificial Intelligence and Lecture Notes in Bioinformatics); Springer: Berlin/Heidelberg, Germany, 2010; pp. 180–187. [CrossRef]
24. Todorova, A.; Gadzhev, G.; Jordanov, G.; Syrakov, D.; Ganev, K.; Miloshev, N.; Prodanova, M. Numerical Study of Some High PM10 Levels Episodes. In *Proceedings of the 7th International Conference on Large-Scale Scientific Computations, Sozopol, Bulgaria, 4–8 June 2009*; Lecture Notes in Computer Science (Including Subseries Lecture Notes in Artificial Intelligence and Lecture Notes in Bioinformatics); Springer: Berlin/Heidelberg, Germany, 2010; pp. 223–230. [CrossRef]
25. Todorova, A.; Syrakov, D.; Gadjhev, G.; Georgiev, G.; Ganev, K.G.; Prodanova, M.; Miloshev, N.; Spiridonov, V.; Bogatchev, A.; Slavov, K. Grid Computing for Atmospheric Composition Studies in Bulgaria. *Earth Sci. Inform.* **2010**, *3*, 259–282. [CrossRef]
26. San, J.R.; Pérez, J.L.; Gonzalez-Barras, R.M. The Use of LES CFD Urban Models and Mesoscale Air Quality Models for Urban Air Quality Simulations. In *Proceedings of the 1st International Conference on Environmental Protection and Disaster Risks EnviroRISK 2020, Sofia, Bulgaria, 29–30 September 2020*; Dobrinkova, N., Gadzhev, G., Eds.; Studies in Systems, Decision and Control; Springer: Cham, Switzerland, 2021; pp. 185–189. [CrossRef]
27. Gryning, S.E.; Batchvarova, E. Advances in Urban Dispersion Modelling. In *Advances in Air Pollution Modeling for Environmental Security*; NATO Science Series (Series IV: Earth and Environmental, Series); Faragó, I., Georgiev, K., Havasi, Á., Eds.; Springer: Dordrecht, The Netherlands, 2005; Volume 54. [CrossRef]
28. Karagiannidis, A.; Poupkou, A.; Giannaros, T.; Giannaros, C.; Melas, D.; Argiriou, A. The Air Quality of a Mediterranean Urban Environment Area and Its Relation to Major Meteorological Parameters. *Water Air Soil Pollut.* **2015**, *226*, 2239. [CrossRef]
29. Zanis, P.; Katragkou, E.; Markakis, K.; Lysaridis, I.; Poupkou, A.; Melas, D.; Kanakidou, M.; Psiloglou, B.; Gerasopoulos, E.; Zerefos, C.; et al. Effects on Surface Atmospheric Photo-Oxidants over Greece during the Total Solar Eclipse Event of 29 March 2006. *Atmos. Chem. Phys.* **2007**, *7*, 6061–6073. [CrossRef]
30. Poupkou, A.; Symeonidis, P.; Lisaridis, I.; Melas, D.; Ziomas, I.; Yay, O.D.; Balis, D. Effects of Anthropogenic Emission Sources on Maximum Ozone Concentrations over Greece. *Atmos. Res.* **2008**, *89*, 374–381. [CrossRef]
31. Katragkou, E.; Zanis, P.; Tegoulias, I.; Melas, D.; Krüger, B.C.; Huszar, P.; Halenka, T.; Rauscher, S. Decadal Regional Air Quality Simulations over Europe in Present Climate: Near Surface Ozone Sensitivity to External Meteorological Forcing. *Atmos. Chem. Phys.* **2010**, *9*, 10675–10710. [CrossRef]
32. Kanakidou, M.; Mihalopoulos, N.; Kindap, T.; Im, U.; Vrekoussis, M.; Gerasopoulos, E.; Dermitzaki, E.; Unal, A.; Kocak, M.; Markakis, K.; et al. Megacities as Hot Spots of Air Pollution in the East Mediterranean. *Atmos. Environ.* **2010**, *45*, 1223–1235. [CrossRef]

33. Im, U.; Poupkou, A.; Incecik, S.; Markakis, K.; Kindap, T.; Unal, A.; Melas, D.; Yenigun, O.; Topcu, S.; Odman, M.T.; et al. The Impact of Anthropogenic and Biogenic Emissions on Surface Ozone Concentrations in Istanbul. *Sci. Total Environ.* **2011**, *409*, 1255–1265. [CrossRef] [PubMed]
34. Juda-Rezler, K.; Reizer, M.; Huszar, P.; Krüger, B.C.; Zanis, P.; Syrakov, D.; Katragkou, E.; Trapp, W.; Melas, D.; Chervenkov, H.; et al. Modelling the effects of climate change on air quality over Central and Eastern Europe: Concept evaluation and projections. *Clim. Res.* **2012**, *53*, 179–203. [CrossRef]
35. Marécal, V.; Peuch, V.-H.; Andersson, C.; Andersson, S.; Arteta, J.; Beekmann, M.; Benedictow, A.; Bergström, R.; Bessagnet, B.; Cansado, A.; et al. A Regional Air Quality Forecasting System over Europe: The MACC-II Daily Ensemble Production. *Geosci. Model Dev.* **2015**, *8*, 2777–2813. [CrossRef]
36. Shamarock, W.C.; Klemp, J.B.; Dudhia, J.; Gill, D.O.; Barker, D.M.; Duda, M.G.; Huang, X.-y.; Wang, W.; Powers, J.G. A Description of the Advanced Research WRF Version 3. 2007. Available online: https://opensky.ucar.edu/islandora/object/technotes%3A500/datastream/PDF/view (accessed on 20 October 2021).
37. Byun, D.; Young, J.; Gipson, G.; Godowitch, J.; Binkowski, F.S.; Roselle, S.; Benjey, B.; Pleim, J.; Ching, J.; Novak, J.; et al. Description of the Models-3 Community Multiscale Air Quality (CMAQ) Modeling System. In *Proceedings of the 10th Joint Conference on the Applications of Air Pollution Meteorology with the A&WMA*; Phoenix, AZ, USA, 11–16 January 1998, pp. 264–268. Available online: http://www.epa.gov/asmdnerl/models3/doc/science/science.html (accessed on 20 October 2021).
38. Byun, D.; Ching, J. Science Algorithms of the EPA Models-3 Community Multiscale Air Quality (CMAQ) Modeling System. Available online: https://cfpub.epa.gov/si/si_public_record_report.cfm?dirEntryId=63400&Lab=NERL (accessed on 20 October 2021).
39. *SMOKE v3.5.1 User's Manual*; The University of North Carolina at Chapel Hill: Chapel Hill, NC, USA, 2013. Available online: https://www.cmascenter.org/smoke/documentation/3.5.1/manual_smokev351.pdf (accessed on 20 October 2021).
40. Vestreng, V. *Emission Data Reported to UNECE/EMEP: Evaluation of the Spatial Distribution of Emissions. Meteorological Synthesizing Centre-West*; The Norwegian Meteorological Institute: Oslo, Norway, 2001; Research Note 56, EMEP/MSC-W Note 1.
41. Visschedijk, A.J.H.; Zandveld, P.Y.J.; Denier van der Gon, H.A.C. A High Resolution Gridded European Emission Database for the EU Integrate Project GEMS. Available online: http://lap.physics.auth.gr/gems/docu/TNO%20Short%20Emissions%20Report.pdf (accessed on 20 October 2020).
42. Denier van der Gon, H.; Visschedijk, A.; van de Brugh, H.; Droge, R. *A High Resolution European Emission Data Base for the Year 2005*; TNO-Report TNO-034-UT-2010-01895 RPT-ML; Environment, Health and Safety: Utrecht, The Netherlands, 2010. Available online: https://webdosya.csb.gov.tr/db/necen/editordosya/file/NEC/MACC_Training/TNO_PAREST.pdf (accessed on 20 October 2020).

MDPI
St. Alban-Anlage 66
4052 Basel
Switzerland
www.mdpi.com

Atmosphere Editorial Office
E-mail: atmosphere@mdpi.com
www.mdpi.com/journal/atmosphere

www.ingramcontent.com/pod-product-compliance
Lightning Source LLC
Chambersburg PA
CBHW061556210926
51850CB00009B/2123